INTRODUCTION TO FEEDBACK CONTROL SYSTEMS

McGRAW-HILL SERIES IN ELECTRICAL ENGINEERING

Consulting Editor
Stephen W. Director, Carnegie-Mellon University

NETWORKS AND SYSTEMS
COMMUNICATIONS AND INFORMATION THEORY
CONTROL THEORY
ELECTRONICS AND ELECTRONIC CIRCUITS
POWER AND ENERGY
ELECTROMAGNETICS
COMPUTER ENGINEERING AND SWITCHING THEORY
INTRODUCTORY AND SURVEY
RADIO, TELEVISION, RADAR, AND ANTENNAS

Previous Consulting Editors

Ronald M. Bracewell, Colin Cherry, James F. Gibbons, Willis W. Harman, Hubert Heffner, Edward W. Herold, John G. Linvill, Simon Ramo, Ronald A. Rohrer, Anthony E. Siegman, Charles Susskind, Frederick E. Terman, John G. Truxal, Ernst Weber, and John R. Whinnery

CONTROL THEORY

Consulting Editor
Stephen W. Director, Carnegie-Mellon University

Athans and Falb OPTIMAL CONTROLS
Auslander, Rabins, and Takahashi INTRODUCING SYSTEMS AND CONTROL
Canon, Cullum, and Polak THEORY OF OPTIMAL CONTROL AND MATHEMATICAL PROGRAMMING
Cruz FEEDBACK SYSTEMS
D'Azzo and Houpis FEEDBACK CONTROL SYSTEM ANALYSIS AND SYNTHESIS
D'Azzo and Houpis LINEAR CONTROL SYSTEM ANALYSIS AND DESIGN: Conventional and Modern
Elgerd CONTROL SYSTEMS THEORY
Emanuel and Leff INTRODUCTION TO FEEDBACK CONTROL SYSTEMS
Eveleigh INTRODUCTION TO CONTROL SYSTEMS DESIGN
Gibson NONLINEAR AUTOMATIC CONTROL
Leondes MODERN CONTROL SYSTEMS THEORY
Meditch STOCHASTIC OPTIMAL LINEAR ESTIMATION AND CONTROL
Melsa and Schultz COMPUTER PROGRAMS FOR COMPUTATIONAL ASSISTANCE IN THE STUDY OF LINEAR CONTROL THEORY
Minorsky THEORY OF NONLINEAR CONTROL SYSTEMS
Raven AUTOMATIC CONTROL ENGINEERING
Sage and Melsa ESTIMATION THEORY WITH APPLICATIONS TO COMMUNICATIONS CONTROL
Schultz and Melsa STATE FUNCTIONS AND LINEAR CONTROL SYSTEMS

INTRODUCTION TO FEEDBACK CONTROL SYSTEMS

Pericles Emanuel
Edward Leff
Professors of Electrical Technology
Queensborough Community College

INTERNATIONAL STUDENT EDITION

McGRAW-HILL KOGAKUSHA, LTD.

Tokyo Auckland Bogota Guatemala Hamburg Johannesburg
Lisbon London Madrid Mexico New Delhi Panama
Paris San Juan São Paulo Singapore Sydney

Library of Congress Cataloging in Publication Data

Emanuel, Pericles, date
Introduction to feedback control systems.

(Electrical engineering series)
Includes index.
1. Feedback control systems. I. Leff, Edward, date joint author. II. Title.
TJ216.E4 629.8'3 78-11264
ISBN 0-07-019310-X

INTRODUCTION TO FEEDBACK CONTROL SYSTEMS

INTERNATIONAL STUDENT EDITION

This book was set in Electra by Bi-Comp, Incorporated.
The editors were Frank J. Cerra and James W. Bradley;
the designer was Charles A. Carson;
the production supervisor was Dominick Petrellese.
The drawings were done by J & R Services, Inc.

KOSAIDO PRINTING CO., LTD. TOKYO, JAPAN

to
SANDRA,
MELANIE, and MICHAEL

P.E.

to
HERMINE,
GAYE, NANCY, and GREGG

E.L.

Contents

Preface

This book has been designed for use in two-year associate degree and four-year baccalaureate degree programs in electrical, electronic, and computer technology. The systems approach using block diagrams has been employed throughout the text. Students with only a fundamental calculus background should be able to understand the Laplace transforms described in Chapter 3. However, an instructor may elect to use Chapter 3 to teach all the mathematics necessary to understand the approach to control system analysis used in this book.

Students without a background in electrical machinery should be able to use the material in Chapters 2 to 5 to understand the dc motor as a fundamental building block in control systems.

Emphasis has been placed on the digital control system. The coverage of A/D and D/A converters as well as other digital components will enable the student to bridge the gap between digital computers and linear control systems.

The basic material in this text has been used for 5 years in courses in servomechanisms and computer control systems at Queensborough Community College. A one-semester course in control systems at the two-year level should include the chapters on components, mathematics, block diagrams, second-order servos, and frequency response (Chapters 1 to 6). A full-year course should also include compensation, other types of controls, digital servos and components, computer controls, motor controls, and analog computers (Chapters 1 to 13). Most of the material can be covered in a one-semester course for fourth-year students in a baccalaureate program, where a much broader mathematical background is expected.

Control-systems technologists already in industry may find the chapter on motor controls and digital servos of particular interest. Permanent-magnet stepper motors, which are usually ignored in the classroom, have been covered in Chapter 9. It is hoped that this topic will be included in all control systems or servomechanism courses. Any student who eventually finds employment in a computer-related industry will come in contact with computer peripherals, which invariably contain a stepper motor.

A one-semester course on digital control systems like the one offered at Queensborough should include Chapters 1 to 5, 9, 10, and 11. Chapter 10 is devoted to interface components (A/D, D/A, MUX, S/H, etc.). Emphasis has been placed not only on use of these components for interfacing computers and controls but also on data acquisition. Although it is a sophisticated topic, data acquisition has been presented so that both two- and four-year schools can cover it in the classroom.

In Chapter 11 the control system is interfaced to a microprocessor. The chapter includes some practical worked out examples.

More advanced topics, e.g., nonlinearities in control systems, root locus, Routh criterion, Nyquist criterion, and the increasingly popular phase-locked loop, are presented in the appendixes. This eliminates the need for a second (reference) text when any of these topics is to be covered.

The authors wish to thank those members of the Department of Electrical Technology at Queensborough Community College who offered their valuable comments. In addition, we wish to thank Sandra Emanuel for typing the entire manuscript.

Pericles Emanuel
Edward Leff

INTRODUCTION TO FEEDBACK CONTROL SYSTEMS

chapter 1

Introduction to Feedback Control Systems

A *control system* can be defined as an interconnection of several components all working together to perform a certain function. A more complete description of these components will be given in Chap. 2. More specifically a control system can be described as a group of components which respond to a signal. The response of these components is the performance of a function. In most cases this function is the control of a physical variable, such as, speed, temperature, position, voltage, or pressure. The signal which causes these components to operate and perform their individual functions is called an *actuating signal*.

1-1 OPEN- AND CLOSED-LOOP SYSTEMS

The broad concept of control systems can be subdivided into two basic categories, open-loop control systems and closed-loop control systems. An *open-loop control system* is one in which the actuating signal is made up solely of an input signal which represents the command to the system. A *closed-loop control system* is one in which the actuating signal is made up of the difference between the input signal and a feedback signal representing the output or controlled variable. The functional diagrams in Figs. 1-1 and 1-2 will help clarify the difference between open- and closed-loop systems. In this text the emphasis will be on closed-loop control systems of more notable feedback control systems.

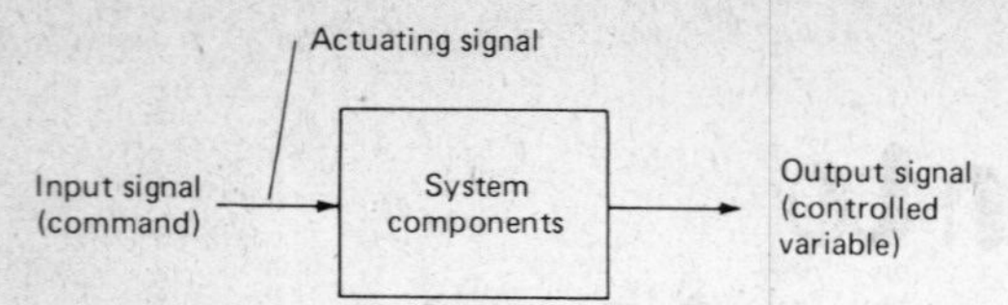

Fig. 1-1 Functional diagram of an open-loop system.

1-2 THE NEED FOR CONTROL SYSTEMS

Why do we need control systems? In modern technology, the greatest use for control systems has been found when a large amount of energy or power is to be controlled with a smaller amount of energy or power. This underlying principle can be illustrated in many fields.

1. In a simple vacuum-tube or transistor amplifier a low-level signal applied to the grid or base will control a relatively higher-level signal on the plate or collector.
2. The pilot of a 350,000-lb aircraft, e.g., the McDonnell-Douglas DC-10, by simply turning a dial on the autopilot control panel can cause the aircraft to climb 20,000 ft in altitude.
3. By turning a key the driver of an automobile can start a 300-hp engine.
4. A person can lower the temperature in the room by 20°F simply by turning a knob on the air conditioner.
5. A rocket ship can be flown to the moon by the manipulation of controls here on earth. (This exhibits another important aspect of control systems, the ability to control something a great distance from the controller or command to the system.)

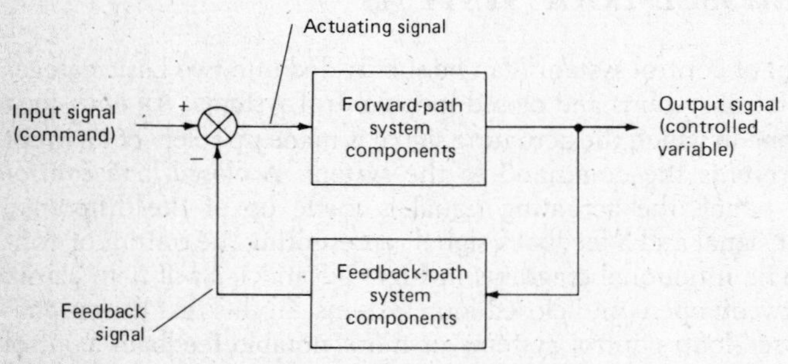

Fig. 1-2 Functional diagram of a closed-loop system.

6. The driver of a 2-ton automobile can control its motion by the simple use of a steering wheel, accelerator pedal, and brake pedal.
7. An antenna-rotor system controls the direction of the antenna on the roof by the simple dial adjustment on top of the television.

1-3 EFFECTS OF FEEDBACK

In each case listed in Sec. 1-2 the performance of the system is vastly improved by making the system operate closed-loop, i.e., by the addition of feedback.

1. A relatively poor amplifier with low input impedance, high output impedance, poor frequency response, and poor gain stability can have each of these factors improved with the addition of feedback.
2. The jumbo jet can attain its selected altitude comfortably and accurately without any overshoot or oscillation with the use of feedback.
3. A person starting a car hears the sound of the engine starting. This sound fed back by the ears to the brain signals the driver to release the key. (While driving the car the driver gets visual feedback on the speed and position of the car by looking at a speedometer and checking through the window.)
4. The air conditioner has a thermostat which provides feedback on the temperature of a room. When the room reaches the selected temperature, the thermostat senses it and switches the compressor off.
5. Were it not for position and velocity feedback, the rocket ship would have a difficult time reaching the moon.

In summary, feedback improves system accuracy, frequency response, and stability and provides for self-monitoring; however, although feedback can stabilize a system, in some instances it may cause a system to become unstable, as will be shown in a later chapter.

1-4 EXAMPLES OF FEEDBACK CONTROL SYSTEMS

One of the most common applications of feedback occurs in everyday life. The toilet-bowl flushing mechanism is a perfect example of feedback. After every flush, the tank automatically fills up with water to a predetermined level and then shuts off without overflowing. Figure 1-3 is a schematic diagram of the mechanism. The command to fill the tank comes from the main water-inlet valve and is always present. When the tank is empty, as shown in Fig. 1-3*a*,

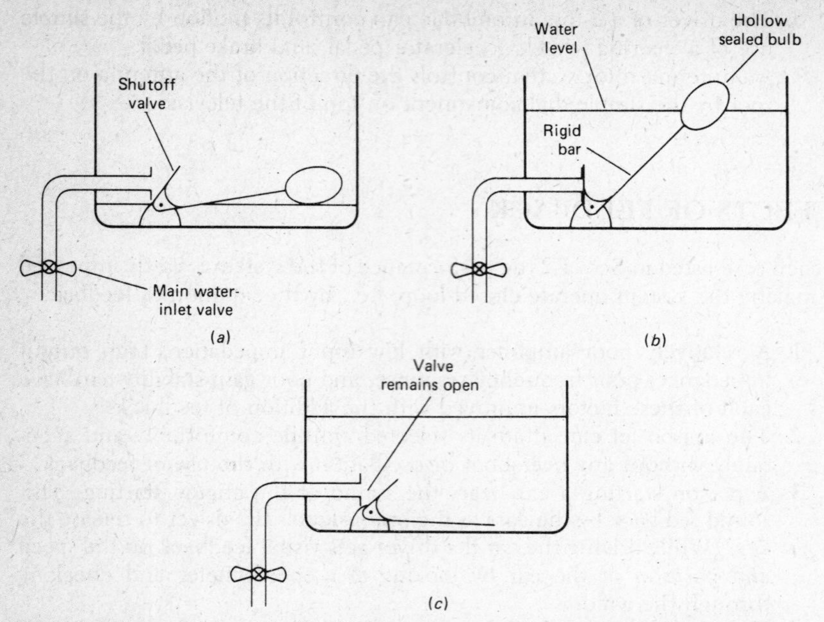

Fig. 1-3 Toilet-tank schematic simplified: (*a*) empty, (*b*) full, (*c*) with bulb (feedback mechanism) removed.

water flows freely into the tank. As the water fills the tank, its level is sensed and fed back to the shutoff valve by the floating bulb and rigid bar. When the tank is full (Fig. 1-3*b*), the rising bulb causes the shutoff valve to seal off the inlet pipe, thus preventing any more water from entering the tank. If the feedback (bulb) is removed, as in Fig. 1-3*c*, the rigid bar will not rise, the shutoff valve will not close, and water will continue to flow indefinitely, flooding the surrounding area.

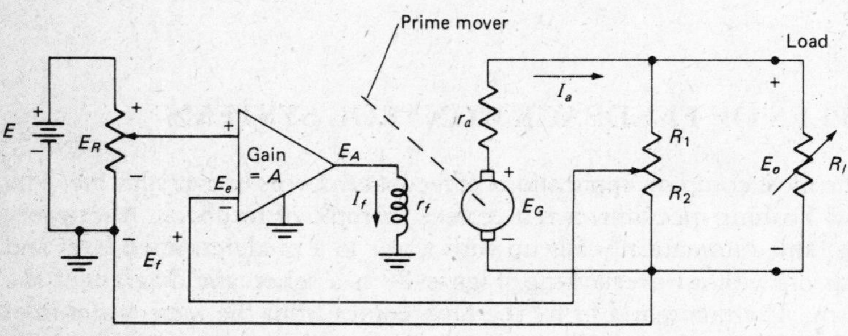

Fig. 1-4 Voltage-regulator system.

Figure 1-4 shows a simplified voltage-regulator control system. As the reference voltage E_R is dialed in with a potentiometer, an error signal E_e appears at the input of the amplifier. After the error is amplified, the output of the amplifier is applied across the field winding of a dc generator. Since the generator is being driven by a prime mover and due to E_A has a field current supplied, it generates a voltage E_G. A feedback signal E_f is supplied as a means of sensing the output E_o. In order to examine the operation more accurately, let us look at the system equations

$$E_e = E_R - E_f \tag{1-1}$$

$$E_A = AE_e = A(E_R - E_f) \tag{1-2}$$

$$E_G = KI_f = \frac{KE_A}{r_f} = K'E_A \tag{1-3}$$

where K is a generator constant

$$E_o = \frac{R_l}{R_l + r_a} E_G \qquad \text{where } R_1 + R_2 \gg R_l \tag{1-4}$$

Since r_a is generally small compared with R_l,

$$E_o \approx E_G \tag{1-5}$$

$$E_f = \frac{R_2}{R_1 + R_2} E_o = hE_o \qquad h = \frac{R_2}{R_1 + R_2} \tag{1-6}$$

When Eqs. (1-1) to (1-6) are combined, we get the system equation

$$E_o = K'A(E_R - E_f) \tag{1-7}$$

and $E_o = K'A(E_R - hE_o)$ (1-8)

Solving for E_o gives

$$E_o = \frac{K'AE_R}{1 + K'Ah} \tag{1-9}$$

If $K'Ah \gg 1$, then

$$E_o = \frac{E_R}{h} = \frac{\text{input}}{\text{feedback gain}} \tag{1-10}$$

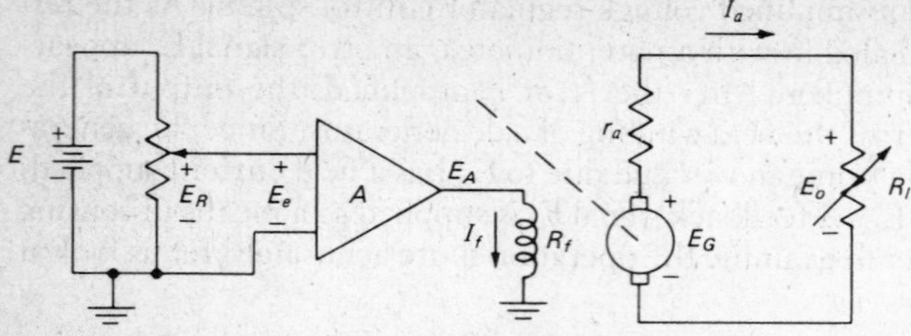

Fig. 1-5 Voltage-regulator system with feedback removed.

Equation (1-10) states that the output is strictly a function of the input or reference voltage and the feedback gain h. It can be expressed as a general equation simply stating that the output of any control system is equal to the input to the system divided by the feedback gain. This is an approximate equation and is true as long as the system loop gain (defined in Chap. 4) is much larger than 1.

If the feedback were not present, the system would be open-loop and the output would be given by

$$E_o = K'AE_R \tag{1-11}$$

which means that the output is now dependent on the generator constant and amplifier gain, which have higher tolerance variations than the feedback factor h (see Fig. 1-5).

A simple position control system is shown in Fig. 1-6. Electrical connections are drawn with solid lines, and mechanical connections are shown as dashed lines. A position command is dialed into the system. The input potentiometer converts this mechanical motion into an electric signal $V_{\theta c}$ representative of the command. A gear connection on the output shaft transfers the output angle θ_o to a shaft feeding back the sensed output θ_f. The feedback shaft is connected to a potentiometer; this converts θ_f into an electric signal $V_{\theta f}$, which thus represents the output variable θ_o. The error amplifier amplifies

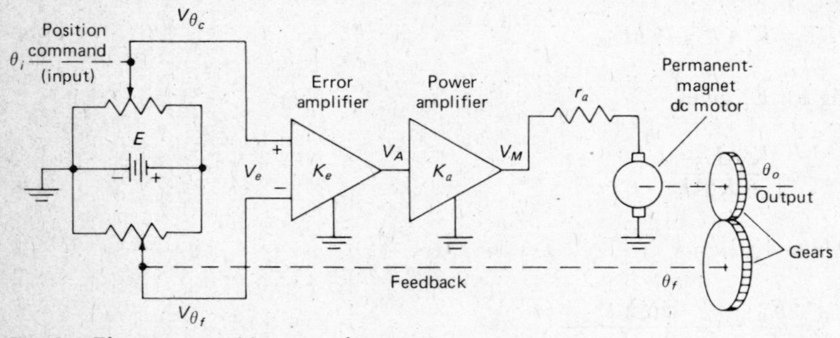

Fig. 1-6 Elementary position control system.

the difference V_e between V_{θ_c} and V_{θ_f}. The amplified error signal V_A is fed into a power amplifier which drives a dc motor. The motor turns until the output shaft is at a position such that V_{θ_f} is equal to V_{θ_c}. When this situation is reached, the error signal V_e, or the difference between V_{θ_c} and V_{θ_f}, is zero. The outputs of the amplifiers are thus zero, and the motor stops turning. Writing the system equations is left as an exercise for the student (Prob. 1-5).

In reality the error signal may not be driven to exactly zero, or the output shaft may overshoot the commanded position. Depending on the system gains and the rate at which the input is applied, different system performance is attained. This text presents various techniques for determining a system's performance and improving it.

1-5 FEEDBACK CONTROL SYSTEMS AND THE TECHNOLOGIST

The question may be asked: Why should a technologist be taught a sophisticated subject like feedback control systems? The answer is "to train a more complete technologist." In practically any job technologists come in contact with feedback control systems and should be able to

1. Understand the basic concept of control systems
2. Read and understand system block diagrams
3. Relate the block diagrams to system components
4. Be able to determine whether a system is operating as specified
5. Be able to isolate the malfunctioning components, if the system is not operating properly

PROBLEMS

1-1. What are the two basic types of control systems, and what is the difference between them?

1-2. Name any physical variables that might be controlled by the use of a feedback control system.

1-3. What benefits are gained by the use of feedback?

1-4. In Fig. 1-4 what is the purpose of the voltage divider determined by R_1 and R_2?

1-5. Derive an equation which relates θ_o to θ_i for the system in Fig. 1-6. *Hint:*

$$V_{\theta_c} = K_p\theta_i \qquad V_{\theta_f} = K_p\theta_f \qquad \theta_f = a\theta_o \qquad \text{and} \qquad \theta_o = K_T V_M$$

where K_p = the potentiometer gain
a = gear ratio
K_T = steady-state gain of motor

1-6. What is the feedback gain in Fig. 1-6?

chapter 2

Servo Components

Since a servo system in many cases controls energy in one form by using energy in another form, it becomes necessary to use special servo components called *transducers*. A transducer is a device which converts a signal from one form of energy into another form of energy; i.e., a transducer takes an input signal (electrical, mechanical, temperature, pressure, etc.) and produces a signal in another form. Since control systems must be accurate, the components that constitute them, including transducers, must also be highly accurate. A transducer must also be reliable.

2-1 POTENTIOMETER

A potentiometer is a device which converts a position signal, usually rotational, into an electrical signal. Figure 2-1 shows a single-turn potentiometer. As the shaft of the potentiometer is turned, the arm, or wiper, (shown as an arrow in Fig. 2-1) makes contact with a resistive element. As the resistance between the arm and one of the other terminals changes, the voltage between the arm and that terminal also changes.

Figure 2-2 is a schematic of the potentiometer. The input to the potentiometer is shown as a dashed line. The output is the voltage e_o. If the potentiometer is of the single-turn type, θ_i can vary from 0 to θ_{max}, where θ_{max} is slightly less than 360° due to mechanical limitations. The input-output relationship can be written

$$e_o = \frac{\theta_i}{\theta_{max}} E_R \tag{2-1}$$

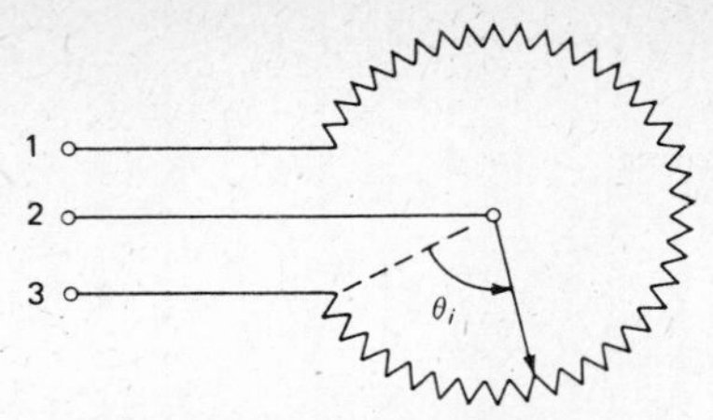

Fig. 2-1 Single-turn potentiometer.

$$\text{or} \quad e_o = \frac{E_R}{\theta_{max}} \theta_i \tag{2-2}$$

$$\text{Output} = \text{gain} \times \text{input}$$

A dimensional analysis of Eq. (2-2) shows that the units of the input θ_i are degrees or radians, the gain E_R/θ_{max} has units of volts per degree (or volts per radian), and the output e_o has units of volts. For a given reference voltage E_R, the gain can be varied by changing θ_{max}. That is, a two-turn potentiometer would have a θ_{max} slightly less than 720°, a three-turn potentiometer would have a θ_{max} less than 1080°, etc.

Because of the way potentiometers are made, the resistance does not vary precisely with the input angle θ_i but approximates a linear relationship. This effect can be seen on a plot of output voltage vs. input angle. Figure 2-3 shows the ideal characteristic, the actual one, and dashed lines which represent the tolerance limits. The tolerance band shown is specified by the linearity of a potentiometer. *Linearity* is the maximum deviation in resistance that a potentiometer will have at any setting from the theoretical value and is given as a percent of the maximum value.

The slope of the graph in Fig. 2-3 represents the *gain* of the potentiometer

$$\text{Gain} = \frac{\text{change in output}}{\text{change in input}} = \frac{\Delta e_o}{\Delta \theta_i} \tag{2-3}$$

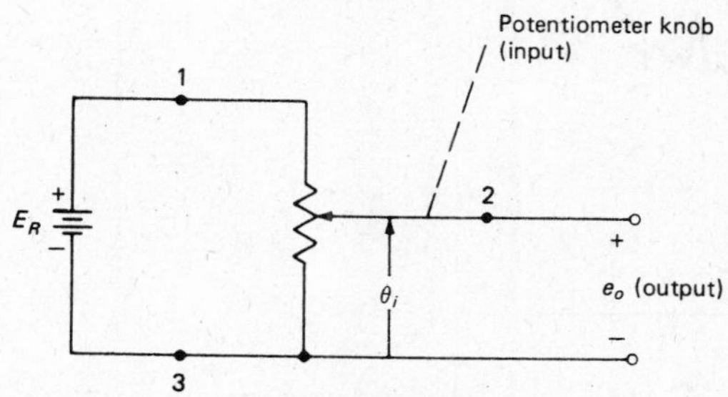

Fig. 2-2 Schematic of potentiometer.

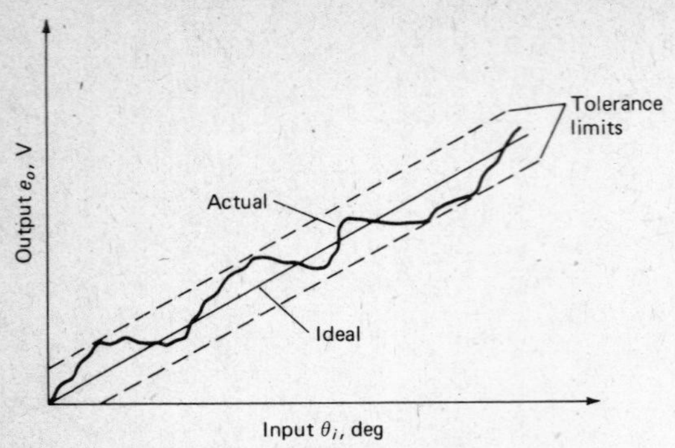

Fig. 2-3 Potentiometer output vs. input.

With increased use, low-quality materials, and poor construction, the characteristic shown in Fig. 2-3 can deteriorate into the one shown in Fig. 2-4, which is noisy and has discontinuities. Points in the characteristic marked "open" represent positions where the arm loses contact with the resistive element, resulting in no output signal. An "open" in a potentiometer means that at that setting the potentiometer has zero gain. If the potentiometer is in the feedback path, a loss of feedback will occur. In some systems this can cause a nonlinear oscillation called a *limit cycle*.

An enlarged view of the characteristic in Fig. 2-3 appears as a stepping pattern in some potentiometers. In wire-wound potentiometers this is caused by

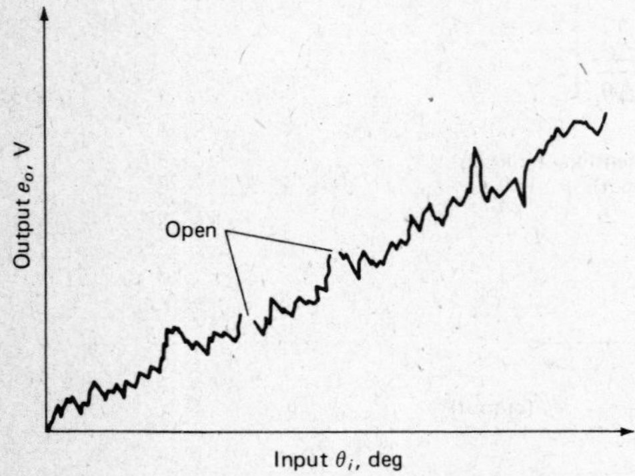

Fig. 2-4 Poor potentiometer characteristic.

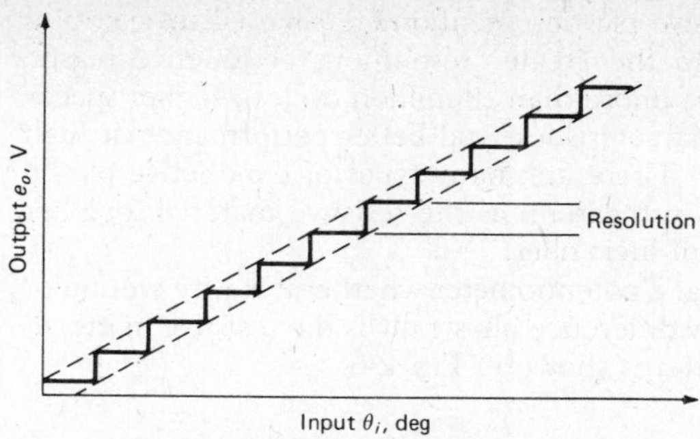

Fig. 2-5 Resolution of potentiometer.

the arm making contact with one or more turns at a time.* This stepping characteristic is shown in Fig. 2-5.

Obviously a multiturn potentiometer with many turns of wire would have a much finer resolution than a single-turn potentiometer. If an almost infinite

* Although this phenomenon does not occur in conductive plastic potentiometers, wire-wound potentiometers are preferred for some applications where resistance stability is very important or where nonlinear functions are desired. Cost reductions are achieved in many of the nonlinear functions desired in wire-wound potentiometers.

Fig. 2-6 A wire-wound potentiometer. (*Courtesy New England Instrument Corporation.*)

resolution is wanted, a conductive plastic potentiometer is used instead of a wire-wound one. In addition to the greater resolution, conductive plastic potentiometers offer a longer life (more than 10 million cycles), higher speeds (up to 1000 r/min), better ac characteristics, and better performance at high frequencies (megahertz region). There are many types of conductive plastic potentiometers, but all of them use carbon as the resistive material together with a plastic resin binder and an inert filler.

One cannot tell by looking at a potentiometer whether it is wire-wound or conductive plastic since the only difference physically is the resistive material, which is internal. A potentiometer is shown in Fig. 2-6.

Example 2-1

A two-turn, 200-winding, 10-kΩ wire-wound potentiometer has a reference of 20 V impressed across its terminals. Assume that it turns a full 720°. Find the potentiometer constant or gain in

(a) Volts per degree
(b) Volts per winding (resolution)
(c) Volts per radian
(d) Volts per turn

Solution

(a) $\text{Gain} = \dfrac{E_R}{\theta_{\max}} = \dfrac{20}{720} = 0.028 \text{ V/deg}$

(b) $\text{Gain} = \dfrac{E_R}{200 \text{ windings}} = \dfrac{20 \text{ V}}{200 \text{ windings}} = 0.1 \text{ V/winding}$

(c) $\text{Gain} = \dfrac{E_R}{(2 \text{ turns})(2\pi \text{ rad/turn})} = \dfrac{20}{2(2\pi)}$

$= \dfrac{20}{4\pi} = 1.59 \text{ V/rad}$

(d) $\text{Gain} = \dfrac{E_R}{2 \text{ turns}} = \dfrac{20}{2} = 10 \text{ V/turn}$

The gain of a potentiometer can vary slightly due to its linearity; however, large gain variations will be obtained when loading occurs. *Loading* is caused by connecting to the arm of the potentiometer a load whose resistance is the same order of magnitude as that of the potentiometer (See Example 2-2).

Example 2-2

A 10-turn 50-kΩ potentiometer with 1 percent linearity uses a 40-V supply.

(a) Find the potentiometer constant in volts per turn.
(b) Find the range of voltages at the midpoint setting due to the linearity of the potentiometer. Assume that the potentiometer is unloaded.
(c) Assuming that the potentiometer is perfectly linear, find the voltage at the midpoint when the potentiometer is loaded with 500 kΩ.
(d) Repeat part (*c*) for a 25-kΩ load.

Solution

(a) $K_p = \dfrac{40 \text{ V}}{10 \text{ turns}} = 4 \text{ V/turn}$

(b) Due to linearity of 1 percent, at midpoint the resistance between the arm and each end terminal will be 25 kΩ ± [0.01(50 kΩ)]. Therefore, the range of voltage at the midpoint using voltage division is

$$V_{max} = \frac{25.5 \text{ k}\Omega}{50 \text{ k}\Omega}(40 \text{ V}) = 20.4 \text{ V}$$

$$V_{min} = \frac{24.5 \text{ k}\Omega}{50 \text{ k}\Omega}(40 \text{ V}) = 19.6 \text{ V}$$

This result could also be obtained by the following method. The nominal voltage at the midpoint would be $\frac{1}{2}V_{max}$:

$$V_{nom} = \tfrac{1}{2}V_{max} = \tfrac{1}{2}(40) = 20 \text{ V}$$

$$\text{Range} = V_{nom} \pm 0.01V_{max} = 20 \pm 0.01(40) = 20 \pm 0.4 \text{ V}$$

(c) Since 500 kΩ ≫ 50 kΩ, the voltage at the midpoint would be one-half of 40 V or 20 V. At the midpoint the potentiometer has gone through five turns; therefore, using the gain found in part (a), we get

$$V = (5 \text{ turns})(4 \text{ V/turn}) = 20 \text{ V}$$

(d) If a 25-kΩ load is connected to the potentiometer, the equivalent circuit is as shown in Fig. 2-7. The effective resistance between the arm and ground is now 25 kΩ‖25 kΩ, or 12.5 kΩ. Using voltage division, we find the output voltage to be

$$e_o = \frac{12.5 \text{ k}\Omega}{12.5 \text{ k}\Omega + 25 \text{ k}\Omega}(40 \text{ V}) = 13.33 \text{ V}$$

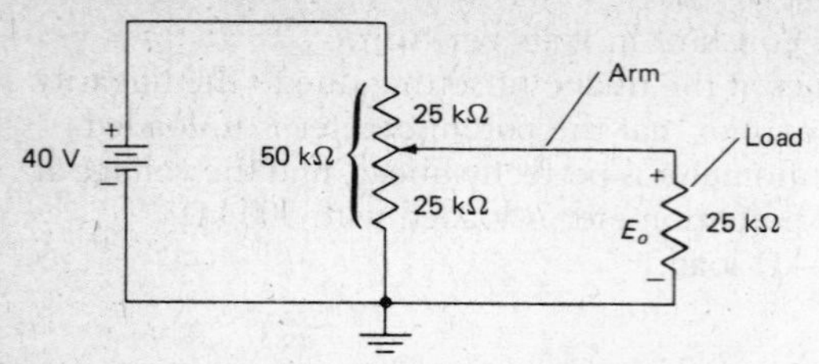

Fig. 2-7 Equivalent circuit for part (*d*) of Example 2-2.

This means that the gain of the potentiometer is now

$$\frac{13.33 \text{ V}}{5 \text{ turns}} = 2.67 \text{ V/turn}$$

as opposed to its unloaded value of 4 V/turn.

Example 2-3

A potentiometer is a 1-turn pot with a resistance of 1 kΩ.

(a) If the measured resistance at its midpoint is 510 Ω, what is its linearity?
(b) If the measured resistance at its quarter point is 225 Ω, what is its linearity?

Solution

$$\text{(a) Linearity} = \frac{\text{deviation from nominal}}{\text{maximum}} \times 100\%$$

$$= \frac{10}{1000} \times 100\% = 1\%$$

$$\text{(b) Linearity} = \frac{25}{10{,}000} \times 100\% = 2.5\%$$

2-2 SYNCHROS

Synchro is a general term (the trade names Selsyn, Autosyn, Telesyn, and others are also used) which encompasses a family of several components, all fairly similar in appearance but differing in construction and operation.

Synchro Generator (Transmitter). This device is different from a normal dc generator (Sec. 2-3) in that it produces a voltage which is a function of its rotor angle and not rotor speed. Schematically it can be represented as in Fig.

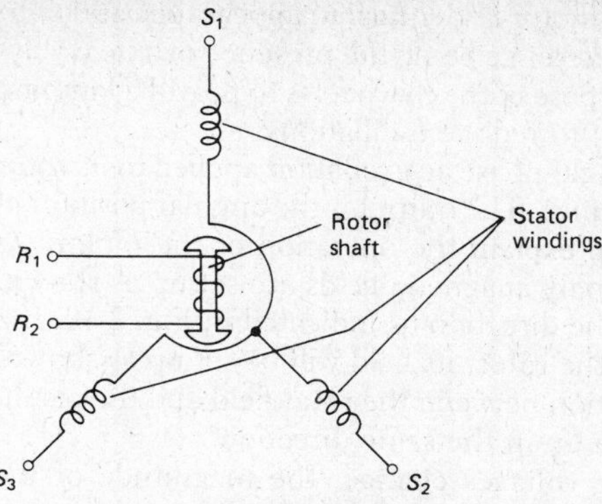

Fig. 2-8 Schematic of synchro generator.

2-8. The generator has two inputs and one output. One input is the rotor shaft position, which is varied to control the output. The other input is an ac voltage applied to the rotor terminals and is generally kept constant. The output consists of voltages induced on the stator windings S_1, S_2, and S_3. Functionally it works as follows. The ac excitation applied to the rotor sets up an alternating magnetic field whose direction is determined by the rotor's position. This rotor field induces a voltage in each of the stator windings through transformer action. The magnitude of the induced voltages depends on the position of the rotor with respect to the stator windings. Thus, a different voltage is induced in each stator winding. If a load is connected to the stator windings, permitting a current to flow, a magnetic field due to the stator windings will always be equal and opposite to the rotor field. Figure 2-9 may help clarify the magnetic field orientations.

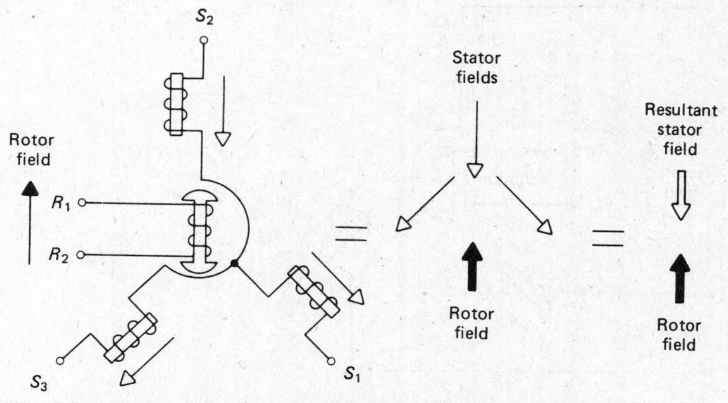

Fig. 2-9 Structure of the magnetic field of synchro generator.

Synchro Motor (Receiver). The motor is identical in appearance and construction to the generator, the sole difference being the presence of a heavy flywheel mounted on the shaft. The purpose of the flywheel is to provide damping to the motion of the rotor shaft, i.e., to reduce oscillations.

The inputs to the motor are a single-phase ac excitation applied to its rotor and voltages applied to its stator windings. The output is the angular position of the rotor shaft. Figure 2-10 will help explain the operation of the motor. As voltage is applied to the stator terminals, magnetic fields are set up as shown. The resultant stator field will be in the direction as indicated in Fig. 2-10. If a single-phase excitation is applied to the rotor, its field will be set up as shown. There will be a strong force of attraction between the rotor field and the resultant stator field, causing them to line up in the same direction.

If the magnitudes of the stator voltages change, the magnitude of the stator currents will change. The resultant stator field will change direction, causing the rotor field to follow it. As the rotor field follows the resultant stator field, it causes the rotor or motor shaft to turn. The rotor's motion is damped by the flywheel.

Differential Synchro. The differential generator and differential motor are similar to the synchro generator and motor except that the differential synchros have three windings on their rotors as opposed to the single winding on the rotors of the generator and motor described above.

The schematic of Fig. 2-11 applies to both differential generator and motor. In a differential generator voltage is applied to the stator windings. As the rotor is turned, a varying voltage is generated at the rotor terminals. In the differential motor voltages are applied to both stator and rotor. The motor's

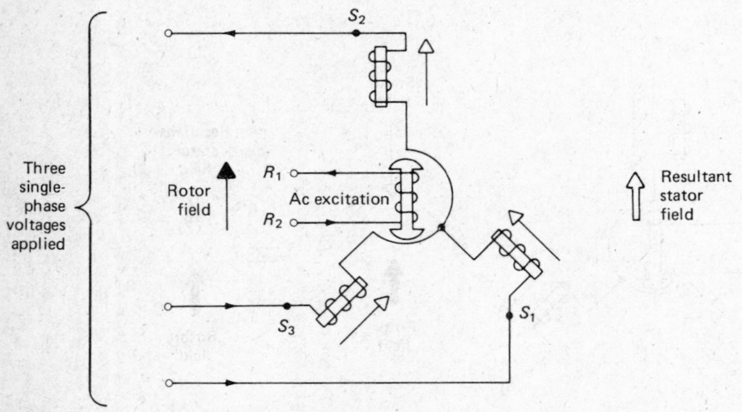

Fig. 2-10 Schematic of a synchro motor.

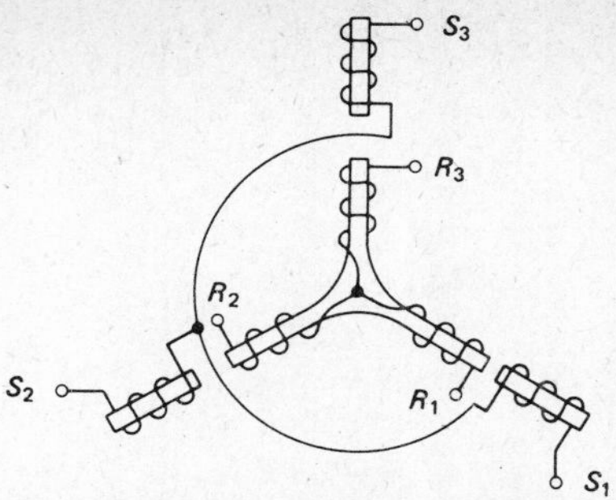

Fig. 2-11 Schematic of a differential synchro.

rotor takes a position depending on the resultant magnetic fields of its stator and rotor.

Control Transformer. The control transformer is a synchro whose output is an electrical signal taken from the rotor. The two inputs are the voltages applied to the stator and the angular position of the rotor. It is similar in appearance to the generator with two exceptions. The stator windings are made up of many more turns of finer wire than in the generator. This gives the control transformer a high input impedance, and so it draws very little current. The other difference is the shape of the rotor winding. The rotor is wound completely around the shaft, so that when a current passes through the rotor winding, there is no net magnetic field developed to interact with the stator field. Hence there is no torque present to oppose the motion of the rotor. The

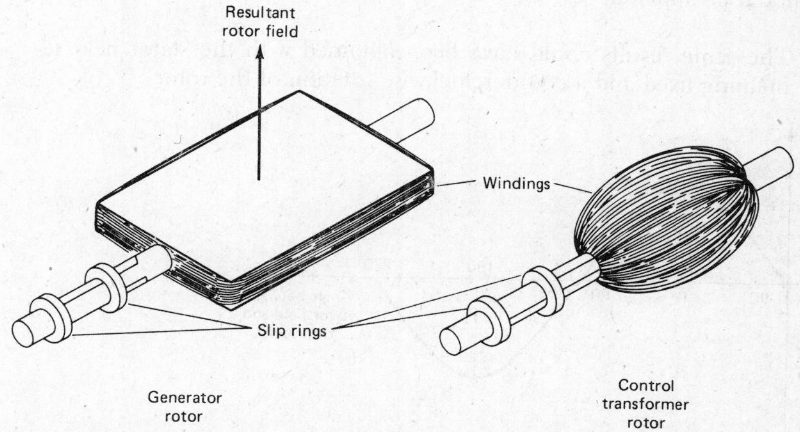

Fig. 2-12 Comparison of rotors of a synchro generator and control transformer.

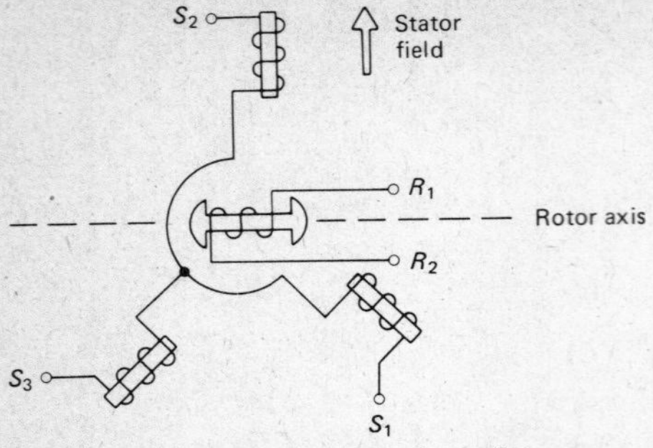

Fig. 2-13 Schematic of a control transformer.

only torque is that which is applied to the rotor by the mechanical system connected to it. Figure 2-12 compares the rotors of a generator and a control transformer.

While the synchro generator is used primarily as a transducer, i.e., output proportional to input, the control transformer can give the sum or difference of a mechanical and electric signal. Assume in Fig. 2-13 that the resultant stator field is vertical, as shown. In the position shown, the rotor axis is horizontal, and so it lies at right angles to the stator field and there is no induced voltage in the rotor. If the stator field rotates, say 30° clockwise, a voltage will be induced across the rotor terminals. As the stator field rotates farther clockwise, the induced voltage at the rotor terminals also increases to a maximum value when the angle between the stator field and rotor axis is 0°. Figure 2-14 shows the output as a function of the angle between stator field and rotor axis. The following facts should be apparent:

1. The same results could have been obtained with the stator field remaining fixed and a counterclockwise rotation of the rotor.

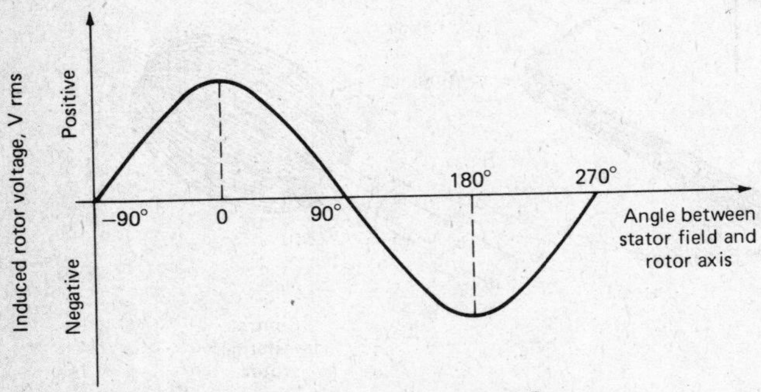

Fig. 2-14 Output characteristic of a control transformer.

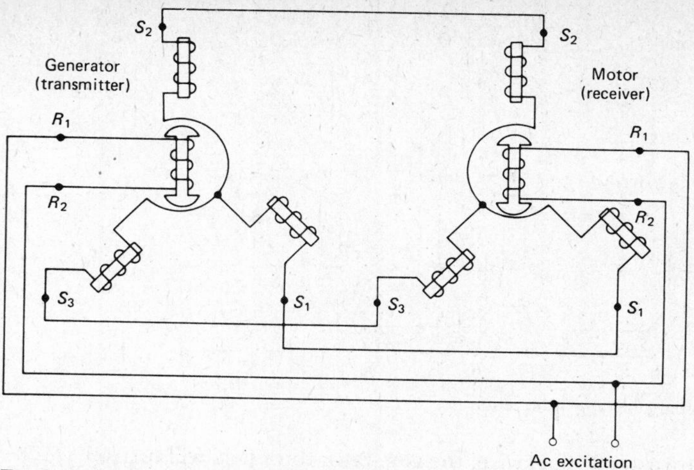

Fig. 2-15 Synchro generator-motor hookup.

2. After rotating the stator field 30° clockwise, the induced rotor voltage could have been brought to zero by rotating the rotor 30° clockwise.
3. The induced rotor voltage is a function of the angle between the stator field and rotor axis.

2-2-1 Interconnection of Synchro Components

Synchro Generator-Motor. Figure 2-15 shows a generator-motor connection. When the ac excitation is applied to the generator rotor, transformer action sets up a field on its stator, inducing a voltage in S_1, S_2, and S_3. A current thus flows through the motor stator windings, setting up a stator field. When the ac excitation is applied to the motor's rotor, a rotor field is set up. This rotor field is attracted by the motor stator field, causing the two to line up. If the gen-

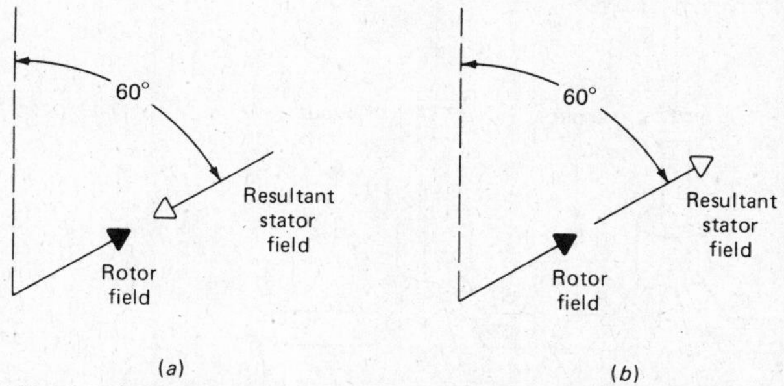

Fig. 2-16 Magnetic field orientation for a 60° clockwise rotation: (*a*) generator and (*b*) motor.

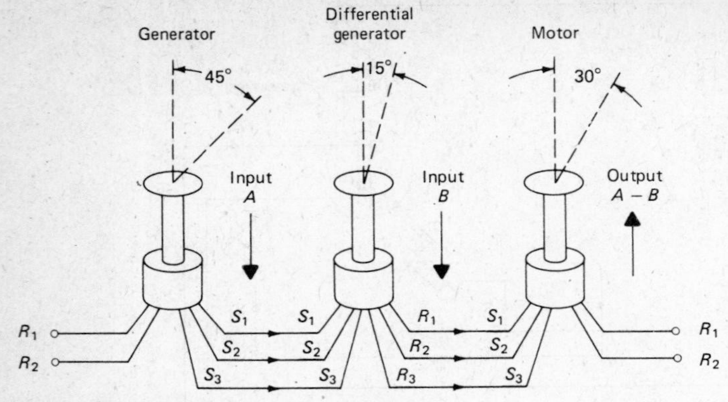

Fig. 2-17 Differential generator subtraction.

erator rotor is turned 60° clockwise, the generator rotor field will turn 60° clockwise. The resultant generator stator field will also rotate 60° clockwise, thus rotating the resultant motor stator field. As the motor stator field rotates, it attracts the motor rotor field, thus turning the rotor. See Fig. 2-16 for the magnetic field diagram.

Synchro Differential. Figure 2-17 shows a schematic of a differential generator used for subtracting two angles. In order to add the two inputs the stator leads S_1 and S_3 and the rotor leads R_1 and R_3 should be interchanged, as shown in Fig. 2-18. Figure 2-19 shows the configuration for differential motor subtraction. In order to add two angles, leads R_1 and R_3 of the differential motor should be interchanged, as shown in Fig. 2-20.

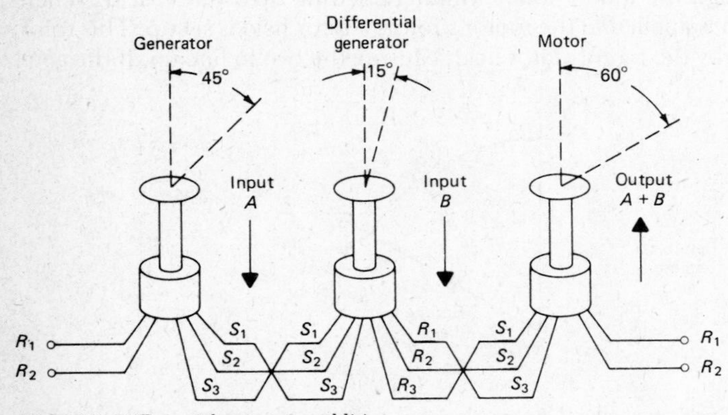

Fig. 2-18 Differential generator addition.

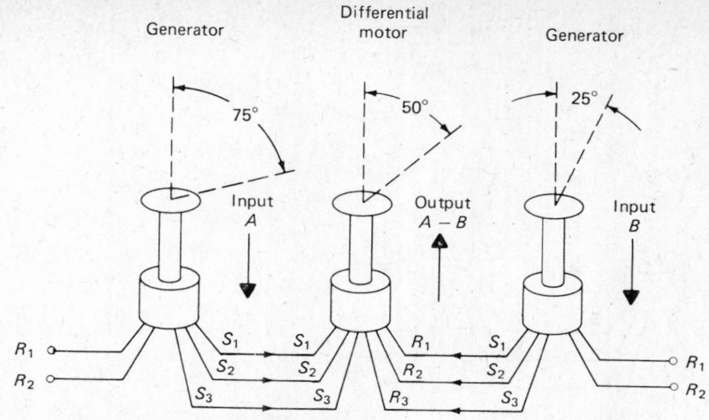

Fig. 2-19 Differential motor subtraction.

Control Transformer-Generator Connection. In the previous discussion, a differential synchro was used to obtain the sum or difference of two angular displacements. The output of the differential was also an angular displacement. In many control-system applications it is necessary to express the difference of two angular displacements as an electrical signal rather than another mechanical signal. Referring to Fig. 2-17, we see that it would be possible to connect a generator to the shaft of the motor and thus provide the needed electrical output. However, it would be necessary to use four synchros to accomplish this feat. The control transformer enables us to do this with the use of two synchros.

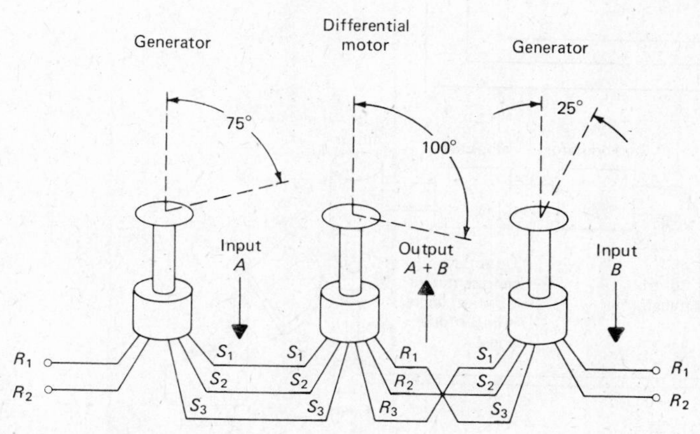

Fig. 2-20 Differential motor addition.

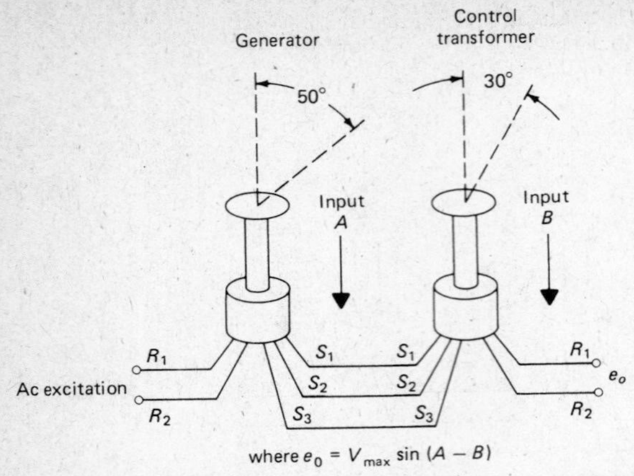

Fig. 2-21 Control transformer-generator hookup.

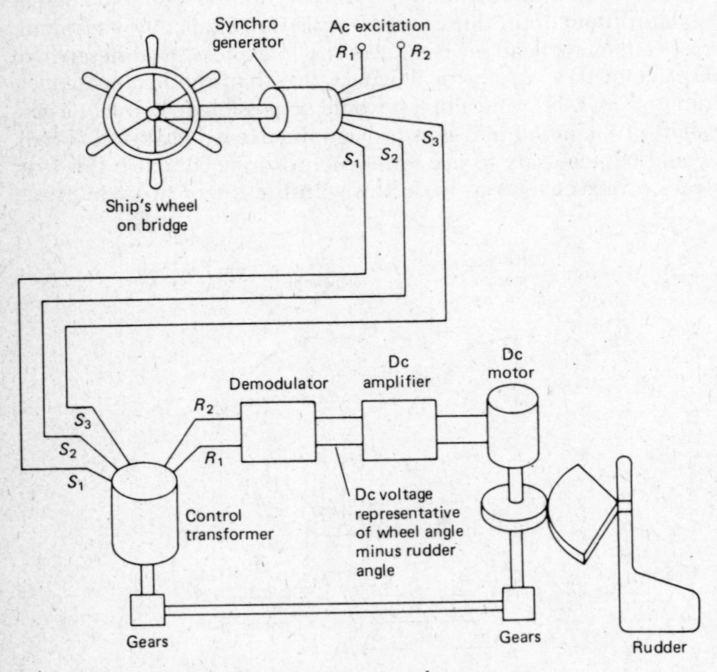

Fig. 2-22 Simplified diagram of rudder-control system.

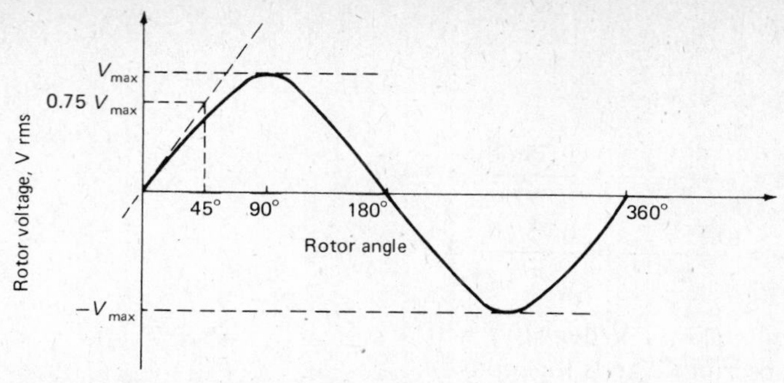

Fig. 2-23 Control-transformer rotor voltage vs. rotor angle.

Figure 2-21 is a schematic of a control transformer-generator hookup. The rms value of the induced voltage e_o is a function of the difference between the two angles A, and B. In many instances, this output signal is demodulated. In effect this is a means of converting the ac to an equivalent dc signal.

Figure 2-22 is a simplified diagram of a ship's rudder-control system using the synchro control transformer-generator team.

Synchros come in many sizes for many torque requirements. They are relatively free of friction and are quite reliable. There is one drawback, but it can be overcome. As can be seen from Fig. 2-14, the output voltage varies sinusoidally rather than linearly with input angle. Therefore in order to maintain linear gains in a control system using synchros, synchro operation is usually kept to the range of angles between -45 and $+45°$, where the sine wave can be approximated by a straight line. (See Fig. 2-23.)

The linear range of operation can be extended through the use of gears, as described in Sec. 2-6. The approximate gain of the synchro in the linear region can be obtained by making a straight-line approximation of the sine wave between 0 and 45°. As can be seen from Fig. 2-23, the gain is given by

$$\text{Gain} = \frac{\text{change in output}}{\text{change in input}} = \frac{0.75V_{max}}{45°}\ \text{V/deg} = \frac{0.75V_{max}}{\pi/4}\ \text{V/rad} \qquad (2\text{-}4)$$

Example 2-4

A control transformer has the characteristic given in Fig. 2-23 with $V_{max} = 60$ V.

(a) What is the approximate gain of the synchro in its linear region?
(b) Using this gain, calculate the output when the rotor is turned to a 10° position.
(c) What is the actual output using the sine wave to determine it?

Solution

(a) Using Eq. (2-4), we have

$$\text{Gain} = \frac{0.75V_{\max}}{45^\circ} = \frac{0.75(60)}{45} = 1 \text{ V/deg}$$

$$\text{or} \quad \text{Gain} = \frac{0.75V_{\max}}{\pi/4} = \frac{0.75(60)}{\pi/4} = 57.3 \text{ V/rad}$$

(b) $e_o = \text{Gain} \times \theta_i = (1 \text{ V/deg})(10^\circ) = 10 \text{ V}$
where subscript i stands for input.

(c) $e_o = V_{\max} \sin\theta = 60 \sin\theta = 60 \sin 10^\circ = 60(0.17) = 10.42 \text{ V}$

2-3 DC GENERATOR

The dc generator is more commonly used as a control element than as a transducer. The generator works on the principle that if a conductor is moved through a magnetic field, a voltage will be induced across the conductor (a simplified statement of Faraday's law). The induced voltage is directly proportional to both the strength of the magnetic field and the speed at which the conductor cuts through the field. The equation describing the generator output is

$$e_g = k\phi\omega \quad \text{V} \tag{2-5}$$

where k = generator constant
ϕ = field strength
ω = speed of conductor

In some applications, the field is constant, in which case the output e_o is strictly a function of the speed ω, which is the input. In other applications the speed is held constant while the field strength is varied. In this case the actual input to the generator would be a voltage. Applied across the field winding of the generator, this voltage gives rise to a current through the winding, which in turn creates and controls the field strength.

Figure 2-24 is a schematic of a separately excited dc generator. The actual output e_o differs from the generated voltage e_g by an amount equal to the voltage drop across the generator's armature, $i_a r_a$.

$$e_o = e_g - i_a r_a \tag{2-6}$$

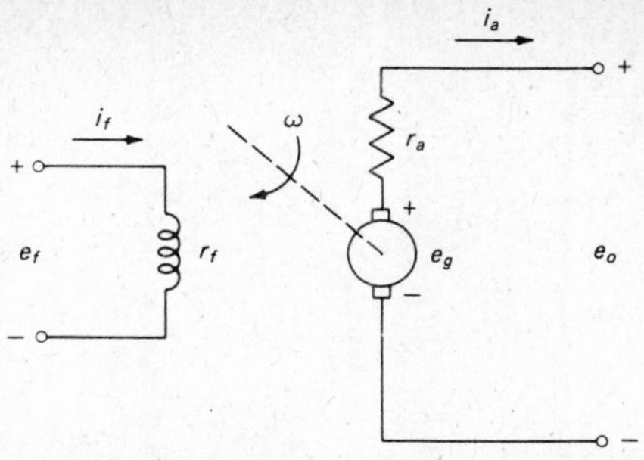

Fig. 2-24 Schematic of a separately excited dc shunt generator.

2-4 TACHOMETER

There are many types and principles of operation of tachometers. Perhaps the most common is the *dc tachometer generator*. A tachometer is a transducer which converts mechanical rotation into an electrical signal. It finds most use in position and velocity control systems. In velocity control systems a tachometer provides the basic feedback signal. In position control systems a tachometer provides rate feedback for velocity damping. A dc tachometer generator is in essence a generator. It has a wound armature and permanent magnet providing a constant field. When its input shaft is rotated, it generates a dc voltage

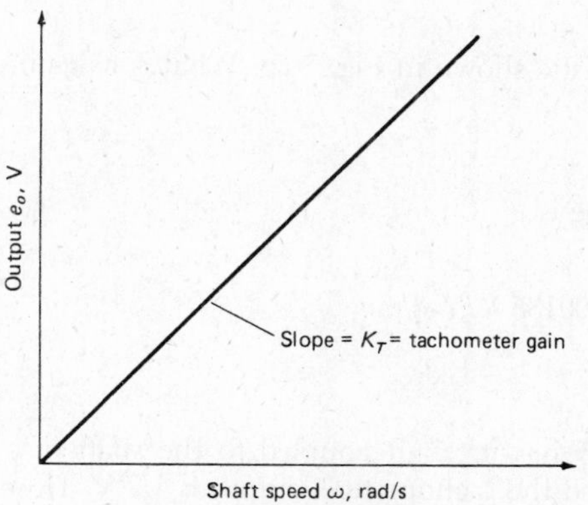

Fig. 2-25 Output characteristic of a dc tachometer.

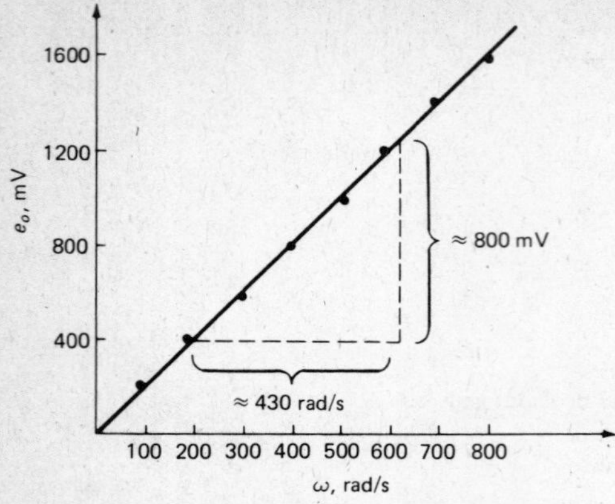

Fig. 2-26 Tachometer characteristic for Example 2-5.

proportional to the shaft speed. The tachometer gain, or *sensitivity*, is defined as the voltage out divided by the shaft speed in radians per second:

$$K_T = \text{V/(rad/s)} \qquad \text{tachometer gain} \tag{2-7}$$

$$e_o = K_T \omega \tag{2-8}$$

where ω is the shaft speed in radians per second. Figure 2-25 shows the output characteristic of a dc tachometer.

Example 2-5

A tachometer has the characteristic shown in Fig. 2-26. What is its gain?

Solution

From the characteristic the slope is

$$K_T = \text{slope} = \frac{800 \text{ mV}}{430 \text{ rad/s}} = 0.00186 \text{ V/(rad/s)}$$

Example 2-6

The tachometer of Example 2-5 has its shaft coupled to the shaft of a motor. The motor is turning, and the tachometer's output is 1.7 V. How fast is the motor turning?

Solution

$$K_T = 0.00186 \text{ V/(rad/s)} \qquad \text{from Example 2-5}$$

$$\text{tachometer output } e_o = K_T\omega$$

$$\text{or} \quad \omega = \frac{e_o}{K_T} = \frac{1.7 \text{ V}}{0.00186 \text{ V/(rad/s)}} = 914 \text{ rad/s}$$

2-5 DC MOTOR

A dc motor is a device which provides a rotational output for an input dc voltage. It is probably the most important control-system component in use. Its operation is just the reverse of that of the dc generator. Because of its importance and wide use, it will be given a more thorough analysis. Remember that a good dc motor can be used either as a motor or as a generator.

The principle on which the motor's operation is based is the following. If a current passes through a conductor which is in a magnetic field, a force is exerted on the wire by the magnetic field.

The simplest and most popular dc motor is the permanent-magnet type, in which the magnetic field is provided by a permanent magnet as part of the motor. The motor also has a wire-wound armature axially mounted in the magnetic field. A simplified diagram is shown in Fig. 2-27 with only one turn of wire on the armature shown. When a voltage v_m, where the subscript stands for motor, is applied as shown, a current i_a flows in the direction shown. As a result of the current, a force is exerted on the wire. The force on the top wire is to the

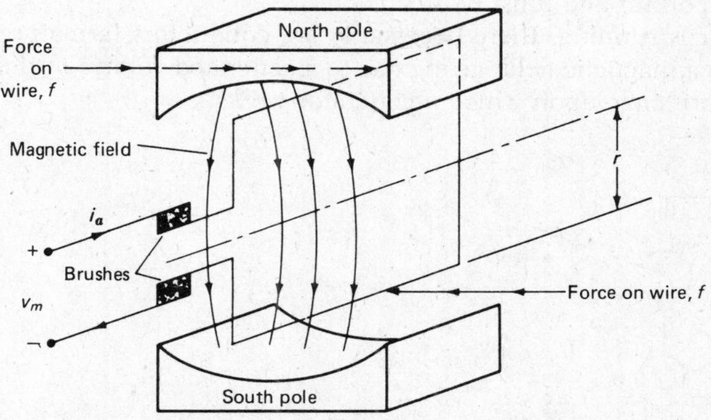

Fig. 2-27 Simplified diagram of a permanent-magnet dc motor.

right, and the force on the bottom wire is to the left. The result will be a clockwise rotation of the armature upon which the wire is mounted. If the applied voltage polarity or the magnetic field polarity were reversed, the rotation would be counterclockwise.

The product of a force and its radial distance is a *torque*. This is a basic definition in physics. Therefore, a torque is produced on the armature which is equal to the force f times the radial distance r.

Another basic law of physics states that a body's angular acceleration is equal to the applied torque divided by the body's inertia. Thus, the behavior of the motor can be described as follows:

1. An applied voltage causes a current to flow through an armature winding.
2. The current coupled with the magnetic field produces a force on the winding.
3. The force creates a torque on the armature which tends to accelerate the armature.

Obviously, larger voltage, larger current, smaller armature resistance, stronger magnetic field, and greater radial distance all result in a larger torque output; however, all these factors tend to increase the size and cost of the motor.

Figure 2-28 is a schematic of a permanent-magnet dc motor. Two parts have not been discussed, the motor's inductance l_m and the voltage generator v_b. The inductance is present because wire is wound around an iron core (the armature). However, since the electrical time constant (l_m/r_a) is very small compared with the mechanical time constant (to be derived in Chap. 4), it can be neglected. It has been introduced here only for completeness. The voltage source v_b is quite important and must be treated.

When the motor is running, there necessarily are conductors (armature windings) rotating in a magnetic field, giving rise to a generated voltage in the winding (this is the principle upon which a generator works).

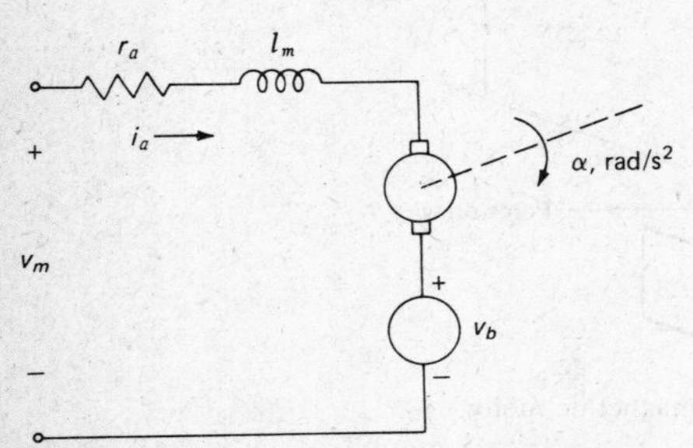

Fig. 2-28 Schematic of a permanent-magnet dc motor.

For a given motor, the generated voltage, called the *back electromotive force* (back emf), is directly proportional to the speed (ω rad/s) at which the motor turns. The polarity is such as to oppose the current which is making the motor turn. The magnitude of the back emf gradient (K_b) is a function of the number of turns of armature winding and the magnetic field strength.

Since the motor current produces a torque, there exists for every dc motor a torque gradient K_t which depends on the physical parameters of the motor. Referring to Fig. 2-28 and neglecting the motor's inductance l_m, we can write

$$v_b = K_b \omega \qquad \text{V} \tag{2-9}$$

$$t_{dev} = K_t i_a \qquad \text{oz} \cdot \text{in} \tag{2-10}$$

$$i_a = \frac{v_m - v_b}{r_a} \qquad \text{A} \tag{2-11}$$

where t_{dev} = motor-developed torque, oz · in
K_b = back emf gradient, V/(rad/s)
K_t = torque gradient, oz · in/A

The units chosen are merely illustrative and may differ in other motors. Substituting Eqs. (2-9) and (2-11) into (2-10) gives

$$t_{dev} = \frac{K_t v_m}{r_a} - \frac{K_t K_b \omega}{r_a} \qquad \text{oz} \cdot \text{in} \tag{2-12}$$

The equation can be simplified by calling K_t/r_a a motor constant K_m and $K_t K_b/r_a$ a new constant f_m. Thus, the motor-developed torque is

$$t_{dev} = K_m v_m - f_m \omega \tag{2-13}$$

The quantity $f_m \omega$ is sometimes referred to as *viscous-friction torque* since in Eq. (2-13) it appears as an opposing torque proportional to motor speed.

In the absence of extraneous torques, the accelerating torque t_{ac} equals the developed torque. If the total moment of inertia of the motor, including the rotor inertia and the shaft it is mounted on, is equal to J_m oz · in · s²/rad, the angular acceleration equals

$$\alpha = \frac{t_{ac}}{J_m} \qquad \text{rad/s}^2 \tag{2-14}$$

If an external load torque t_l is present, the net accelerating torque is equal to

$$t_{ac} = t_{dev} - t_l = K_m v_m - f_m \omega - t_l \tag{2-15a}$$

and if mechanical viscous friction $f_v \omega$ is present, it must also be subtracted from the developed torque before accelerating torque is obtained

Due to back emf — Due to mechanical friction

$$t_{ac} = t_{dev} - t_l - f_{v\omega} = K_m v_m - (f_m + f_v)\omega - t_l \tag{2-15b}$$

or $$t_{ac} = K_m v_m - F_m \omega - t_l \tag{2-15c}$$

where $F_m = f_m + f_v$. Solving for the load torque in the steady state, i.e., when the acceleration and hence t_{ac} is zero, we get

$$t_l = K_m v_m - F_m \omega \tag{2-16}$$

Equation (2-16) can be plotted as a *torque-speed curve*. It is shown in Fig. 2-29. The quantity $K_m v_m$ is called the motor stall torque t_s, that is, the value of load torque for which the motor will stall. The quantity $K_m v_m / F_m$ is the motor's no-load speed ω_{nl}. It is the speed at which the motor will run when there is no opposing load torque. Figure 2-30 shows the effects of different applied motor voltages on the torque-speed curve.

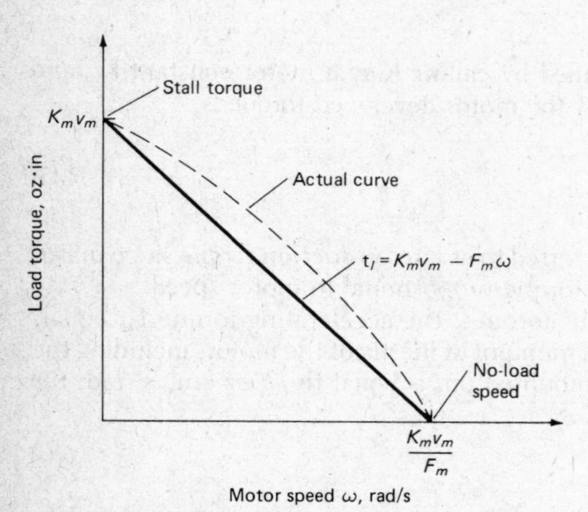

Fig. 2-29 Torque-speed curve of motor.

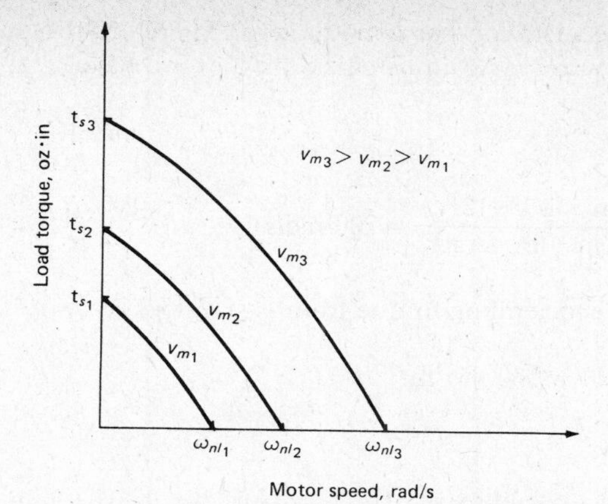

Fig. 2-30 Torque-speed curves for different motor voltages.

Example 2-7

A motor has the following parameters, and v_m is applied at $t = 0$:

$$K_m = 20 \text{ in} \cdot \text{lb/V},\ v_m = 12 \text{ V},\ F_m = 1.2 \text{ in} \cdot \text{lb} \cdot \text{s/rad}$$
$$J_m = 2 \text{ in} \cdot \text{lb} \cdot \text{s}^2/\text{rad},\ t_l = 0 \text{ (no load)}$$

Find:

(a) The acceleration at $t = 0^+$
(b) The motor's steady-state speed
(c) The acceleration when the speed is 50 rad/s
(d) Repeat parts (a) and (b) for a constant-load torque t_l of 80 in · lb connected to the shaft

Solution

(a) At $t = 0$ the motor's speed is zero; therefore from Eq. (2-15c)

$$\begin{aligned} t_{ac} &= K_m v_m - F_m(0) - 0 \\ &= (20 \text{ in} \cdot \text{lb/V})(12 \text{ V}) = 240 \text{ in} \cdot \text{lb} \end{aligned}$$

Substituting this into Eq. (2-14) gives

$$\alpha = \frac{240 \text{ in} \cdot \text{lb}}{2 \text{ in} \cdot \text{lb} \cdot \text{s}^2/\text{rad}} = 120 \text{ rad/s}^2$$

(b) In the steady state the acceleration has gone to zero. Therefore the accelerating torque is now zero. Again using Eq. (2-15*c*), we have

$$0 = K_m v_m - F_m \omega$$
$$F_m \omega = K_m v_m$$
$$\omega = \frac{K_m v_m}{F_m} = \frac{(20 \text{ in} \cdot \text{lb/V})(12 \text{ V})}{1.2 \text{ in} \cdot \text{lb} \cdot \text{s/rad}} = 200 \text{ rad/s}$$

(c) When $\omega = 50$ rad/s, the accelerating torque from Eq. (2-15*c*) is

$$t_{ac} = 20(12) - 1.2(50) = 180 \text{ in} \cdot \text{lb}$$

and from Eq. (2-14)

$$\alpha = \frac{180}{2} = 90 \text{ rad/s}^2$$

(d) Parts (a) to (c) will be repeated using the above method; however, t_l is no longer zero: $t_l = 80$ in · lb. At $t = 0$

$$t_{ac} = 20(12) - 1.2(0) - 80 = 160$$
$$\alpha = \frac{160}{2} = 80 \text{ rad/s}^2$$

in the steady state

$$0 = 20(12) - 1.2(\omega) - 80$$
$$1.2\omega = 160$$
$$\omega = \frac{160}{1.2} = 133.33 \text{ rad/s}$$

When $\omega = 50$ rad/s,

$$t_{ac} = 20(12) - 1.2(50) - 80 = 240 - 60 - 80 = 100 \text{ in} \cdot \text{lb}$$
$$\alpha = \frac{100}{2} = 50 \text{ rad/s}^2$$

2-6 GEARS

In almost every control system which involves rotational motion gears are necessary. Gears can be used for any one or all of the following reasons:

1. It is often necessary to match the motor to the load it is driving. A motor which usually runs at high speed and low torque output may be required to drive a load at low speed and high torque. This is the principle behind the operation of an automobile. In starting, a small gear ratio is used between motor and drive shaft to provide low wheel speed and high torque. Since acceleration is directly proportional to torque, this provides a large acceleration. Once the automobile has accelerated, a large gear ratio is used to provide a high wheel speed for a low motor speed. Torque, which provides acceleration, is no longer needed.
2. A gear can be used to adjust a gain in a control system mechanically.
3. A gear can be used to reverse the direction of rotation.
4. Gears can be used to extend the range of transducers. For example, the rotation of a shaft is to be sensed with a synchro. The shaft will rotate a total of 300°. This presents a problem since the linear range of a synchro is only ±45°, or a total of 90°. The problem can be solved by placing a set of gears between the shaft and the synchro with a gear ratio of 0.25. This means that the synchro will turn only a total of 75° (0.25 × 300), which keeps it within its linear range. The loss of gain introduced by the gears can be picked up electrically after the synchro.

The most popular type of gear is the spur gear shown in Fig. 2-31. Figure 2-32 shows the spur gears with the gear teeth removed. N_i, r_i, ω_i, and θ_i represent the number of teeth, radius, speed, and angular displacement of the input gear. The corresponding parameters of the output gear have the subscript o.

The gear ratio a can be defined in any of the following ways:

$$a = \frac{\theta_o}{\theta_i} = \frac{\omega_o}{\omega_i} = \frac{r_i}{r_o} = \frac{N_i}{N_o} \tag{2-17}$$

A gear pass is very much like a transformer in an electric circuit (Fig. 2-33). The voltage source V does not see the resistance R_L but R_L reflected through the transformer. The source sees a resistance equal to a^2R_L.

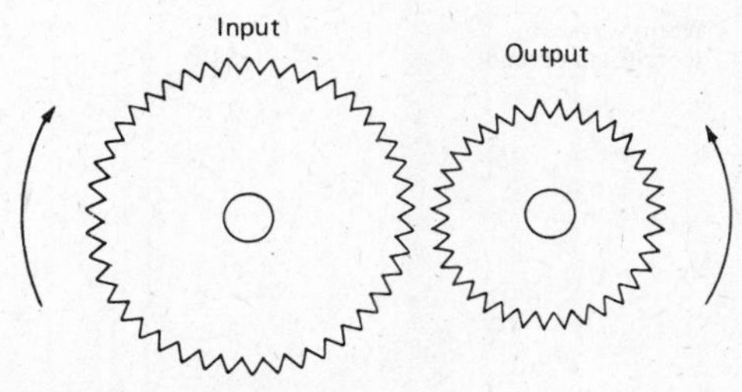

Fig. 2-31 Spur gears.

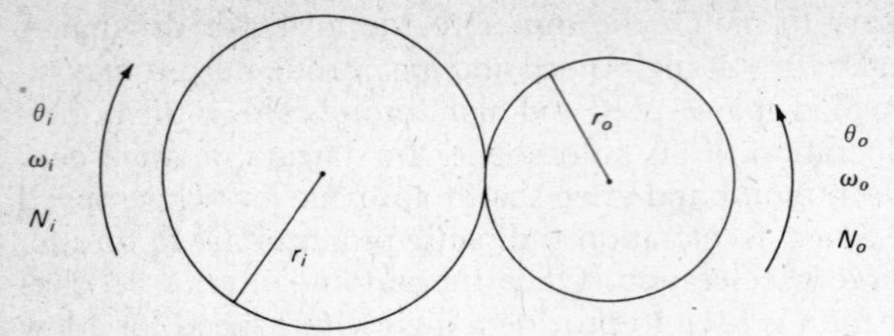

Fig. 2-32 Simplified spur gears.

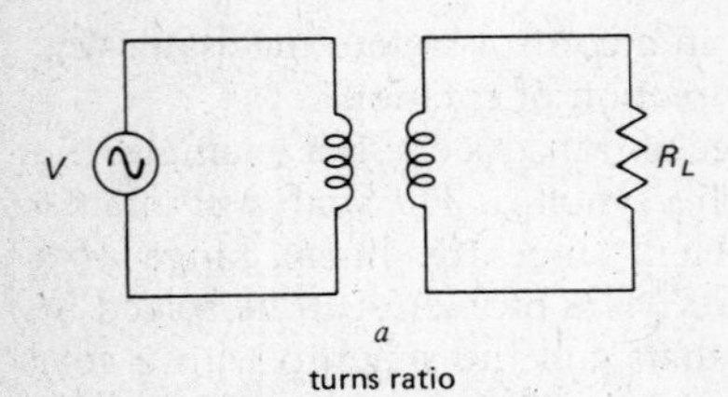

Fig. 2-33 Simple transformer circuit.

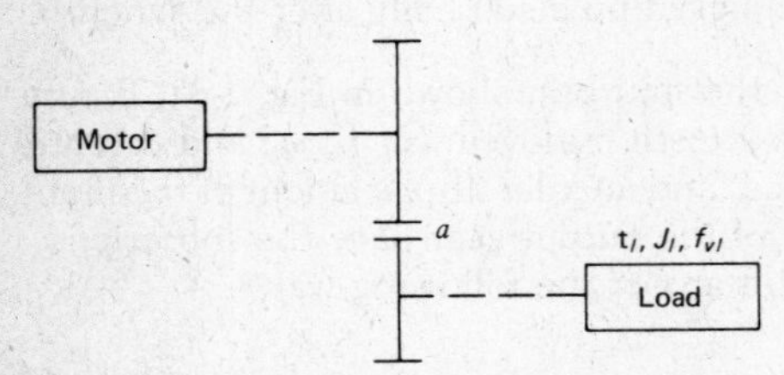

Fig. 2-34 Motor-load connection through gear pass.

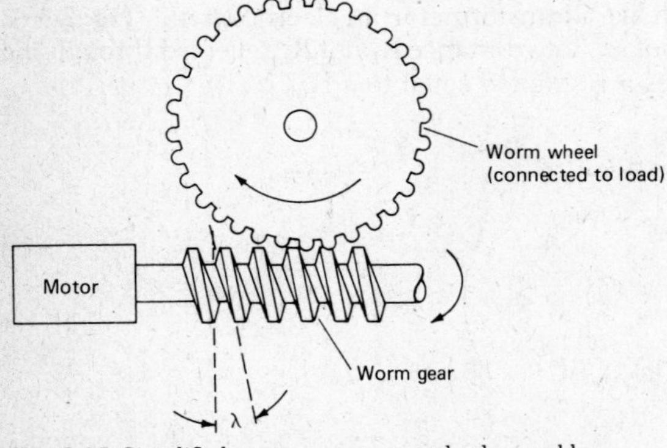

Fig. 2-35 Simplified worm-gear–worm-wheel assembly.

In the same manner quantities can be reflected through a gear pass. A motor which is connected to a load through a gear pass does not see the opposing load torque or load inertia but these quantities reflected through the gear pass. The following equations along with Fig. 2-34 explain the rules for reflecting quantities through a gear pass:

$$\mathrm{t}_r = a\mathrm{t}_l \tag{2-18}$$

$$J_r = a^2 J_l \tag{2-19}$$

$$f_{v_r} = a^2 f_{v_l} \tag{2-20}$$

where t_l = load torque
t_r = load torque as seen by motor
J_l = load inertia
J_r = load inertia as seen by motor
f_{v_l} = viscous friction coefficient of load
f_{v_r} = viscous friction coefficient of load as seen by motor

Another type of gear which is extremely important is the worm gear (Fig. 2-35), which can be used to obtain extremely small gear ratios (0.001 and smaller). It also transmits rotation through 90°. An important feature of the worm gear is that it can provide irreversibility. As long as the pitch angle λ is less than a given amount, the load cannot backdrive the motor. This is both economical and a safety feature. Power can be turned off after a motor has moved a load to a commanded position, and in the event of a power failure, the load will not backdrive the motor. The value of this feature can readily be seen in the operation of an elevator.

All the equations given for the spur gears apply to worm gears with the exception of those referring to the number of teeth and the radii of the gears.

Example 2-8

A motor is connected to a load as shown in Fig. 2-36. (Units have been omitted for simplicity.)

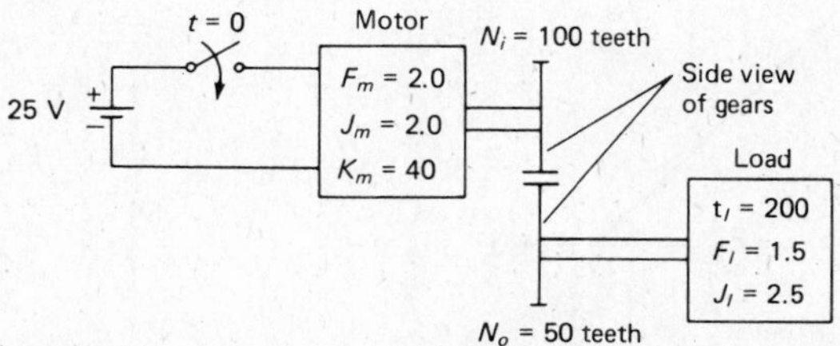

Fig. 2-36 Motor-load configuration for Example 2-8.

(a) Find the starting acceleration of the motor.
(b) Draw the torque-speed curve for the system.
(c) Using the torque-speed curve drawn in part (b), find the steady-state speed of the motor and the load.

Solution

(a) In order to use Eqs. (2-14) and (2-15c) we must first find all quantities reflected to the motor. From Eq. (2-17) the gear ratio is

$$a = \frac{N_i}{N_o} = \frac{100}{50} = 2$$

$$\text{At the motor:} \begin{cases} J_{\text{tot}} = J_m + a^2 J_l = 2 + (2)^2(2.5) = 12 \\ F_{\text{tot}} = F_m + a^2 F_l = 2 + (2)^2(1.5) = 8 \\ \mathrm{t}_r = a\mathrm{t}_l = 2(200) = 400 \end{cases}$$

From Eq. (2-15c), rewritten here with total quantities instead of just motor quantities, noting that $\omega = 0$ at starting, we get

$$\begin{aligned} \mathrm{t}_{ac} &= K_m v_m - F_{\text{tot}}\,\omega - \mathrm{t}_r \\ &= 40(25) - 8(0) - 400 \\ &= 1000 - 400 = 600 \end{aligned}$$

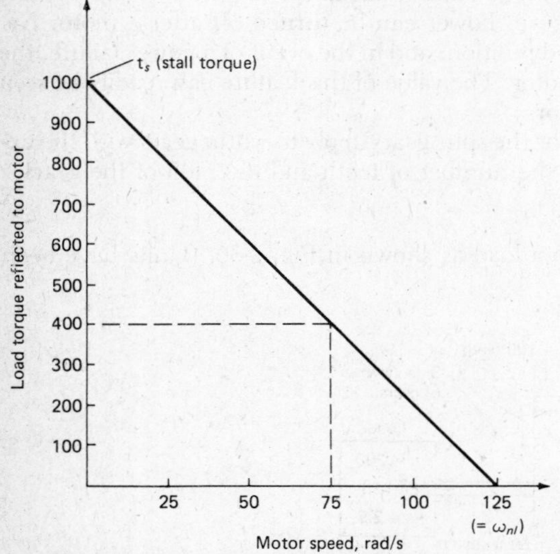

Fig. 2-37 Torque-speed curve for Example 2-8.

From Eq. (2-14)

$$\alpha = \frac{t_{ac}}{J_{tot}} = \frac{600}{12} = 50$$

(b) The torque-speed curve is plotted by finding the stall torque and no-load speed and joining them with a straight line (Fig. 2-37):

$$t_s = K_m v_m = 40(25) = 1000$$

$$\omega_{nl} = \frac{K_m v_m}{F_{tot}} = \frac{1000}{8} = 125$$

(c) To find the motor speed, find the reflected load on the ordinate, move horizontally across to the curve, and then vertically down to read the steady-state speed

$$\omega_m = 75 \qquad \text{from Fig. 2-37}$$

The load speed is found from Eq. (2-17)

$$\omega_l = \omega_o = a\omega_i = a\omega_m = 2(75) = 150$$

2-7 AC SERVO MOTOR

In control-system applications where the output is in the fractional-horsepower range, up to about $\frac{1}{8}$hp, the two-phase ac motor is quite popular. Above that range, the efficiency begins to fall off significantly. The principal advantages of the ac motor are

1. Simple circuitry
2. Absence of brushes or slip rings, which introduce friction and must be replaced after extensive use
3. Stalling without causing damage
4. Almost instantaneous starting, stopping, and reversal
5. Long life
6. High reliability
7. Elimination of the need to demodulate ac signals (as is the case with the dc motor)

The motor is made up of a rotor, which can have several forms, and a stator with two windings, the axes of which are spaced 90° apart. One winding,

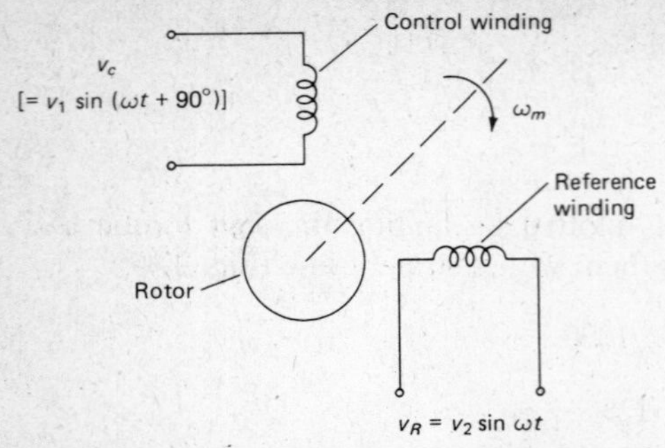

Fig. 2-38 Schematic of a two-phase ac servomotor.

with a fixed ac voltage applied, is called the *reference winding*. The other winding, with an ac voltage of variable amplitude applied, is called the *control winding* since the voltage applied to it is used to control the motor's speed and torque. Not only must the two windings be 90° apart in space, but the two voltages applied to them also must differ in time phase by 90° in order to produce torque efficiently. This is usually accomplished either with an external resistor and capacitor or with a phase-shifting circuit in the servo amplifier whose output is applied to the control winding.

A schematic of the motor is shown in Fig. 2-38 and a typical set of torque-speed curves in Fig. 2-39. Of particular interest is the fact that although the curves are similar to those of a dc motor, the values of stall torque and

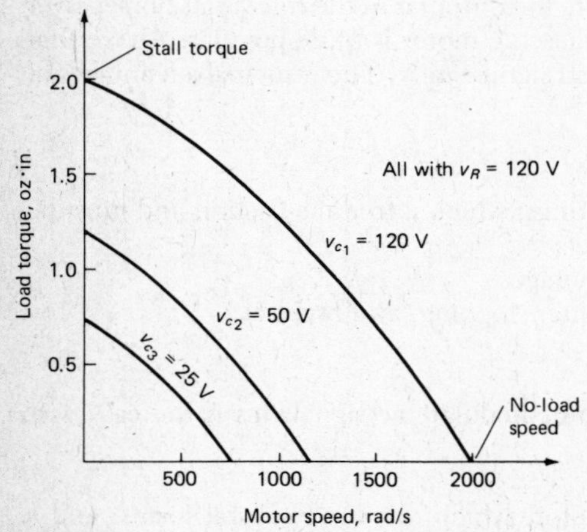

Fig. 2-39 Typical torque-speed curves for an ac servomotor.

no-load speed are not proportional to control voltage. As indicated in Fig. 2-39, the stall torque and no-load speed are given by

$$t_s = K_m v_c \qquad \omega_{nl} = \frac{K_m v_c}{f_m} \tag{2-21}$$

where the value of the motor constant K_m is different for differing values of v_c. If the nonlinear curves of Fig. 2-39 are approximated by straight lines, the torque equation becomes

$$t_{\text{dev}} = K_m v_c - f_m \omega \tag{2-22}$$

The similarity of Eqs. (2-22) and (2-13) should be noted. In the steady state, with external viscous friction and constant load torque applied,

$$t_l = K_m v_c - F_{\text{tot}}\,\omega \tag{2-23}$$

Example 2-9

A 60 cycle servo motor has a stall torque of 1.5 oz · in and no-load speed of 2000 r/min when both control and reference windings are excited with 115 V. Find the speed of the motor when it is connected directly to a constant load of 0.8 oz · in and viscous-friction coefficient of 3×10^{-3} oz · in · s.

Solution

First using Eq. (2-21), we determine the motor's viscous friction

$$\omega_{nl} = \frac{K_m v_c}{f_m}$$

$$f_m = \frac{K_m v_c}{\omega_{nl}} = \frac{t_s}{\omega_{nl}} = \frac{1.5 \text{ oz} \cdot \text{in}}{(2000 \text{ r/min})(2\pi \text{ rad/r}) \dfrac{1 \text{ min}}{60 \text{ s}}}$$

$$= 0.007 \text{ oz} \cdot \text{in} \cdot \text{s}$$

Next, the total viscous friction under load conditions will be determined. Since the load is connected directly to the motor, no gears are being used. Thus, the total viscous friction is that of the motor plus that of the load

$$F_{\text{tot}} = f_m + a^2 f_l$$

where $a = 1$ since direct coupling is the same as unity gear ratio

$$F_{\text{tot}} = 0.007 + 0.003 = 0.01 \text{ oz} \cdot \text{in} \cdot \text{s}$$

Applying Eq. (2-23) and solving for ω, we get

$$\begin{aligned} t_l &= K_m v_c - F_{\text{tot}}\omega \\ 0.8 &= 1.5 - 0.01\omega \\ 0.01\omega &= 0.7 \\ \omega &= 70 \text{ rad/s} \end{aligned}$$

2-8 LINEAR VARIABLE DIFFERENTIAL TRANSFORMER (LVDT)

An LVDT is a transducer whose input is linear motion (as opposed to rotational) and whose output is a voltage proportional to the input displacement. Figure 2-40 shows schematics of two types of LVDTs. In both the principle of operation is the same. The input to the device is the linear motion x of a magnetic material. This linear motion varies the reluctance of a magnetic circuit, which in turn changes the induced voltages e_1 and e_2. The output is given by

$$e_o = e_1 - e_2 \tag{2-24}$$

$$e_o = kx \tag{2-25}$$

Equation (2-25) expresses the output as a function of input x. The gain k (with units of volts per unit length) is a function of the physical constants, turns ratios, and the magnitude of the excitation voltage. Since this output also is an ac signal whose frequency is equal to the frequency of the excitation signal, it too must be demodulated in control systems making use of dc motors and controls.

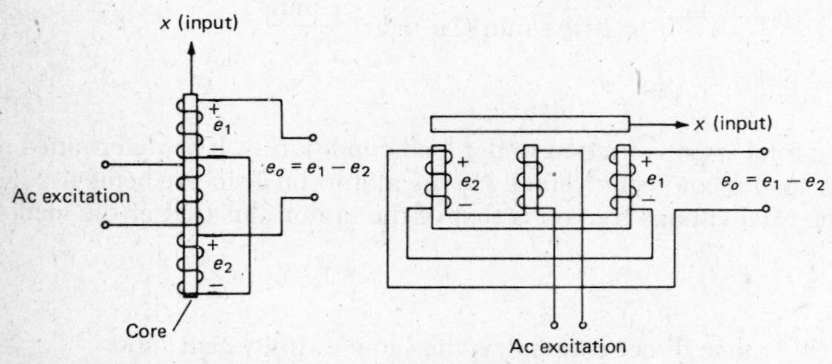

Fig. 2-40 Linear variable differential transformers (LVDTs).

2-9 PRESSURE TRANSDUCER

A pressure transducer is a device whose input is pressure and whose output is a voltage proportional to the input pressure. There are two basic types; one uses a diaphragm, and the other type uses a bourdon tube, i.e., a metallic tube designed to deflect proportionally to a change in pressure. In general, for pressures ranging from 0 to 30 lb/in² a diaphragm type is used. For pressures ranging from 0 to 10,000 lb/in² a bourdon tube is used.

Figure 2-41 is a schematic of a typical diaphragm pressure transducer. The pressure p deflects the diaphragm, which is connected to an LVDT. The LVDT produces a voltage proportional to its shaft displacement, which is in turn proportional to the pressure being sensed. The output e_o is given by

$$e_o = kp \tag{2-26}$$

where k is a constant depending on the LVDT gain and the physical constants of the diaphragm.

Figure 2-42 shows a C-type bourdon-tube pressure transducer. As the pressure increases, the tube begins to straighten out (moving upward as shown). The end of the tube is flexibly connected to the shaft of a rotary transducer. The output is given by

$$e_o = kp \tag{2-27}$$

where k is a constant depending on the gain of the transducer, the ac excitation, and physical constants pertaining to the bourdon tube. In both transduc-

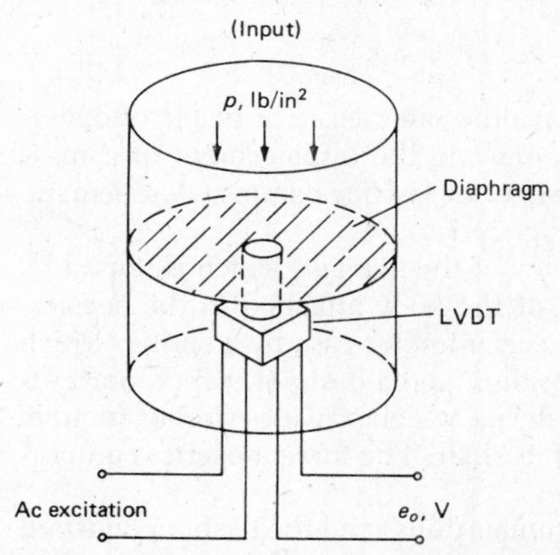

Fig. 2-41 Schematic of diaphragm pressure transducer.

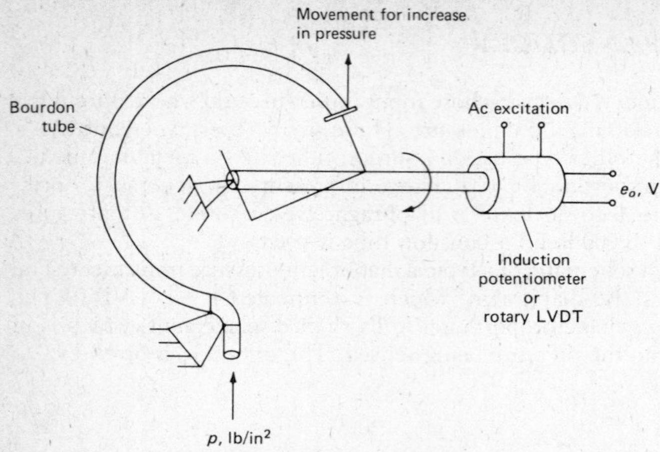

Fig. 2-42 C-type bourdon-tube pressure transducer.

ers mentioned above, the output is an ac signal and must be demodulated to provide dc control.

Pressure transducers are essential in automatic piloted aircraft. It is by sensing barometric air pressure that the altitude of a plane can be determined in flight. Since aircraft have generators which produce 400 Hz ac (this is standard for aircraft), the excitation indicated in Figs. 2-41 and 2-42 would be 400 Hz.

2-10 ACCELEROMETER

The many types of accelerometers all have the same basic principle of operation. The device senses acceleration by sensing the displacement of a mass which is restrained by a spring and a dashpot or viscous damper. A schematic of a typical accelerometer is shown in Fig. 2-43.

It will now be shown that the motion y of the mass m, which is sensed by the LVDT, represents the acceleration of the body attached to the accelerometer. The mass is connected to the accelerometer case by a spring, which tends to restore the mass to its initial position, and a dashpot, which prevents the mass from oscillating. A dashpot is a device which provides viscous friction; i.e., it exerts a force to oppose motion of its shaft. The force it exerts is proportional to the velocity of its shaft.

The spring constant is k_x oz/in (or similar units) and the dashpot constant is f_v oz/(in/s) (or similar units). Newton's second law states that the sum of the

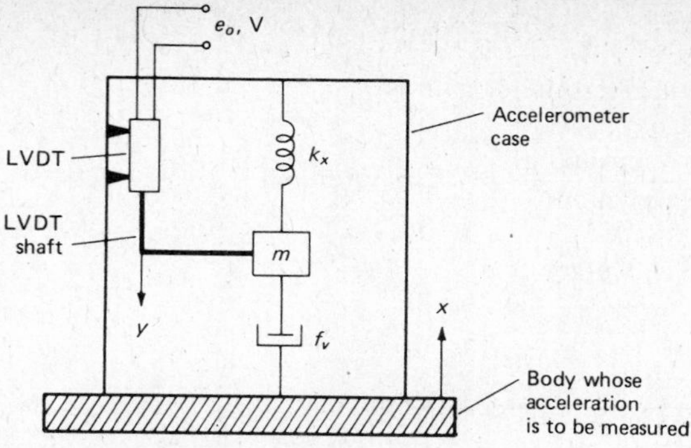

Fig. 2-43 Typical accelerometer schematic.

forces on a body is equal to its mass m times its acceleration d^2y/dt^2. The displacement of mass m with respect to inertial space is $y - x$. Applying Newton's law, we can now write

$$m \frac{d^2(y - x)}{dt^2} = -k_x y - f_v \frac{dy}{dt} \tag{2-28}$$

$$m \frac{d^2x}{dt^2} = m \frac{d^2y}{dt^2} + f_v \frac{dy}{dt} + k_x y \tag{2-29}$$

$$ma = m \frac{d^2y}{dt^2} + f_v \frac{dy}{dt} + k_x y \tag{2-30}$$

where $a = d^2x/dt^2$ is the acceleration of the body. Equation (2-30) has a solution which has a gain, neglecting transients, given by

$$e_o = \frac{k_l m}{k_x} \frac{d^2x}{dt^2} = k'a \tag{2-31}$$

where k_l is the gain of the LVDT. Note that the viscous damper (f_v) is absent from Eq. (2-31) since it affects only the transient response.

Example 2-10

An accelerometer uses an LVDT with a gain of 1.8 V/cm, a spring with a force constant of 500 dyn/cm and a mass of 60 g. Find:

(a) The accelerometer constant

(b) The acceleration of the body on which the accelerometer is mounted when the output voltage is 4.5 V rms

Solution

(a) From Eq. (2-31) the accelerometer gain is

$$k' = \frac{k_l m}{k_x} = \frac{(1.8\ \text{V/cm})(60\ \text{g})}{500\ \text{dyn/cm}} = 0.216\ \text{V/(cm/s}^2\text{)}$$

(b) Again using Eq. (2-31), we get

$$e_o = k'a$$

$$a = e_o/k' = \frac{4.5\ \text{V}}{0.216\ \text{V/cm/s}^2} = 20.83\ \text{cm/s}^2$$

2-11 GYROSCOPE

A gyroscope is a device which senses angles and angular rates in inertial space. In the flight of aircraft it is essential to know the plane's position relative to the earth and the rate of change of its position. The property of a gyroscope which enables it to determine these variables is its tendency to maintain its axis in a fixed direction in space.

2-12 RESOLVER

In many control applications it becomes necessary to break a vector quantity down into its two rectangular components. A resolver is a transducer which

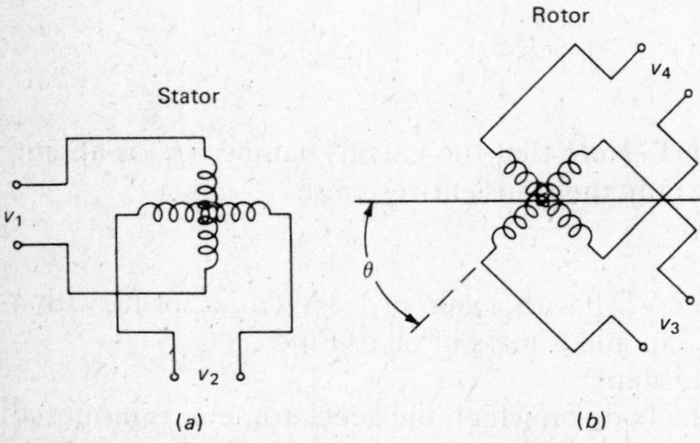

Fig. 2-44 Schematic of a resolver: (a) stator and (b) rotor.

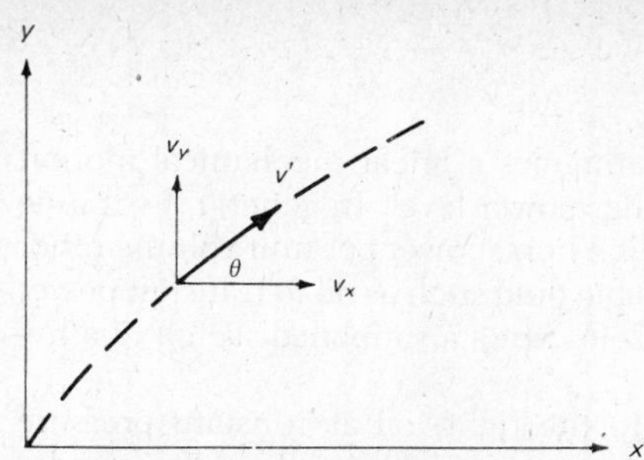

Fig. 2-45 Vector diagram of missile velocity.

performs this function. The resolver works on the principle of magnetic induction. It consists of a rotor and stator, each with two windings 90° apart. Figure 2-44 is a schematic of a resolver. The input-output relationship is described by

$$v_3 = k_r(v_1 \cos\theta - v_2 \sin\theta) \qquad v_4 = k_r(v_1 \sin\theta + v_2 \cos\theta) \tag{2-32}$$

where k_r is a resolver constant.

If one stator winding is shorted, making v_2 equal to zero, the equations become

$$v_3 = v' \cos\theta = v_x \qquad v_4 = v' \sin\theta = v_y \tag{2-33}$$

where $v' = k_r v_1$.

It should be evident from Eqs. (2-33) that if v' is a voltage representative of the velocity of a missile and θ is the missile's flight-path angle relative to the earth, voltage v_3 is the component of velocity parallel to the earth while voltage v_4 is the component of velocity perpendicular to the earth. Figure 2-45 may help clarify the preceding statement. Resolvers can be built to be highly accurate.

2-13 AMPLIDYNE

The amplidyne is a special type of dc generator, also called a *rotary amplifier*. It is special in that, by design, it has a very high gain-bandwidth product. It also is available for handling high power levels up to 50 kW and more. It is commonly used as the power stage driving a dc motor and has therefore found much use in position and speed control systems. When used, it is driven at a constant speed, and its input is generally the output voltage of an electronic amplifier.

2-14 HYDRAULIC ACTUATOR

A hydraulic actuator is a device which amplifies a linear mechanical motion into a mechanical motion at a much higher power level. In general, hydraulic actuators are economical and provide a high horsepower per unit volume ratio. The device uses a relatively noncompressible fluid such as oil to transmit power through the action of fluid flow. Figure 2-46 shows a simplified sketch of a hydraulic actuator.

When the input shaft is displaced to the right, oil at constant pressure flows from the center valve up into the actuator filling the left cylinder chamber. This causes the piston to move to the right, moving the load. As the piston moves to the right, it expels the oil in the right cylinder chamber down through the right-side return valve, which opened up when the input shaft moved right originally. The oil thus returns to the main supply. Input motion to the left provides load motion to the left in a similar manner.

Hydraulic actuators can be used for precise linear motion more easily than electrically operated actuators. They are used in automobiles for power steering and power brakes, in ships for controlling the rudder, in large machine equipment and in airplanes for moving the control surfaces, namely, the rudder, ailerons, elevators, horizontal stabilizer, flaps, and spoilers.

In airplanes, the input shaft movement of the hydraulic actuator is controlled by the autopilot through the use of an electrohydraulic servo valve. This is similar to a hydraulic actuator except that the input is an electrical signal. The electrical signal through magnetic induction moves a valve which ports oil into the main valve, providing output-shaft movement. This output shaft is linked to the input shaft of the hydraulic actuator as shown in Fig. 2-47.

Provision is made through cables for the pilot to operate the hydraulic actuator manually, also shown in Fig. 2-47.

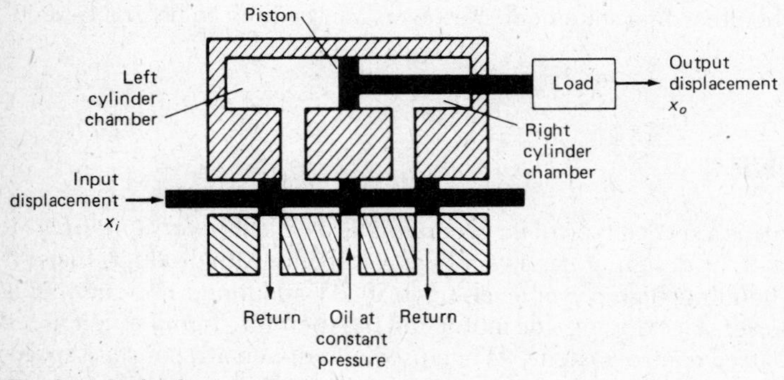

Fig. 2-46 Hydraulic actuator.

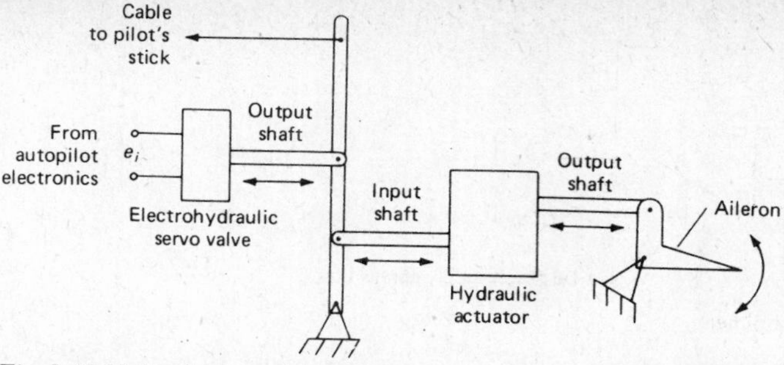

Fig. 2-47 Hydraulic control system used to position an airplane's control surface.

2-15 OPERATIONAL AMPLIFIER

Several years ago when analog computers were first being designed and built, a component was needed that could be used for the simulation of many devices (electrical, mechanical, etc.). The operational amplifier (OP-AMP) was developed for just that purpose. Today, however, it has become so widely used as an electronic circuit component that its original intent represents only a small portion of its use.

The OP-AMP is an electronic amplifier characterized by an almost infinite gain (50,000 to 500,000), extremely high input impedance, and practically zero output impedance. The device has three terminals to which passive components (resistors, capacitors, etc.) are connected.

These terminals are known as the inverting input, the noninverting input, and the output. Depending on the magnitude of the components and the configuration used, different characteristics are obtained. The derivations will not be given here. More details on the use of OP-AMPs will be covered in Chap. 13.

2-15-1 The Inverting Amplifier

Figure 2-48 shows the configuration for an inverting amplifier. Note that the noninverting input is grounded. The input-output relationship is

$$e_o = -\frac{R_f}{R_1} e_1 \tag{2-34}$$

Example 2-11

The circuit of Fig. 2-48 has $R_1 = 10\ \text{k}\Omega$, $R_f = 100\ \text{k}\Omega$, and $e_1 = 0.25$ V. Find e_o.

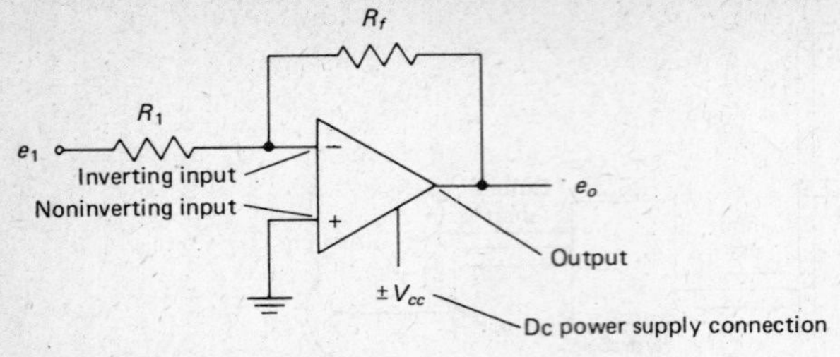

Fig. 2-48 Inverting amplifier.

Solution

Using Eq. (2-34), we get

$$e_o = -\frac{100\ \text{k}\Omega}{10\ \text{k}\Omega}(0.25) = -10(0.25) = -2.5\ \text{V}$$

2-15-2 The Noninverting Amplifier

Figure 2-49 shows a configuration for a noninverting amplifier. The input-output relationship is

$$e_o = \left(1 + \frac{R_f}{R_1}\right) e_1 \tag{2-35}$$

Example 2-12

The circuit of Fig. 2-49 is to be used to obtain a voltage gain of 5. Determine the values of R_f and R_1.

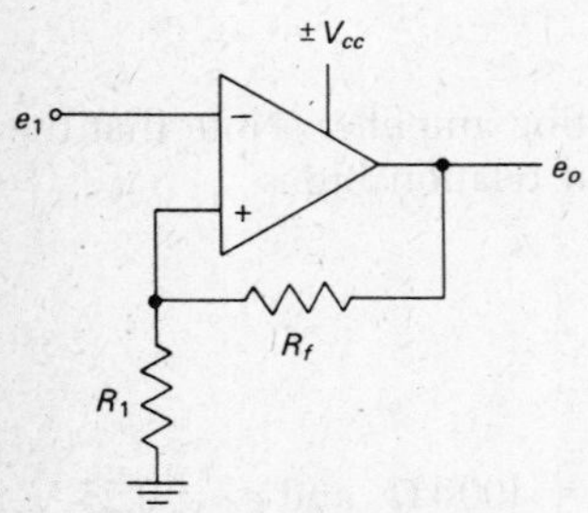

Fig. 2-49 Noninverting amplifier.

Solution

From Eq. (2-35) the gain (output divided by input) is

$$\text{Gain} = \frac{e_o}{e_1} = \left(1 + \frac{R_f}{R_1}\right) = 5 \qquad \frac{R_f}{R_1} = 5 - 1 = 4$$

Since the ratio of $R_f/R_1 = 4$, if we select one of the resistors, the other can be calculated. If $R_1 = 10\ \text{k}\Omega$ is selected, $R_f = 4(R_1) = 40\ \text{k}\Omega$ should be used, or if $R_f = 200\ \text{k}\Omega$ is selected, $R_1 = R_f/4 = 50\ \text{k}\Omega$ should be used.

There are many values of R_f and R_1 that could satisfy the above example.

2-15-3 The Summing Amplifier

This is a configuration in which an output is obtained by summing several inputs with their corresponding negative gains. The circuit is shown in Fig. 2-50, and the input-output relationship is

$$e_o = -\frac{R_f}{R_1}(e_1) - \frac{R_f}{R_2}(e_2) - \frac{R_f}{R_3}(e_3) - \cdots \tag{2-36}$$

Example 2-13

The circuit of Fig. 2-50 has $R_f = 220\ \text{k}\Omega$, $R_1 = 40\ \text{k}\Omega$, $R_2 = 100\ \text{k}\Omega$, and $R_3 = 18\ \text{k}\Omega$. If $e_1 = -1.2\ \text{V}$, $e_2 = 5\ \text{V}$, and $e_3 = -1\ \text{V}$, determine the output e_o.

Solution

Using Eq. (2-36), we get

$$\begin{aligned} e_o &= -\frac{220\ \text{k}\Omega}{40\ \text{k}\Omega}(-1.2) - \frac{220\ \text{k}\Omega}{100\ \text{k}\Omega}(5) - \frac{220\ \text{k}\Omega}{18\ \text{k}\Omega}(-1) \\ &= -5.5(-1.2) - 2.2(5) - 12.22(-1) \\ &= 6.6 - 11 + 12.22 = 7.82\ \text{V} \end{aligned}$$

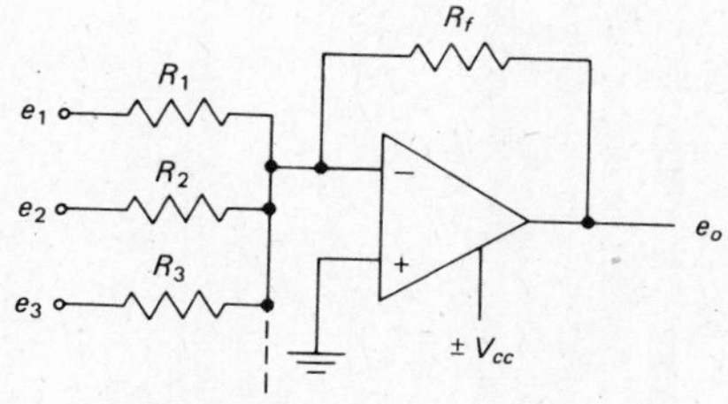

Fig. 2-50 Summing amplifier.

2-15-4 The Integrator

This configuration performs the mathematical operation of integration. The output is equal to the integral of the input, for as long as the input is present, multiplied by a gain. The input-output relationship is given by

$$e_o(t) = -\frac{1}{RC}\int_0^t e_1(t)\,dt \tag{2-37}$$

For dc (constant) inputs, Eq. (2-37) becomes

$$e_o(t) = -\frac{1}{RC}e_1 t \tag{2-38}$$

The circuit is shown in Fig. 2-51.

Example 2-14

The circuit of Fig. 2-51 has $e_1 = -2$ V, $R = 1$ MΩ, and $C = 5$ μF. The supply V_{cc} is 12 V, and the input is applied at $t = 0$. Determine the output at $t = 30$ s, and plot the output for all time.

Solution

Since the input is constant, Eq. (2-38) is used to find the output

$$e_o = -\frac{1}{(2\ \text{M}\Omega)(5\ \mu\text{F})}(-2)t = 0.2t \tag{2-39}$$

At $t = 30$ s, the output is found by substituting $t = 30$ into Eq. (2-39)

$$e_o = 0.2(30) = 6\ \text{V}$$

If the output equation (2-39) is plotted versus time, the graph in Fig. 2-52 is obtained.

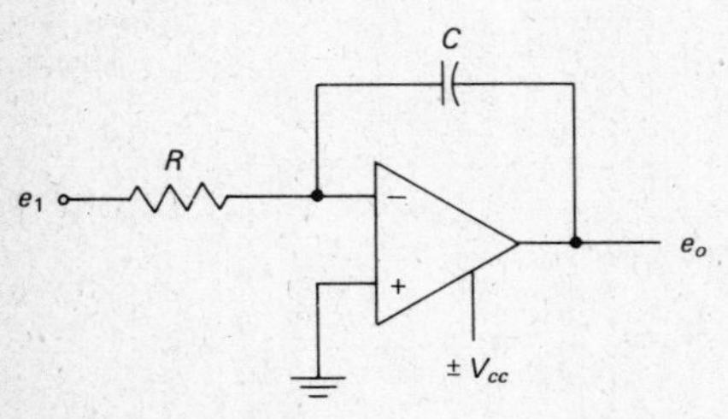

Fig. 2-51 Integrator.

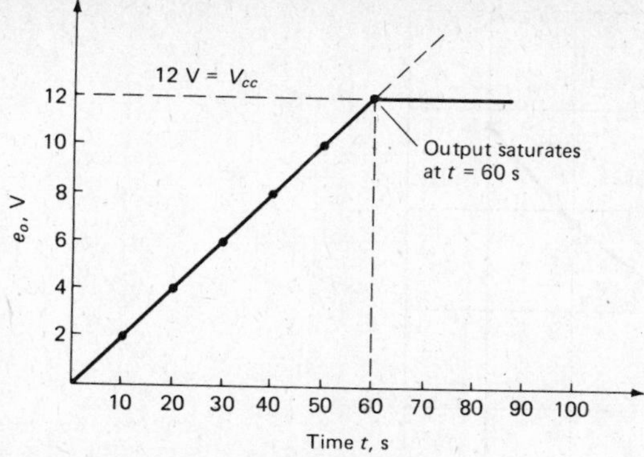

Fig. 2-52 Plot of integrator output for all time.

It should be noted that the output does not increase forever. The output increases until the input is removed or until $e_o = V_{cc}$, at which point the amplifier will saturate. The time at which saturation occurs is found by letting $e_o = V_{cc}$ in Eq. (2-39) and solving for t

$$e_o = V_{cc} = 0.2t \qquad \text{or} \qquad t = \frac{V_{cc}}{0.2} = \frac{12}{0.2} = 60 \text{ s}$$

The point of saturation is indicated on Fig. 2-52.

The circuits shown here are just samples of what can be done with the OP-AMP, but they are the configurations most commonly used in control systems.

PROBLEMS

2-1. A helical 5-turn potentiometer has a resistance of 10 kΩ and 9000 winding turns.
  (a) If the measured resistance at its midpoint setting is 5050 Ω, what is its linearity?
  (b) If the measured resistance at its quarter-point setting is 2600 Ω, what is its linearity?

2-2. (a) What is the resolution of the potentiometer of Prob. 2-1 in volts per winding turn (volts per step) if (a) 90 V and (b) 45 V is impressed across it?

2-3. (a) For the pot of Prob. 2-1, what is the potentiometer constant K_p in volts per radian and volts per degree if the input voltage is (a) 90 V, (b) 45 V, and (c) 22.5 V?

2-4. The calibration curve for a three-turn 100-kΩ potentiometer with no load is shown in Fig. P2-4.

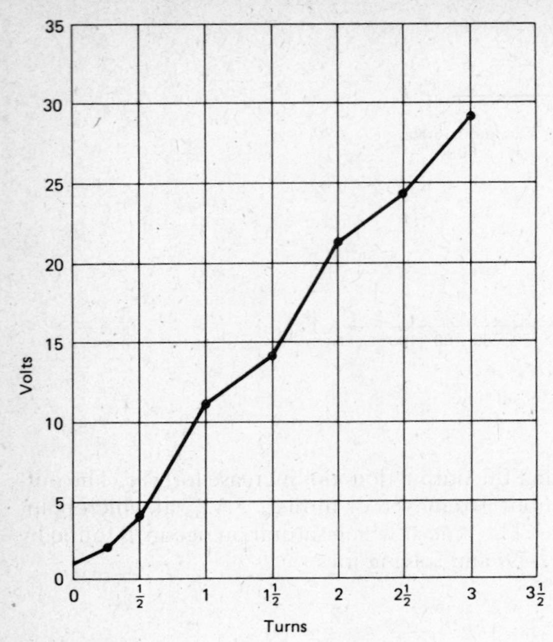

Fig. P2-4

(a) What is the approximate potentiometer constant in volts per turn?
(b) Draw the approximate calibration curve when the potentiometer is loaded with 500 kΩ. Over what turns range would you use the potentiometer to achieve linear operation and what is the approximate potentiometer constant over this range?
(c) Repeat part (*b*) for a 50-kΩ load.

2-5. A 3-turn 100-kΩ potentiometer with 1 percent linearity uses a 30-V supply.
(a) Find the potentiometer constant in volts per turn.
(b) Find the range of voltages at the midpoint setting.
(c) Assuming that the potentiometer is perfectly linear, find the voltage at the midpoint when the potentiometer is loaded with 500-kΩ.
(d) Repeat part (c) for a 50-kΩ load.

2-6. A 50-kΩ, 10-turn potentiometer has a ± 12-V supply connected to it as shown in Fig. P2-6.
(a) Find the potentiometer constant K_p in volts per radian.
(b) Find the output at the arm (unloaded) when the shaft is rotated 70° from its midpoint toward +12 V.
(c) Repeat part (b) when the potentiometer is connected to a 40-kΩ load.

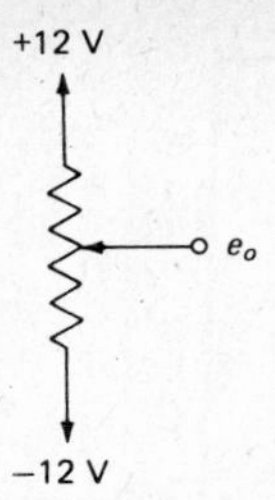

Fig. P2-6

2-7. A 10-turn 10-kΩ potentiometer is connected as in Fig. P2-6 to a ± 24-V supply. It has a linearity of 0.5 percent. Find:
(a) K_p in volts per radian
(b) K_p in volts per degree
(c) the range of voltage at the midpoint (unloaded)

2-8. Describe the difference between a synchro torque transmitter (TX) and a control transformer (CT).

2-9. Describe two types of pressure transducers.

2-10. A tachometer has a gain of 0.05 V/(rad/s). Find:
(a) The output voltage when the shaft speed is 40 deg/s
(b) The output voltage when the shaft speed is 20 rad/s
(c) The shaft speed in radians per second and degrees per second when the output voltage is 1.8 V.

2-11. A tachometer has the experimental input-output characteristic shown in Fig. P2-11. What is its approximate gain in volts per radian per second, in volts per degree per second?

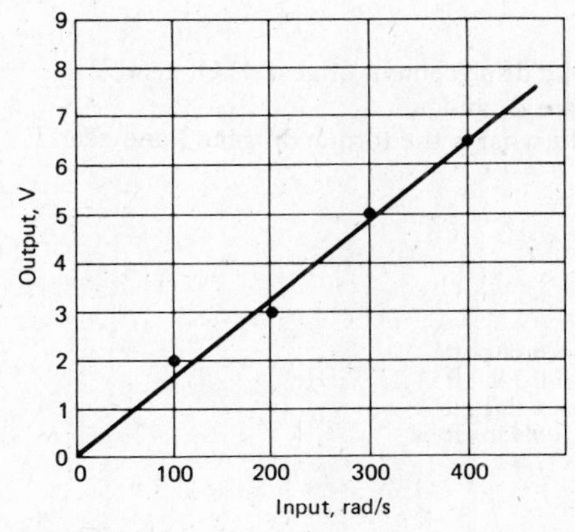

Fig. P2-11

2-12. Referring to Fig. P2-12:

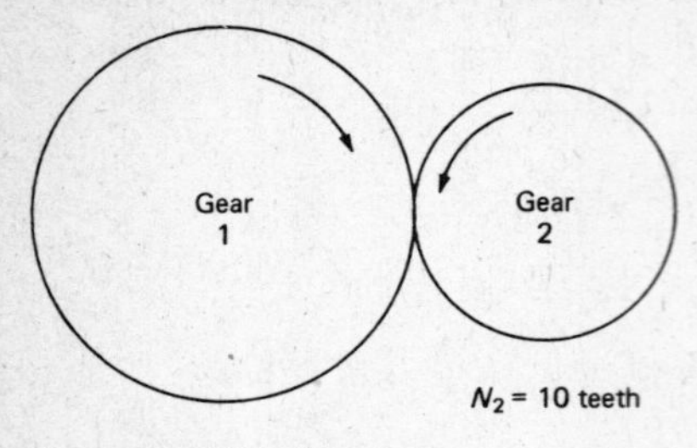

Fig. P2-12

(a) What is ratio of the diameters D_1/D_2?
(b) If gear 1 is displaced 40°, what is the displacement of gear 2?
(c) If ω_1 is 30 rad/s, what is the value of ω_2?
(d) If α_2 is 4 rad/s², what is the value of α_1, the angular acceleration of gear 1?
(e) If the torque of gear 1 is 5 lb · ft, what is the torque of gear 2?

2-13. Referring to Fig. P2-13:

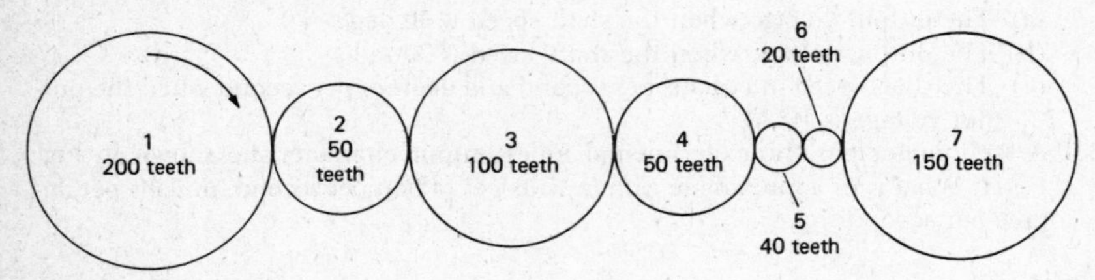

Fig. P2-13

(a) If $\theta_1 = 2$ rad clockwise, what is the displacement of gear 4? Of gear 7?
(b) If ω_6 is 20 rad/s clockwise, what are ω_1 and ω_2?
(c) If the torque of gear 1 is 10 lb · ft, what is the torque on gear 3 and gear 7?

2-14. Referring to Fig. P2-14:

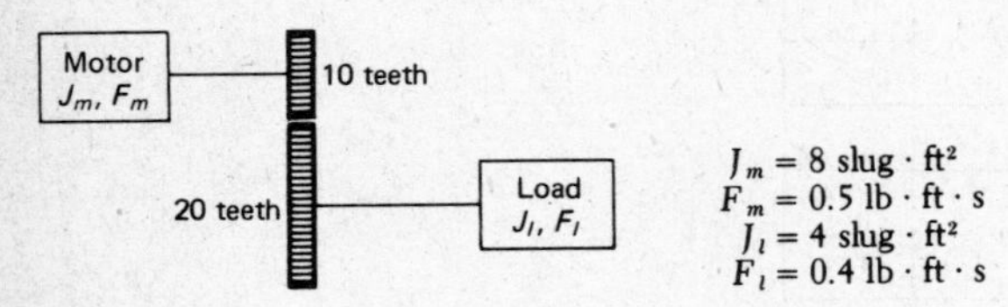

$J_m = 8$ slug · ft²
$F_m = 0.5$ lb · ft · s
$J_l = 4$ slug · ft²
$F_l = 0.4$ lb · ft · s

Fig. P2-14

(a) Calculate the total moment of inertia, J_{tot} seen by the motor. Repeat for F_{tot} the coefficient of viscous friction.

(b) Calculate the motor stall torque if $\omega_m = 20$ rad/s and $\alpha_m = 2$ rad/s².
(c) Repeat part (b) if the load is removed.

2-15. A motor has the following constants

$$K_m = 2 \qquad F_m = 0.5 \qquad J_m = 1$$

For an input of 10 V find:
(a) The steady state speed in radians per second.
(b) Sketch the speed vs. time if the motor time constant is given by J_m/F_m. *Hint:* The motor speed starts at zero and exponentially approaches the steady state speed.

2-16. (a) Sketch the linear torque-speed characteristics of a motor for $v = 2$, 4, and 6 V if the friction coefficient is 0.04 lb · ft · s and the stall torque at $v = 2$ is 0.4 lb · ft.
(b) Find the motor constant K_m.

2-17. An ac servo motor has both windings excited with 115 V ac. It has a stall torque of 2 lb · ft. Its coefficient of viscous friction is 0.2 lb · ft · s.
(a) Find its no-load speed.
(b) It is connected to a constant load of 0.9 lb · ft and coefficient of viscous friction of 0.05 lb · ft · s through a gear pass with a ratio of 4. Find the speed at which the motor will run.

2-18. Find e_o in Fig. P2-18.

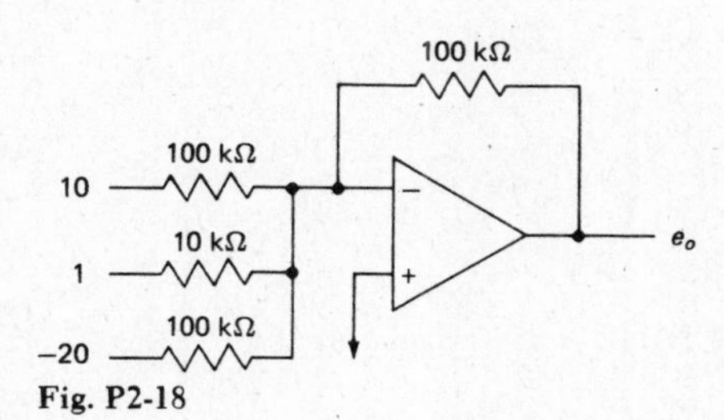

Fig. P2-18

2-19. Find e_o in Fig. P2-19.

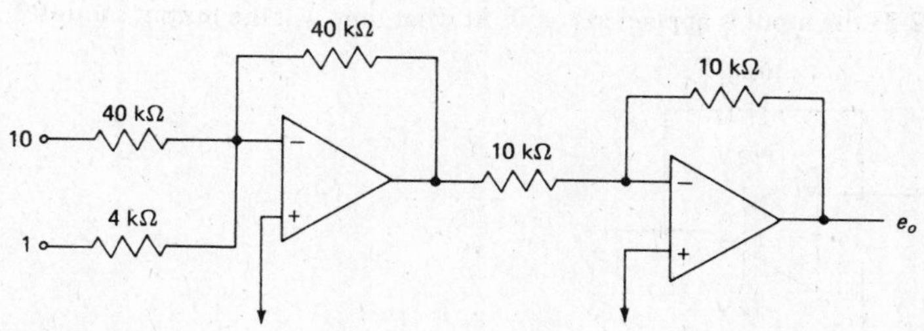

Fig. P2-19

2-20. Sketch e_o vs. time and find the value of e_o at $t = 0.2$ s in Fig. P2-20.

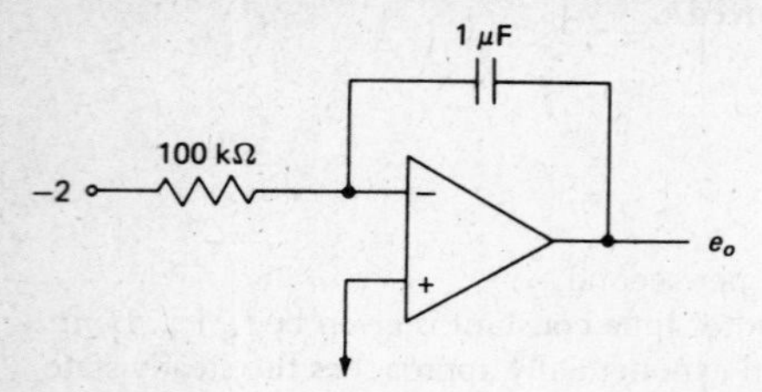

Fig. P2-20

2-21. Sketch e_o in Fig. P2-21.

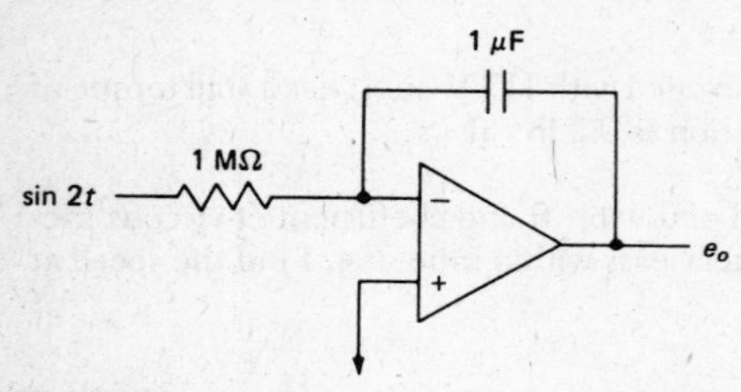

Fig. P2-21

2-22. Find e_o in Fig. P2-22.

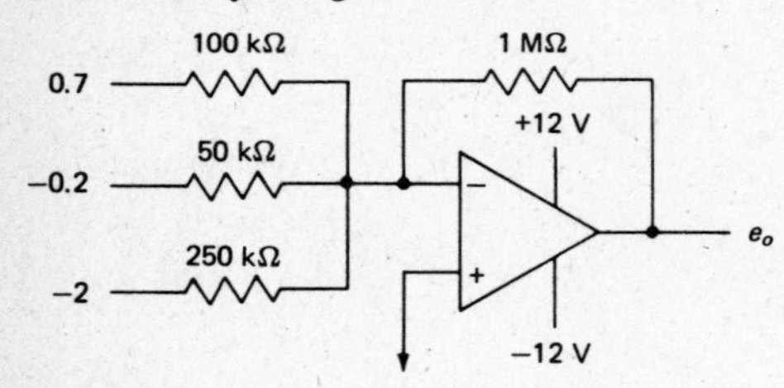

Fig. P2-22

2-23. In Fig. P2-23 the input is applied at $t = 0$. At what time will the output saturate?

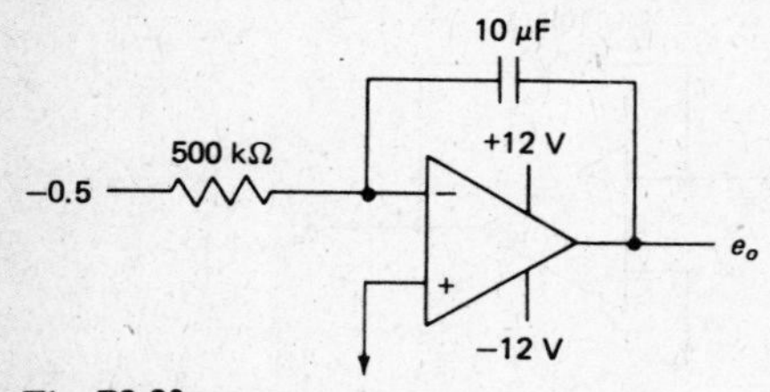

Fig. P2-23

2-24. Find e_o in Fig. P2-24.

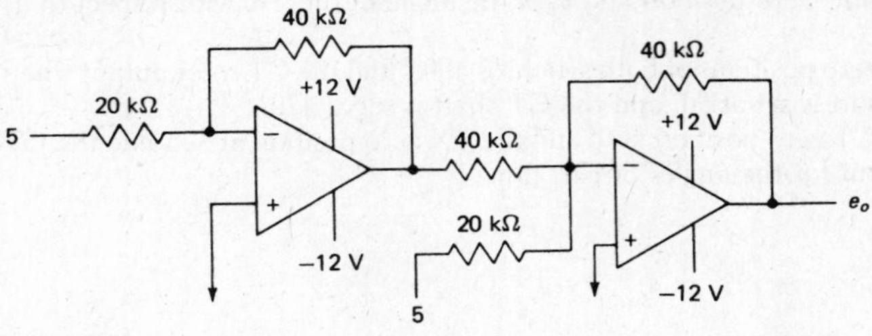

Fig. P2-24

2-25. Find e_o in Fig. P2-25.

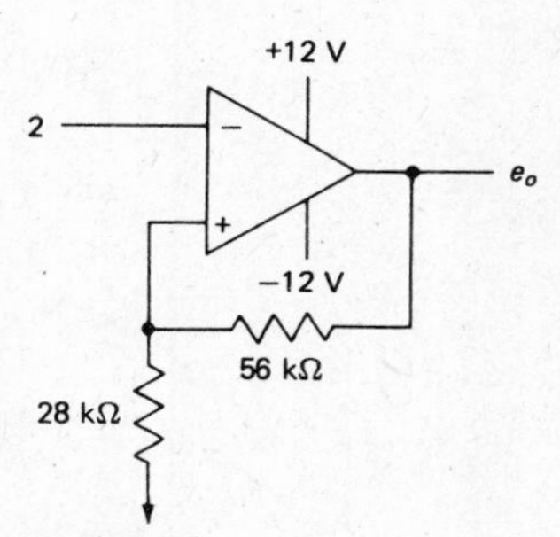

Fig. P2-25

2-26. The switch in Fig. P2-26 is closed for $\frac{1}{2}$ min and then opened. The output which initially was 0 V is at -5.4 V after the switch is opened. What is the input signal?

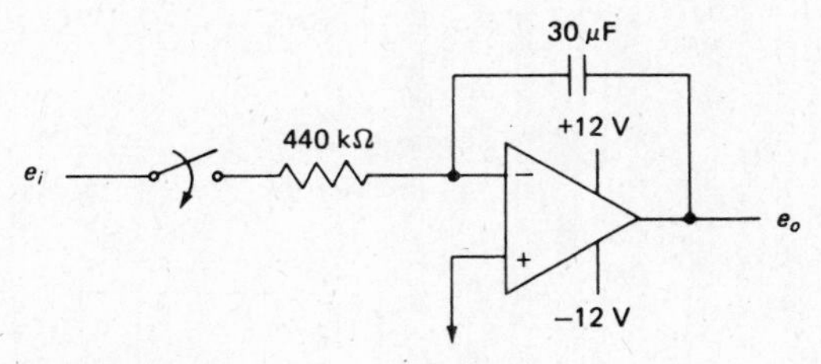

Fig. P2-26

2-27. A synchro control transformer has an output voltage equal to 75 sin θ, where θ is the rotor angle. What is its approximate gain in the linear region?

2-28. A transmitter (TX) and control transformer (CT) have their stators connected. The CT rotor output is given by 90 sin $(\theta_C - \theta_T)$, where θ_C is the angle of the CT with respect to its zero position and θ_T is the angle of the TX with respect to its zero position.

(a) With the zero positions of both synchros at 0°, find the CT rotor output when the TX shaft is set at 50° and the CT shaft is set at 120°.

(b) With the CT zero position at 0° and the TX zero position at 30°, find the CT rotor output for the angles of part (a).

chapter 3

Mathematical Techniques

Mathematicians have developed formal techniques to solve the differential equations which are characteristic of all physical processes. One of the problems facing engineers is the development of equations that describe the dynamic operation of many components and systems. These equations can generally be obtained using basic laws of physics and electricity. In some cases, the equations are derived by a *model* which is obtained from experimental observation.

Once the equations are determined, they must be solved to determine the response of the system to various inputs relevant to the way the system will be used. The differential equations of the linear type can be solved using a formal *classical* method. The technician and engineer, however, are more comfortable with an automated algebraic technique such as the Laplace transform, which lends itself to computerization.

3-1 DIFFERENTIAL EQUATIONS

A differential equation is an equation involving the derivatives of a function. For example, the equation

$$2x = 8 \tag{3-1}$$

is an algebraic equation whose solution is

$$x = 4 \tag{3-2}$$

$$\text{while} \quad \frac{dx(t)}{dt} + 2x(t) = 8 \tag{3-3}$$

is a differential equation whose solution is

$$x(t) = 4 + Ke^{-2t} \tag{3-4}$$

where K depends on the initial conditions. This book deals only with the solution of linear differential equations with constant coefficients. This means that all coefficients will be constant and no square, cube, etc., or product terms will be found.

Examples of linear differential equations with constant coefficients are

$$\frac{dx(t)}{dt} + 2x(t) = 8 \tag{1}$$

$$\frac{d^2x(t)}{dt^2} + \frac{2dx(t)}{dt} + 4x(t) = 10 \tag{2}$$

$$\frac{d^3x(t)}{dt^2} + \frac{3d^2x(t)}{dt^2} + \frac{4dx(t)}{dt} + 2x(t) = 7 \tag{3}$$

Examples which are *not* linear differential equations with constant coefficients are

$$\frac{dx(t)}{dt} + 2x(t)^2 = 8 \tag{1}$$

$$\frac{dx(t)^2}{dt} + 5x(t) = 9 \tag{2}$$

$$\frac{dx(t)}{dt} + 2tx(t) = 7 \tag{3}$$

$$\frac{d^2x(t)}{dt^2}\frac{dx(t)}{dt} + 2x(t) = 8 \tag{4}$$

To save time in writing differential equations, the following notation will be used:

$$\dot{x}(t) = \frac{dx(t)}{dt} \qquad \ddot{x}(t) = \frac{d^2x(t)}{dt^2} \qquad \dddot{x}(t) = \frac{d^3x(t)}{dt^3} \tag{3-5}$$

and so forth. Furthermore, if it is assumed that all variables are functions of time, time will be implied and need not be written. As an example, write

$$x \text{ instead of } x(t) \qquad \text{or} \qquad \dot{x} \text{ instead of } \frac{dx(t)}{dt} \tag{3-6}$$

The order of a differential equation is defined as the highest derivative present. Therefore

$$\begin{aligned} \dot{x} + 2x &= 8 && \text{first order} \\ \ddot{x} + 2\dot{x} + 4x &= 10 && \text{second order} \\ \ddot{x} + 2x &= 8 && \text{second order} \\ \dddot{x} + 3\ddot{x} + 2\dot{x} + x &= 12 && \text{third order} \end{aligned}$$

In order to solve a differential equation, a set of initial conditions is required where the number of initial conditions must equal the order of the differential equation. The initial conditions are normally given at the starting time $t = 0$, whereas, in electric circuits the initial conditions are usually quoted at $t = 0+$ (an infinitesimal time after the switch is closed). As an example, suppose a car is moving around a track at a constant velocity of 100 ft/s and the position of the vehicle is required after 10 s. If x is position, then $\dot{x}$ is velocity, and the equation can be written as

$$\dot{x} = 100 \tag{3-7}$$

Without thinking, it is easy to write

$$\begin{aligned} & x = 100t \\ \text{or} \quad & x = 1000 \text{ ft} \qquad \text{when } t = 10 \text{ s} \end{aligned} \tag{3-8}$$

But this solution only tells us the car has traveled 1000 ft, not where it is. One initial condition [Eq. (3-7) is a first-order equation] is required. Suppose that the initial condition is $x(0) = 10$; that is, at $t = 0$, the vehicle is located 10 ft beyond a marker on the track. Then the true answer to the problem is

$$x = 1010 \qquad \text{when } t = 10 \text{ s}$$

Example 3-1

Since

$$\ddot{x} + 2\dot{x} + x = 7 \qquad \begin{aligned} x(0) &= 0 \\ \dot{x}(0) &= 0 \end{aligned}$$

is a second-order equation, two initial conditions are required.

Since

$$\dot{x} + 2x = 5 \qquad x(0) = 5$$

is first order, one initial condition is required.
Since

$$\ddot{x} + 5x = 8 \qquad \begin{aligned} x(0) &= 5 \\ \dot{x}(0) &= -2 \end{aligned}$$

is second order, two initial conditions are required.

3-2 CLASSICAL METHOD

This method is based on an observation that the solution of a differential equation consists of two parts. One part is called the *forced* or *steady-state response*. The other part is called the *natural* or *transient response*. If the variable in the differential equation is x, the solution is written as

$$x = x_f + x_n \tag{3-9}$$

where x_n is the natural response and x_f is the forced response. For example, if we have

$$\dot{x} + 2x = 8 \qquad x(0) = 0 \tag{3-10}$$

we substitute (3-9) into (3-10) and get

$$(\dot{x}_f + \dot{x}_n) + 2(x_f + x_n) = 8 \tag{3-11}$$

3-2-1 The Forced Response

The right-hand side of the differential equation is called the *input* or *driving function*. The forced or steady-state response always resembles the input. Since the input is a constant, the conclusion is that

$$x_f = \text{const}$$

Assuming that the natural response has died away when the forced response exists, Eq. (3-11) can be rewritten as

$$\dot{x}_f + 2x_f = 8 \tag{3-12}$$

but if x_f = constant, then

$$\dot{x}_f = 0$$

and Eq. (3-12) becomes

$$2x_f = 8 \tag{3-13}$$

or $$x_f = 4 \tag{3-14}$$

3-2-2 The Transient Response

The natural response is the solution of the differential equation with the input equal to zero. The equation that results with the right-hand side set to zero is called the *homogeneous equation*. For this example

$$\dot{x}_n + 2x_n = 0 \tag{3-15}$$

In order for a derivative plus a multiple of a function to add to zero, the derivative and the function must "look alike." One function in which all the derivatives have the same form as the original function is the exponential. Assume that the answer is

$$x_n = Ae^{st} \tag{3-16}$$

Taking derivatives, where A and s are constants, gives

$$\dot{x}_n = sAe^{st} \tag{3-17}$$

Inserting (3-16) and (3-17) into (3-15) leads to

$$sAe^{st} + 2Ae^{st} = 0 \tag{3-18}$$

or $$Ae^{st}(s + 2) = 0 \tag{3-19}$$

The solution to (3-19) is

$$s + 2 = 0 \tag{3-20}$$

Otherwise, if $Ae^{st} = 0$, the solution would always be zero. Solving (3-20) gives

$$s = -2 \tag{3-21}$$

Then (3-16) becomes

$$x_n = Ae^{-2t} \tag{3-22}$$

and the solution $x = x_f + x_n$ becomes

$$x = 4 + Ae^{-2t} \tag{3-23}$$

Since Eq. (3-23) is the solution, it must hold for all time, including $t = 0$, but the initial condition is

$$x = 0 \qquad \text{when } t = 0$$

Letting $t = 0$ in (3-23), and knowing that $e^0 = 1$, we have

$$x = 4 + A \tag{3-24}$$

$$\text{or} \quad 0 = 4 + A$$
$$\text{and} \quad A = -4 \tag{3-25}$$

The complete solution (3-23) becomes

$$x = 4 - 4e^{-2t} \tag{3-26}$$

$$\text{or} \quad x = 4(1 - e^{-2t}) \tag{3-27}$$

As another example, consider the second-order differential equation

$$\ddot{x} + 3\dot{x} + 2x = 6 \qquad \begin{array}{l} x(0) = 1 \\ \dot{x}(0) = 0 \end{array} \tag{3-28}$$

The forced response is obtained by letting

$$\ddot{x}_f = 0 \qquad \dot{x}_f = 0$$
$$\text{or} \quad 2x_f = 6$$
$$x_f = 3 \tag{3-29}$$

Assuming that the natural response is

$$x_n = Ae^{st} \tag{3-30}$$

$$\text{we get} \quad \dot{x}_n = sAe^{st} \tag{3-31}$$

$$\text{and} \quad \ddot{x}_n = s^2Ae^{st} \tag{3-32}$$

and the homogeneous equation is

$$\ddot{x}_n + 3\dot{x}_n + 2x_n = 0 \tag{3-33}$$

or $$s^2Ae^{st} + 3sAe^{st} + 2Ae^{st} = 0 \tag{3-34}$$

which becomes

$$(s^2 + 3s + 2)Ae^{st} = 0 \tag{3-35}$$

Since Ae^{st} is not zero, (3-35) becomes

$$s^2 + 3s + 2 = 0 \tag{3-36}$$

Equation (3-36), called the *characteristic equation*, can easily be obtained from the homogeneous equation using the following rule. Let

$$\begin{aligned} x_n &= 1 \\ \dot{x}_n &= s \\ \ddot{x}_n &= s^2 \\ \dddot{x}_n &= s^3 \\ &\cdot\ \cdot\ \cdot\ \cdot\ \cdot \end{aligned}$$

Factoring Eq. (3-36) as

$$s^2 + 3s + 2 = (s + 1)(s + 2)$$

gives

$$(s + 1)(s + 2) = 0 \tag{3-37}$$

which has two solutions

$$s = -1 \qquad \text{and} \qquad s = -2 \tag{3-38}$$

The true natural response has *two* parts (second-order equation)

$$x_n = A_1e^{s_1t} + A_2e^{s_2t} \tag{3-39}$$

or $$x_n = A_1e^{-t} + A_2e^{-2t} \tag{3-40}$$

There are two unknown coefficients, A_1 and A_2, to be found, so two equations are required. First write the total solution

$$x = x_f + x_n$$

or $$x = 3 + A_1e^{-t} + A_2e^{-2t} \tag{3-41}$$

Let $t = 0$ and substitute the initial condition

$$x = 1 \qquad \text{when } t = 0 \tag{3-42}$$

Then $1 = 3 + A_1 + A_2$

Now differentiate both sides of Eq. (3-41)

$$\dot{x} = 0 - A_1e^{-t} - 2A_2e^{-2t} \tag{3-43}$$

Substitute the initial condition $\dot{x} = 0$ when $t = 0$

$$0 = -A_1 - 2A_2 \tag{3-44}$$

Solving (3-44) gives

$$A_1 = -2A_2 \tag{3-45}$$

Substituting (3-45) into (3-42), we get

$$1 = 3 - 2A_2 + A_2 \tag{3-46}$$

or $1 - 3 = -2A_2 + A_2$ or $-2 = -A_2$

or $$A_2 = 2 \tag{3-47}$$

Equation (3-45) results in

$$A_1 = -2A_2 = -4 \tag{3-48}$$

The total solution (3-41) becomes

$$x = 3 - 4e^{-t} + 2e^{-2t} \tag{3-49}$$

It is apparent that the classical method becomes very unwieldy as the order of the differential equation increases. Therefore, it would be much better to use an algebraic method which combines the natural response, forced response, and determination of the coefficients into one procedure. The Laplace-transform method does just that.

3-3 LAPLACE TRANSFORMS

Suppose you only understand the English language and someone used the Spanish word *mañana*. If you consulted a Spanish-English dictionary you would find that *mañana* translated into "tomorrow." It is important to note that *mañana* has no conceptual meaning to you; only the translation into your language brings you the concept of tomorrow or the following day. The Laplace transform is basically the same as a new language, a mathematical language. The translation from the real-world language of mathematics (called the *time domain*) to the Laplace language (called the *frequency domain*) is accomplished with mathematical manipulation. Just as there are Spanish-English and English-Spanish dictionaries, there are also tables of Laplace-transform pairs (see Appendix A). These tables allow a translation from the time domain into the frequency domain (known as the *transform*) or from the frequency domain into the time domain (known as the *inverse transform*).

3-3-1 Mathematical Derivation of Transforms

Defining $x(t)$ as a function in the time domain and $X(s)$ as the equivalent function in the frequency domain, by definition we get

$$X(s) = \int_0^\infty x(t)e^{-st}\,dt \tag{3-50}$$

where $X(s)$ is the Laplace transform of $x(t)$. In shorthand notation

$$X(s) = \mathscr{L}[x(t)] \tag{3-51}$$

Given $X(s)$ and translating into the inverse transform $x(t)$, we have

$$x(t) = \mathscr{L}^{-1}[X(s)] \tag{3-52}$$

As an example, suppose $x(t) = 1$. To be mathematically precise, write

$$x(t) = \begin{cases} 0 & \text{for } t < 0 \\ 1 & \text{for } t > 0 \end{cases}$$

This is called a *unit step function*, written

$$x(t) = u(t) \tag{3-53}$$

Substituting (3-53) into (3-50) gives

$$X(s) = \int_0^\infty u(t)e^{-st}\,dt \tag{3-54}$$

but since $u(t) = 1$ for $t > 0$,

$$X(s) = \int_0^\infty e^{-st}\,dt \tag{3-55}$$

$$X(s) = -\frac{1}{s}e^{-st}\Big]_0^\infty = -\frac{1}{s}e^{-\infty} - \left(-\frac{1}{s}e^{-0}\right) = \frac{1}{s} \tag{3-56}$$

In other words

$$\mathscr{L}[u(t)] = \mathscr{L}[1] = \frac{1}{s} \tag{3-57}$$

As a second example let

$$x(t) = ku(t) \qquad \text{where } k = \text{const}$$

and $ku(t) = \begin{cases} 0 & \text{for } t < 0 \\ k & \text{for } t > 0 \end{cases}$

Then $X(s) = \mathscr{L}[ku(t)] = \int_0^\infty ku(t)e^{-st}\,dt = k\int_0^\infty u(t)e^{-st}\,dt$

$$= k\int_0^\infty e^{-st}\,dt = \frac{k}{s}$$

or $\mathscr{L}[k] = \frac{k}{s}$ (3-58)

As another example, let

$$x(t) = ke^{-at} \qquad \text{where } k, a = \text{const}$$

Then

$$X(s) = \int_0^\infty ke^{-at}e^{-st}\,dt = k\int_0^\infty e^{-(s+a)t}\,dt = -\frac{k}{s+a}e^{-(s+a)t}\Big]_0^\infty$$

$$= -\frac{k}{s+a}e^{-\infty} - \left(-\frac{k}{s+a}e^{-0}\right) = \frac{k}{s+a}$$

or $\mathscr{L}[ke^{-at}] = \frac{k}{s+a}$ (3-59)

As another example

$$x(t) = k \cos \omega t$$

Using Euler's theorem

$$e^{j\omega t} = \cos \omega t + j \sin \omega t \qquad \text{and} \qquad e^{-j\omega t} = \cos \omega t - j \sin \omega t$$

we get

$$k \cos \omega t = k \frac{e^{j\omega t} + e^{-j\omega t}}{2}$$

Then

$$\begin{aligned} X(s) &= \int_0^\infty k \cos \omega t e^{-st}\, dt = \frac{k}{2} \int_0^\infty (e^{j\omega t} + e^{-j\omega t}) e^{-st}\, dt \\ &= \frac{k}{2} \left(\int_0^\infty e^{j\omega t} e^{-st}\, dt + \int_0^\infty e^{-j\omega t} e^{-st}\, dt \right) \\ &= \frac{k}{2} \left(\int_0^\infty e^{-(s-j\omega)t}\, dt + \int_0^\infty e^{-(s+j\omega)t}\, dt \right) \end{aligned}$$

or $$X(s) = \frac{k}{2} \left(\frac{1}{s - j\omega} + \frac{1}{s + j\omega} \right) = k \frac{s}{s^2 + \omega^2}$$

or $$\mathcal{L}[k \cos \omega t] = k \frac{s}{s^2 + \omega^2} \tag{3-60}$$

3-3-2 Laplace-Theorem Applications

There are many theorems which make taking the transforms easier. The important ones are listed in Appendix B. The examples below illustrate the use of the theorems.

Example 3-1

If

$$x_1(t) = 1 \qquad x_2(t) = e^{-2t}$$

find $\mathcal{L}[4x_1(t) + 3x_2(t)]$.

Solution

$$X_1(s) = \frac{1}{s} \qquad X_2(s) = \frac{1}{s + 2}$$

$$\mathcal{L}[4x_1(t) + 3x_2(t)] = 4X_1(s) + 3X_2(s) = \frac{4}{s} + \frac{3}{s + 2}$$

Example 3-2

If $x(t) = e^{-2t}$, find $\mathscr{L}[\dot{x}(t)]$.

Solution

Let

$$t = 0+ \qquad x(0+) = e^0 = 1$$

Find X(s) from the transform pairs in Appendix B

$$X(s) = \frac{1}{s+2}$$

Use the theorem that

$$\mathscr{L}[\dot{x}(t)] = sX(s) - x(0+)$$

$$\mathscr{L}[\dot{x}(t)] = \frac{s}{s+2} - 1 = \frac{s-(s+2)}{s+2} = \frac{-2}{s+2}$$

As a check, since $x(t) = e^{-2t}$ and $\dot{x}(t) = -2e^{-2t}$,

$$\mathscr{L}[-2e^{-2t}] = -\frac{2}{s+2}$$

Example 3-3

If $x(t) = e^{-2t}\,dt$, find $\mathscr{L}\left[\int x(t)\,dt\right]$.

Solution

Let $x^{-1}(t) = \int e^{-2t}\,dt = \dfrac{e^{-2t}}{-2}$

so that $x^{-1}(0+) = -\frac{1}{2}$

But since $X(s) = \dfrac{1}{s+2}$

we have $\mathscr{L}\left[\int x(t)\,dt\right] = \dfrac{X(s)}{s} + \dfrac{x^{-1}(0+)}{s}$

$$= \frac{1}{s(s+2)} - \frac{\frac{1}{2}}{s} = \frac{1-\frac{1}{2}(s+2)}{s(s+2)}$$

$$= \frac{-s/2}{s(s+2)} = \frac{-\frac{1}{2}}{s+2}$$

Example 3-4

If $X(s) = 1/(s + 2)$, find $x(t)$ when $t = 0+$.

Solution

$$x(0+) = \lim_{s\to\infty} sF(s) = \lim_{s\to\infty} \frac{s}{s+2}$$

$$= \frac{\infty^*}{\infty + 2} = \frac{\infty}{\infty} = 1$$

As a check, from the transform-pair tables

$$x(t) = e^{-2t} \qquad \text{and} \qquad x(0+) = e^{-0} = 1$$

Example 3-5

If $X(s) = 1/(s + 2)$, find $x(t)$ after a very long time.

Solution

$$x(\infty) = \lim_{s\to 0} sX(s) = \lim_{s\to 0} \frac{s}{s+2} = \frac{0}{2} = 0$$

As a check, $x(t) = e^{-2t}$

$$x(\infty) = e^{-\infty} = 0$$

Example 3-6

If $x(t) = e^{-2t}$, find $\mathscr{L}[\ddot{x}(t)]$.

Solution

Let $t = 0+ \qquad x(0+) = e^{-0} = 1$

Find $\dot{x}(t)$:

$$\dot{x}(t) = -2e^{-2t}$$

Let $t = 0+$

$$\dot{x}(0+) = -2e^{-0} = -2$$

* This is not the proper mathematical way of treating limits, but it illustrates the point.

Determine

$$X(s) = \frac{1}{s+2}$$

$$\mathscr{L}[\ddot{x}(t)] = s^2X(s) - sx(0+) - \dot{x}(0+)$$

$$= \frac{s^2}{s+2} - s + 2 = \frac{s^2 - s(s+2) + 2(s+2)}{s+2}$$

$$= \frac{s^2 - s^2 - 2s + 2s + 4}{s+2} = \frac{4}{s+2}$$

3-3-3 Partial Fractions

Suppose you are given

$$X(s) = \frac{1}{s(s+2)}$$

and asked to find $x(t)$, but your table of transform pairs does not show $X(s)$. If $X(s)$ can be broken down into fractions

$$X(s) = \frac{k_1}{s} + \frac{k_2}{s+2}$$

you can use the Laplace tables to find

$$x(t) = k_1 + k_2e^{-2t}$$

The partial-fractions technique provides a method of finding k_1 and k_2.

Step 1. Write $X(s)$ as the sum of fractions

$$\frac{1}{s(s+2)} = \frac{k_1}{s} + \frac{k_2}{s+2} \tag{3-61}$$

Step 2. To find k_1 eliminate the second term

$$\frac{1}{s(s+2)} = \frac{k_1}{s}$$

and cancel s on both sides

$$k_1 = \frac{1}{s+2}$$

Now let s be that value which would make the denominator of k_1 (in the original partial-fraction expansion) equal to zero.

In this case, let $s = 0$

$$k_1 = \left.\frac{1}{s+2}\right|_{s=0} = \frac{1}{2}$$

Step 3. To find k_2 eliminate the first term in (3-61)

$$\frac{1}{s(s+2)} = \frac{k_2}{s+2}$$

then cancel $s + 2$

$$k_2 = \frac{1}{s}$$

Let $s = -2$ (the denominator of k_2 was $s + 2$; let $s + 2 = 0$, or $s = -2$) and get

$$k_2 = \left.\frac{1}{s}\right|_{s=-2} = -\frac{1}{2}$$

Equation (3-61) becomes

$$\frac{1}{s(s+2)} = \frac{\frac{1}{2}}{s} - \frac{\frac{1}{2}}{s+2}$$

$$\text{or} \quad X(s) = \frac{\frac{1}{2}}{s} - \frac{\frac{1}{2}}{s+2} \tag{3-62}$$

Taking the inverse transform gives

$$x(t) = \mathscr{L}^{-1}[X(s)] = \mathscr{L}^{-1}\left[\frac{\frac{1}{2}}{s} - \frac{\frac{1}{2}}{s+2}\right]$$

$$= \mathscr{L}^{-1}\left[\frac{\frac{1}{2}}{s}\right] - \mathscr{L}^{-1}\left[\frac{\frac{1}{2}}{s+2}\right]$$

and from the tables

$$x(t) = \tfrac{1}{2} - \tfrac{1}{2}e^{-2t}$$

Example 3-7

Solve the Laplace transform

$$I(s) = \frac{10}{(s+1)(s+2)}$$

for the current $i(t)$.

Solution

Step 1

$$\frac{10}{(s+1)(s+2)} = \frac{k_1}{s+1} + \frac{k_2}{s+2}$$

Step 2. Solving for k_1, we get

$$\frac{10}{\cancel{(s+1)}(s+2)} = \frac{k_1}{\cancel{(s+1)}}$$

$$\text{or} \quad k_1 = \left.\frac{10}{s+2}\right|_{s=-1} = \frac{10}{-1+2} = 10$$

Step 3. Solving for k_2, we get

$$\frac{10}{(s+1)\cancel{(s+2)}} = \frac{k_2}{\cancel{(s+2)}}$$

$$\text{or} \quad k_2 = \left.\frac{10}{s+1}\right|_{s=-2} = \frac{10}{-2+1} = -10$$

Then $I(s) = \dfrac{10}{s+1} - \dfrac{10}{s+2}$ or $i(t) = 10e^{-t} - 10e^{-2t}$.

Example 3-8

Solve the Laplace transform

$$V(s) = \frac{s+5}{s(s+1)(s+2)}$$

for $v(t)$.

Solution

Step 1

$$\frac{s+5}{s(s+1)(s+2)} = \frac{k_1}{s} + \frac{k_2}{s+1} + \frac{k_3}{s+2}$$

Step 2. Solving for k_1, we get

$$\frac{s+5}{\cancel{s}(s+1)(s+2)} = \frac{k_1}{\cancel{s}}$$

$$\text{or} \quad k_1 = \left.\frac{s+5}{(s+1)(s+2)}\right|_{s=0} = \frac{5}{(1)(2)} = 2.5$$

Step 3. Solving for k_2, we get

$$\frac{s+5}{s\cancel{(s+1)}(s+2)} = \frac{k_2}{\cancel{s+1}}$$

$$\text{or} \quad k_2 = \left.\frac{s+5}{s(s+2)}\right|_{s=-1} = \frac{-1+5}{(-1)(-1+2)} = \frac{4}{(-1)(1)} = -4$$

Step 4. Solving for k_3, we get

$$\frac{s+5}{s(s+1)\cancel{(s+2)}} = \frac{k_3}{\cancel{s+2}}$$

$$\text{or} \quad k_3 = \left.\frac{s+5}{s(s+1)}\right|_{s=-2} = \frac{-2+5}{(-2)(-2+1)} = \frac{3}{(-2)(-1)} = \frac{3}{2} = 1.5$$

$$\text{Then} \quad V(s) = \frac{2.5}{s} - \frac{4}{s+1} + \frac{1.5}{s+2}$$

$$\text{or} \quad V(t) = 2.5 - 4e^{-t} + 1.5e^{-2t}$$

3-3-4 Solving Differential Equations using Laplace Techniques

Consider the equation previously solved by the classical method

$$\dot{x}(t) + 2x(t) = 8 \qquad x(0+) = 0 \tag{3-63}$$

Since $x(t)$ is the time domain function, let $X(s)$ be the frequency-domain function. Take the Laplace transform of both sides of (3-63)

$$\mathscr{L}[\dot{x}(t) + 2x(t)] = \mathscr{L}[8] \tag{3-64}$$

but $\mathscr{L}[\dot{x}(t) + 2x(t)] = \mathscr{L}[\dot{x}(t)] + 2\mathscr{L}[x(t)]$

and $\mathscr{L}[\dot{x}(t)] = sX(s) - x(0+) = sX(s) - 0$

$2\mathscr{L}[x(t)] = 2X(s)$

Also $\mathscr{L}[8] = \dfrac{8}{s}$

Equation (3-64) becomes

$$sX(s) + 2X(s) = \frac{8}{s} \qquad \text{or} \qquad (s+2)X(s) = \frac{8}{s}$$

Solving for $X(s)$ gives

$$X(s) = \frac{8}{s(s+2)} \tag{3-65}$$

and using partial fractions leads to

$$\frac{8}{s(s+2)} = \frac{k_1}{s} + \frac{k_2}{s+2}$$

$$k_1 = \left.\frac{8}{s+2}\right|_{s=0} = 4$$

$$k_2 = \left.\frac{8}{s}\right|_{s=-2} = -4$$

Equation (3-65) becomes

$$X(s) = \frac{4}{s} - \frac{4}{s+2} \tag{3-66}$$

From the tables of transform pairs (Appendix B) the inverse transform of (3-66) is

$$x(t) = 4 - 4e^{-2t} = 4(1 - e^{-2t}) \tag{3-67}$$

Notice that the Laplace method did not require finding the natural and forced response separately or the coefficient of the natural response.

Solution by Laplace methods is very mechanical and involves the following steps:

1. Take the Laplace of both sides of the differential equation.
2. Solve the resulting algebraic equation for the unknown Laplace variable.

3. Expand the result into partial fractions.
4. Determine the inverse transform.

Example 3-9

This is the same as Eq. (3-63), but the initial condition is *not* zero

$$\dot{x}(t) + 2x(t) = 8 \qquad x(0+) = 1 \tag{3-68}$$

Step 1

$$\mathscr{L}[\dot{x}(t) + 2x(t)] = \mathscr{L}[8] \tag{3-69}$$

or $\mathscr{L}[x(t)] + 2\mathscr{L}[x(t)] = \mathscr{L}[8]$

but $\mathscr{L}[\dot{x}(t)] = sX(s) - x(0+) = sX(s) - 1$

and $2\mathscr{L}[x(t)] = 2X(s)$ and $\mathscr{L}[8] = \dfrac{8}{s}$

Equation (3-69) becomes

$$[sX(s) - 1] + 2X(s) = \frac{8}{s} \tag{3-70}$$

Step 2. Equation (3-70) becomes

$$sX(s) + 2X(s) = \frac{8}{s} + 1$$

or
$$(s + 2)X(s) = \frac{8 + s}{s} \tag{3-71}$$

Then
$$X(s) = \frac{8 + s}{s(s + 2)} \tag{3-72}$$

Step 3. Since

$$\frac{8 + s}{s(s + 2)} = \frac{k_1}{s} + \frac{k_2}{s + 2}$$

and
$$k_1 = \left.\frac{8 + s}{s + 2}\right|_{s=0} = \frac{8}{2} = 4 \qquad k_2 = \left.\frac{8 + s}{s}\right|_{s=-2} = \frac{8 - 2}{-2} = -3$$

we have
$$X(s) = \frac{4}{s} - \frac{3}{s + 2}$$

Step 4

$$x(t) = \mathscr{L}^{-1}\left[\frac{4}{s} - \frac{3}{s+2}\right] = 4 - 3e^{-2t}$$

Example 3-10

Consider Eq. (3-28), which was also solved using the classical method

$$\ddot{x}(t) + 3\dot{x}(t) + 2x(t) = 6 \qquad \begin{aligned} x(0+) &= 1 \\ \dot{x}(0+) &= 0 \end{aligned} \tag{3-73}$$

Step 1

$$\mathscr{L}[\ddot{x}(t) + 3\dot{x}(t) + 2x(t)] = \mathscr{L}[6] \tag{3-74}$$

or $\mathscr{L}[\ddot{x}(t)] + 3\mathscr{L}[\dot{x}(t)] + 2\mathscr{L}[x(t)] = \mathscr{L}[6]$

but $\mathscr{L}[\ddot{x}(t)] = s^2X(s) - sx(0+) - \dot{x}(0+)$

$= s^2X(s) - s \cdot (1) - 0 = s^2X(s) - s$

and $3\mathscr{L}[\dot{x}(t)] = 3[sX(s) - x(0+)] = 3[sX(s) - 1] = 3sX(s) - 3$

$2\mathscr{L}[x(t)] = 2X(s)$

$$\mathscr{L}[6] = \frac{6}{s}$$

Equation (3-74) becomes

$$[s^2X(s) - s] + [3sX(s) - 3] + 2X(s) = \frac{6}{s} \tag{3-75}$$

Step 2. Equation (3-75) becomes

$$s^2X(s) + 3sX(s) + 2X(s) = \frac{6}{s} + s + 3$$

or $$[s^2 + 3s + 2]X(s) = \frac{6 + s^2 + 3s}{s} \tag{3-76}$$

But since

$$s^2 + 3s + 2 = (s+1)(s+2)$$

we have

$$(s+1)(s+2)X(s) = \frac{6 + s^2 + 3s}{s}$$

$$\text{or} \quad X(s) = \frac{6 + s^2 + 3s}{s(s + 1)(s + 2)} \tag{3-77}$$

Step 3

$$\frac{6 + s^2 + 3s}{s(s + 1)(s + 2)} = \frac{k_1}{s} + \frac{k_2}{s + 1} + \frac{k_3}{s + 2}$$

$$\text{and} \quad k_1 = \left.\frac{6 + s^2 + 3s}{(s + 1)(s + 2)}\right|_{s=0} = \frac{6}{2} = 3$$

$$k_2 = \left.\frac{6 + s^2 + 3s}{s(s + 2)}\right|_{s=-1} = \frac{6 + 1 - 3}{(-1)(1)} = -4$$

$$k_3 = \left.\frac{6 + s^2 + 3s}{s(s + 1)}\right|_{s=-2} = \frac{6 + 4 - 6}{(-2)(-1)} = 2$$

$$\text{Then} \quad X(s) = \frac{3}{s} - \frac{4}{s + 1} + \frac{2}{s + 2}$$

Step 4

$$x(t) = \mathcal{L}^{-1}[X(s)] = 3 - 4e^{-t} + 2e^{-2t}$$

3-3-5 Complex Roots

Consider the equation

$$\ddot{x} + 2\dot{x} + 2x = 8 \qquad \begin{aligned} x(0) &= 0 \\ \dot{x}(0) &= 0 \end{aligned} \tag{3-78}$$

Using the abbreviation $X = X(s)$, and taking Laplace transforms, we get

$$\mathcal{L}[\ddot{x}] = s^2X - sx(0) - \dot{x}(0) = s^2X$$
$$\mathcal{L}[\dot{x}] = sX - x(0) = sX$$
$$\mathcal{L}[x] = X$$

$$\text{Then } s^2X + 2sX + 2X = \frac{8}{s}$$

$$\text{or} \quad (s^2 + 2s + 2)X = \frac{8}{s} \tag{3-79}$$

The term $s^2 + 2s + 2$ is only factorable into complex roots, but by completing the square, that is,

$$s^2 + 2as + b = (s + a)^2 + (b - a^2)$$

we get

$$s^2 + 2s + 2 = (s + 1)^2 + 1$$

Then $[(s + 1)^2 + 1]X = \frac{8}{s}$

or $X = \frac{8}{s[(s + 1)^2 + 1]}$ (3-80)

In our transform tables find the form

$$X = \frac{k}{s[(s + a)^2 + \omega^2]} \tag{3-81}$$

Equations (3-80) and (3-81) are the same if

$$k = 8 \qquad a = 1 \qquad \omega = 1$$

but the inverse transform of (3-81) is

$$x = \frac{k}{a^2 + \omega^2} - \frac{ke^{-at} \sin (\omega t + \psi)}{\omega \sqrt{a^2 + \omega^2}} \qquad \psi = \tan^{-1} \frac{\omega}{a}$$

Therefore the inverse transform of (3-80) is

$$x = \frac{8}{1^2 + 1^2} - \frac{8e^{-t} \sin (t + \psi)}{1 \sqrt{1^2 + 1^2}} \qquad \psi = \tan^{-1} \frac{1}{1} = 45^\circ$$

or $x = 4 - \frac{8}{\sqrt{2}} e^{-t} \sin (t + 45^\circ)$

3-3-6 Repeated Roots

When the time-domain solution contains repeated roots, the transform can be solved as follows:

$$\ddot{y} + 2\dot{y} + y = 8 \qquad \begin{matrix} y(0) = 0 \\ \dot{y}(0) = 0 \end{matrix}$$

$$\mathscr{L}[\ddot{y} + 2\dot{y} + y] = \mathscr{L}[8]$$
$$\mathscr{L}[\ddot{y}] = s^2y - sy(0) - \dot{y}(0) = s^2Y$$
$$\mathscr{L}[\dot{y}] = sY - y(0) = sY$$

or $s^2Y + 2sY + Y = \frac{8}{s}$

$$Y(s^2 + 2s + 1) = \frac{8}{s}$$

but since

$$s^2 + 2s + 1 = (s + 1)^2$$

we have

$$Y = \frac{8}{s(s + 1)^2} \tag{3-82}$$

Because of the square term, the partial-fraction expansion is given by

$$\frac{8}{s(s + 1)^2} = \frac{A}{s} + \frac{B}{(s + 1)^2} + \frac{C}{s + 1}$$

$$A = \left.\frac{8}{(s + 1)^2}\right|_{s=0} = 8$$

$$B = \left.\frac{8}{s}\right|_{s=-1} = -8$$

C cannot be found in exactly the same manner, but using differentiation we get

$$C = \lim_{s \to -1} \frac{d}{ds}\left[(s + 1)^2 \frac{8}{s(s + 1)^2}\right]$$

$$= \lim_{s \to -1} \frac{d}{ds}\left(\frac{8}{s}\right)$$

$$= \lim_{s \to -1} \frac{-8}{s^2} = \frac{-8}{(-1)^2} = -8$$

Equation (3-82) can now be written as

$$Y = \frac{8}{s} - \frac{8}{(s + 1)^2} - \frac{8}{s + 1} \tag{3-83}$$

The inverse transform of (3-83) can be determined from the table of transform pairs (Appendix B)

$$y = 8 - 8te^{-t} - 8e^{-t}$$

3-3-7 Imaginary Roots

When the solution contains imaginary roots, the following technique can be used to solve the transform:

$$\begin{aligned} &\ddot{y} + 4y = 8 \qquad y(0) = \dot{y}(0) = 0 \\ &\mathcal{L}[\ddot{y}] + 4\mathcal{L}[y] = \mathcal{L}[8] \\ &\mathcal{L}[\ddot{y}] = s^2Y - sy(0) - \dot{y}(0) = s^2Y \\ &\mathcal{L}[y] = Y \end{aligned} \tag{3-84}$$

Equation (3-84) becomes

$$s^2Y + 4Y = \frac{8}{s}$$

$$\text{or} \quad Y = \frac{8}{s(s^2+4)} = \frac{8}{s(s+2j)(s-2j)}$$

$$\text{or} \quad Y = \frac{A}{s} + \frac{B}{s+2j} + \frac{C}{s-2j} \tag{3-85}$$

$$\text{and} \quad A = \left.\frac{8}{(s-2j)(s+2j)}\right|_{s=0} = \frac{8}{-4j^2} = 2$$

$$B = \left.\frac{8}{s(s-2j)}\right|_{s=-2j} = \frac{8}{-2j(-4j)} = -1$$

$$C = \left.\frac{8}{s(s+2j)}\right|_{s=2j} = \frac{8}{2j(4j)} = -1$$

Equation (3-85) can be written as

$$Y = \frac{2}{s} - \frac{1}{s+2j} - \frac{1}{s-2j} = \frac{2}{s} - \left(\frac{1}{s+2j} + \frac{1}{s-2j}\right) \tag{3-86}$$

$$\text{but} \quad \frac{1}{s+2j} + \frac{1}{s-2j} = \frac{s-2j+s+2j}{(s+2j)(s-2j)} = \frac{2s}{s^2+4}$$

or Eq. (3-86) becomes

$$Y = \frac{2}{s} - \frac{2s}{s^2+4}$$

and the inverse transform is

$$y(t) = 2 - 2\cos 2t$$

PROBLEMS

3-1. Find the Laplace transforms of the following time functions:

(a) $y(t) = 7 + 5t$

(b) $c(t) = 3(1 - e^{-t})$

(c) $f(t) = t^2 + 2 \sin 4t$

(d) $i(t) = -5e^{-3t} \cos 5t$

3-2. Find the partial-fraction expansions (do not solve for the inverse transforms).

(a) $\dfrac{3s + 1}{s(s + 1)}$

(b) $\dfrac{5s + 6}{(s + 2)(s + 3)}$

(c) $\dfrac{6s^2 + 10s + 2}{s^3 + 3s^2 + 2s}$

(d) $\dfrac{1}{4s(s + 4)}$

3-3. Given the following Laplace transforms, obtain the inverse transforms $f(t)$.

(a) $F(s) = \dfrac{6}{s(s + 3)}$

(b) $F(s) = \dfrac{20}{s(s^2 + 10s + 9)}$

(c) $F(s) = \dfrac{3s}{s^2 + 9}$

(d) $F(s) = \dfrac{5}{s^2 + 1}$

(e) $F(s) = \dfrac{4 - 2s}{s^3 + 4s}$

(f) $F(s) = \dfrac{s}{s^2 + 2s + 2}$

3-4. Given

$$Y(s) = \frac{100}{s[(s + 5)^2 + 25]}$$

find $y(t)$.

3-5. Find $F(s)$ for the following differential equations.

(a) $\dfrac{3df(t)}{dt} + 4f(t) = 5 \qquad f(0) = 0$

(b) $\dfrac{2d^2f(t)}{dt^2} + \dfrac{3df(t)}{dt} + 4f(t) = e^{-5t} + 6 \cos 7t \qquad \dfrac{df(0)}{dt} = 1 \quad f(0) = 2$

3-6. Solve the following differential equations using Laplace transforms.

(a) $\dot{x}(t) + 2x(t) = 8 \qquad x(0) = 0$

(b) $\ddot{y}(t) + 4\dot{y}(t) + 3y(t) = 6 \qquad y(0) = 0 \quad \dot{y}(0) = 0$

(c) $\ddot{y}(t) + 9y(t) = 0 \qquad \dot{y}(0) = 3 \quad y(0) = 0$

(d) $\ddot{x}(t) + 2\dot{x}(t) + 10x(t) = 10 \qquad \dot{x}(0) = 0 \quad x(0) = 0$

3-7. If $\mathscr{L}[x(t)] = X(s)$, find:

(a) $\mathscr{L}[\dot{x}(t)]$ when $X(s) = \dfrac{2}{s}$ and $x(0) = 2$

(b) $\mathscr{L}[\ddot{x}(t)]$ when $X(s) = \dfrac{5}{s+1}$ and $x(0) = 5$, $\dot{x}(0) = -5$

(c) $\mathscr{L}[\int x(t)]$ when $X(s) = \dfrac{3}{s^2+9}$ and $x^{-1}(0) = -\frac{1}{3}$

3-8. Find i using Laplace transforms (see Fig. P3-8).

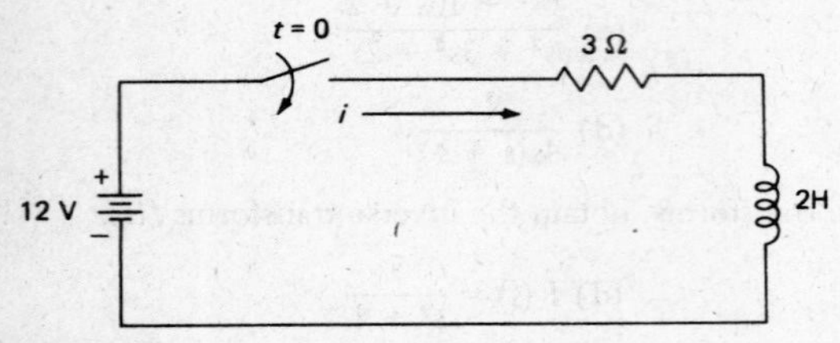

Fig. P3-8

3-9. Solve for i using Laplace transforms (see Fig. P3-9).

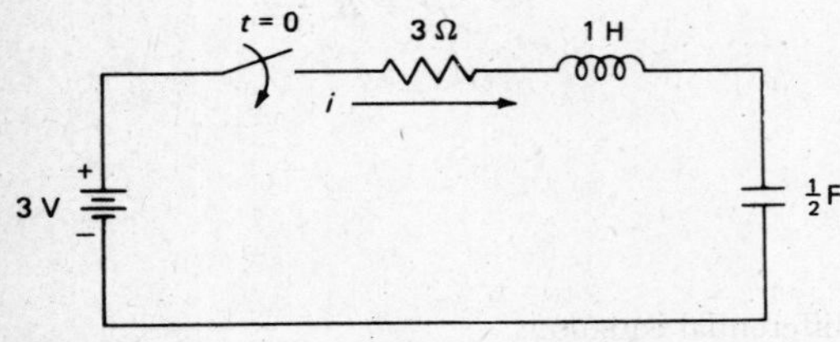

Fig. P3-9

3-10. Repeat Prob. 3-9 for $R = 2\ \Omega$, $L = 1$ H, $C = 0.1$ F.

3-11. The Laplace-transformed error of a servo system is given by

$$E(s) = \frac{17s^3 + 7s^2 + s + 6}{s^5 + 3s^4 + 5s^3 + 4s^2 + 2s}$$

Using the initial- and final-value theorems (Appendix B), find:

(a) The initial-value error $e(0)$

(b) The steady-state or final-value error $e(\infty)$

chapter 4

Transfer Functions of Components and Block Diagrams

Chapter 2 dealt with several control-system components. For each of those components an input-output relationship was obtained called the gain of the component. This gain related the input and output in the time domain. In Chap. 3 a technique was presented which enables us to convert from the time domain into the frequency domain. The gain which relates input to output in the frequency domain is called the *transfer function* of the component. In many instances the gain in the time domain is identical in appearance to the transfer function in the frequency domain. This occurs when the gain is constant for all time.

4-1 TRANSFER FUNCTIONS OF COMPONENTS

4-1-1 Potentiometer

The input-output relationship for the potentiometer was given by Eq. (2-2)

$$e_o(t) = \frac{E_R}{\theta_{\max}} \theta_i(t)$$

which can be rewritten

$$e_o(t) = K_p \theta_i(t) \tag{4-1}$$

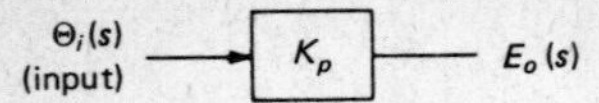

Fig. 4-1 Block diagram of a potentiometer.

where K_p is the gain of the potentiometer in volts per degree. Taking the Laplace transform of Eq. (4-1), we obtain

$$\mathscr{L}[e_o(t)] = \mathscr{L}[K_p\theta_i(t)] \tag{4-2a}$$

$$E_o(s) = K_p\Theta_i(s) \tag{4-2b}$$

where $\Theta_i(s)$ represents the Laplace transform of the input to the potentiometer and $E_o(s)$ represents the transform of its output. When working with control systems it is more practical to represent a component by a simple block containing its transfer function.

From Eq. (4-2*b*) we can obtain the transfer function of the potentiometer

$$K_p = \frac{E_o(s)}{\Theta_i(s)} \tag{4-3}$$

Here we see that the transfer function and the time-domain gain are identical (K_p). The block representing the potentiometer is shown in Fig. 4-1.

4-1-2 Accelerometer

Equation (2-30), repeated here for reference, expressed the relation between the acceleration a of a body and the displacement y of a mass

$$ma = m\frac{d^2y}{dt^2} + f_v\frac{dy}{dt} + k_x y \tag{2-30}$$

Taking the Laplace transform of Eq. (2-30) and assuming zero initial conditions, we obtain

$$mA(s) = ms^2Y(s) + f_v sY(s) + k_x Y(s) \tag{4-4}$$

Solving (4-4) for the transfer function $Y(s)/A(s)$, we get

$$\frac{Y(s)}{A(s)} = \frac{1}{s^2 + (f_v/m)s + k_x/m} \tag{4-5}$$

However the output of the accelerometer is not the displacement y but a voltage e_o which is proportional to y (Fig. 2-43). Since

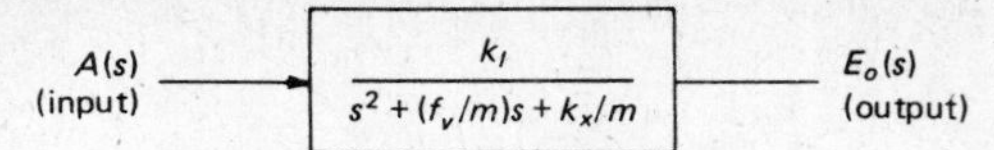

Fig. 4-2 Block diagram of an accelerometer.

$$e_o = k_l y \tag{4-6}$$

for the LVDT, taking the Laplace transform of Eq. (4-6),

$$E_o(s) = k_l Y(s) \tag{4-7}$$

By substituting Eq. (4-7) into (4-5) we obtain

$$\frac{E_o(s)}{A(s)} = \frac{k_l}{s^2 + (f_v/m)s + k_x/m} \tag{4-8}$$

Equation (4-8) is the transfer function of the accelerometer described in Sec. 2-10. The block diagram for the accelerometer is shown in Fig. 4-2. As indicated by its transfer function [Eq. (4-8)], the accelerometer falls into a special category called *second-order systems*. The response of a second-order system is discussed in detail in Chap. 5.

4-1-3 Tachometer

The input-output relationship for the dc Tachometer was given by

$$e_o = K_T \omega \tag{2-8}$$

where e_o = output, V
K_T = tachometer gain, V/(rad/s)
ω = tachometer shaft speed, rad/s

If the shaft displacement is represented by θ rad, the speed will be

$$\omega = \frac{d\theta}{dt} \tag{4-9}$$

Substituting (4-9) into (2-8), we get

$$e_0 = K_T \frac{d\theta}{dt} \tag{4-10}$$

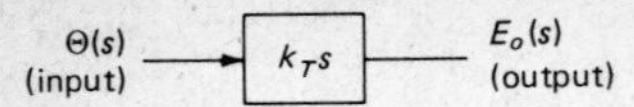

Fig. 4-3 Block diagram of a tachometer.

Taking the Laplace transform of (4-10), we get

$$E_o(s) = K_T s \Theta(s) \tag{4-11}$$

and the transfer function for the tachometer is given by

$$\frac{E_o(s)}{\Theta(s)} = K_T s \tag{4-12}$$

The block diagram for the tachometer is shown in Fig. 4-3.

4-1-4 DC Motor

The equations governing the motor's behavior were derived in Sec. 2-5

$$\mathrm{t}_{\mathrm{dev}} = K_m v_m - f_m \omega \tag{2-13}$$

Since $\omega = d\theta/dt$, (2-15*c*) will be written as

$$\mathrm{t}_{ac} = K_m v_m - F_m \frac{d\theta}{dt} - \mathrm{t}_l \tag{4-13}$$

From Eq. (2-14) the net accelerating torque is given by

$$\mathrm{t}_{ac} = J_m \alpha = J_m \frac{d^2\theta}{dt^2} \tag{4-14}$$

where J_m is the total motor inertia. Equating (4-13) to (4-14), we derive the differential equation

$$K_m v_m - \mathrm{t}_l - F_m \frac{d\theta}{dt} = J_m \frac{d^2\theta}{dt^2} \tag{4-15}$$

Rearranging terms gives

$$K_m v_m - \mathrm{t}_l = J_m \frac{d^2\theta}{dt^2} + F_m \frac{d\theta}{dt} \tag{4-16}$$

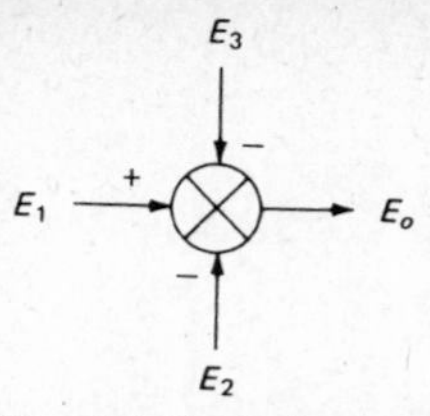

Fig. 4-4 Block-diagram symbol for a summer.

Taking the Laplace transform of Eq. (4-16) and assuming zero initial conditions, we get

$$K_m V_m(s) - T_l(s) = J_m s^2 \Theta(s) + F_m s \Theta(s) \tag{4-17}$$

$$K_m V_m(s) - T_l(s) = (J_m s + F_m) s \Theta(s) \tag{4-18}$$

$$[K_m V_m(s)] - T_l(s)] \frac{1}{J_m s + F_m} = \dot{\Theta}(s) \tag{4-19}$$

Before drawing the block diagram of the motor, the symbol in Fig. 4-4 will be defined. The output E_o of the summer is given by the algebraic sum of all the inputs to the summer taking into account the addition or subtraction symbol assigned to each input.

For example, in Fig. 4-4 if $E_1 = 5$ V, $E_2 = 3$ V, and $E_3 = -7$ V, the output E_o will be

$$E_o = E_1 - E_2 - E_3$$
$$E_o = (5) - (3) - (-7) = 5 - 3 + 7 = 9 \text{ V} \tag{4-20}$$

Figure 4-5 can now be drawn representing the block diagram of the motor. It is a representation of Eq. (4-19).

If the motor is connected to the load T_l through a gear pass with gear ratio a, as in Fig. 4-6, then

$$F_t = f_m + f_v + a^2 F_l \tag{4-21}$$

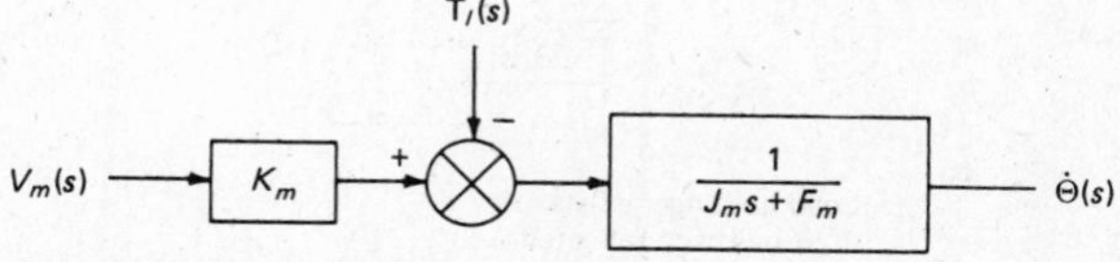

Fig. 4-5 Block diagram of a dc motor.

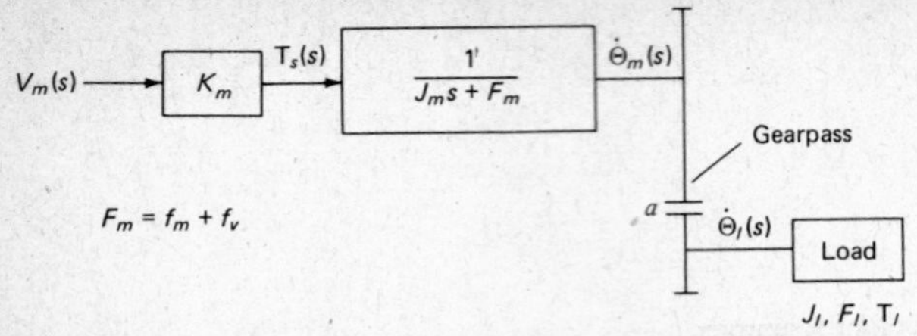

Fig. 4-6 Block diagram of a motor connected to a load through a gearpass.

where f_m = motor viscous friction due to back emf
f_v = motor viscous friction due to bearings
a^2F_l = load viscous friction as seen by motor (Sec. 2-6)

$$\text{and} \quad J_t = J_m + a^2 J_l \tag{4-22}$$

where J_m is the motor, shaft, and gear inertias on the motor side of the gearpass and a^2J_l is the inertia of the load, all its associated shafts, and the load gear as seen by the motor and

$$T_{xm}(s) = aT_l(s)$$

where T_{xm} is the load torque T_l as seen by the motor.

Figure 4-6 is a block diagram showing a motor connected to a load through a gear pass. If the load quantities are reflected to the motor side of the gear pass, Fig. 4-6 can be redrawn as Fig. 4-7, where the quantities F_t and J_t are defined above.

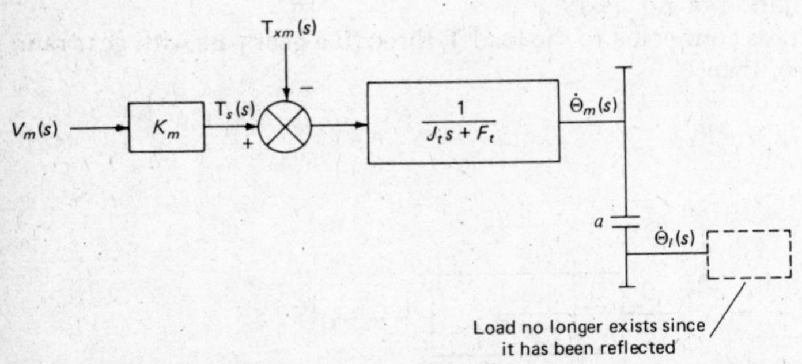

Fig. 4-7 Equivalent representation of Fig. 4-6 with load inertia and friction reflected to the motor.

Referring to Figs. 4-5 to 4-7, we see that if motor A is driving a load with inertia and friction, it is equivalent to a motor B, driving a load without friction or inertia, where motor B has inertia and friction greater than those of motor A by the amount of inertia and friction reflected from the load.

4-2 BLOCK DIAGRAMS

Since the block diagram is to systems analysis what the circuit diagram is to electronic analysis, it is essential to understand and learn the rules for working with block diagrams.

Every feedback control system can be represented by the general block diagram shown in Fig. 4-8, in which $G(s)$ represents all the components in the forward path between input and output. $G(s)$ is sometimes called the *direct transfer function* or *direct transmission gain* (DTG). $H(s)$ represents all the components between the output and input summing point via the feedback path. In many instances $H(s)$ is merely a constant, and quite frequently it is equal to unity. In the case when $H(s) = 1$ (unity feedback) the signal $E(s)$ is the difference between the input and output $[R(s) - C(s)]$. $E(s)$ is commonly called the *error*. It is also called the actuating signal since a signal present at $E(s)$ will actuate or make the system respond. The signal $F(s)$ represents the feedback signal. When $H(s) = 1$, $F(s)$ is equal to the output $C(s)$.

Referring to Fig. 4-8, we can write

$$C(s) = G(s)E(s) \tag{4-23}$$

$$F(s) = H(s)C(s) \tag{4-24}$$

$$E(s) = R(s) - F(s) \tag{4-25}$$

Substituting (4-24) into (4-25) gives

$$E(s) = R(s) - H(s)C(s) \tag{4-26}$$

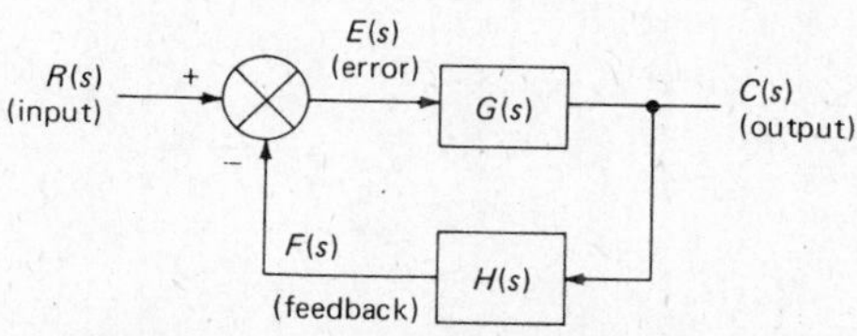

Fig. 4-8 General block diagram of a feedback control system.

and now substituting (4-26) into (4-23) we get

$$C(s) = G(s)[R(s) - H(s)C(s)] = G(s)R(s) - HG(s)C(s)$$
$$C(s)[1 + HG(s)] = G(s)R(s) \tag{4-27}$$

Then $$\frac{C}{R}(s) = \frac{G(s)}{1 + HG(s)} \qquad \textbf{closed-loop transfer function} \tag{4-28}$$

Equation (4-28), which relates output to input, is called the *closed-loop transfer function* of the system. The quantity $HG(s)$ is the product of all the gains in the loop. It is also the ratio of feedback signal to error signal with the feedback loop opened as shown in Fig. 4-9:

$$\frac{F(s)}{E(s)} = HG(s) \qquad \textbf{open-loop transfer function} \tag{4-29}$$

$HG(s)$ is called the *open-loop transfer function* or simply the *loop gain*.

In words, Eq. (4-28) can be stated as

$$\text{Closed-loop transfer function} = \frac{\text{DTG}}{1 + \text{loop gain}} \tag{4-30}$$

Another equation of interest is obtained by substituting (4-23) into (4-28) to get

$$\frac{E}{R}(s) = \frac{1}{1 + HG(s)} \qquad \textbf{actuating-signal ratio} \tag{4-31}$$

called the actuating-signal ratio. Equation (4-31) relates the error to the input.

Referring to Eq. (4-27), if the input $R(s)$ is set to zero, we get

$$1 + HG(s) = 0 \qquad \textbf{characteristic equation} \tag{4-32}$$

called the *characteristic equation* of the system. It is from this equation that information about the dynamic behavior of a system is obtained.

Note that $1 + HG(s)$ is also the denominator of Eq. (4-28). It is important

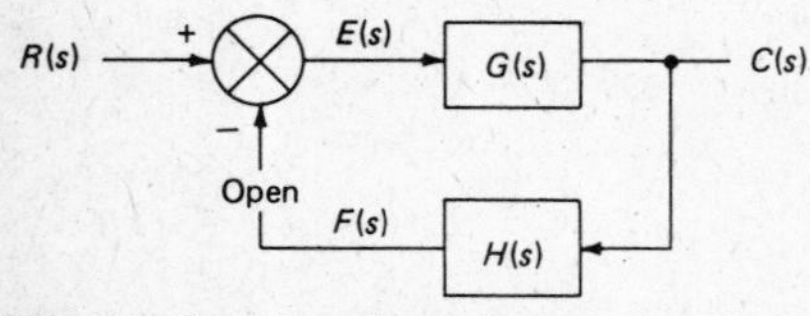

Fig. 4-9 Open-loop block diagram.

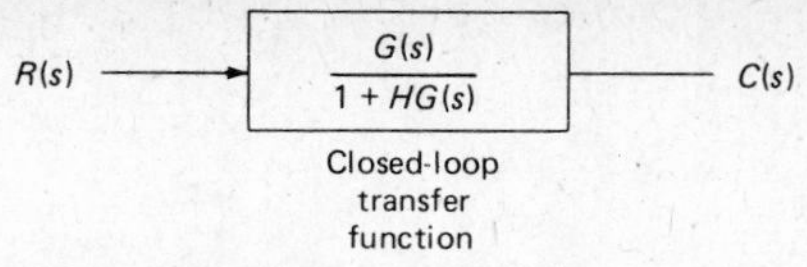

Fig. 4-10 Reduced form of Fig. 4-8.

to realize that all the equations derived above were based on the polarities shown at the summer in Fig. 4-8.

Once the closed-loop transfer function is obtained, Fig. 4-8 can be reduced to Fig. 4-10.

Example 4-1

As an example, consider a system given by the block diagram in Fig. 4-11:

Direct transmission gain:

$$\mathrm{DTG} = G(s) = \frac{10}{s+10}\,\frac{20}{s} = \frac{200}{s(s+10)}$$

Open-loop transfer function:

$$\mathrm{OLTF} = \text{loop gain} = HG(s) = 2\,\frac{200}{s(s+10)}$$

$$HG(s) = \frac{400}{s(s+10)}$$

Closed-loop transfer function:

$$\mathrm{CLTF} = \frac{C}{R}(s) = \frac{\mathrm{DTG}}{1 + \text{loop gain}}$$

$$\frac{C}{R}(s) = \frac{200/s(s+10)}{1 + 400/s(s+10)} = \frac{200}{s(s+10) + 400}$$

$$= \frac{200}{s^2 + 10s + 400}$$

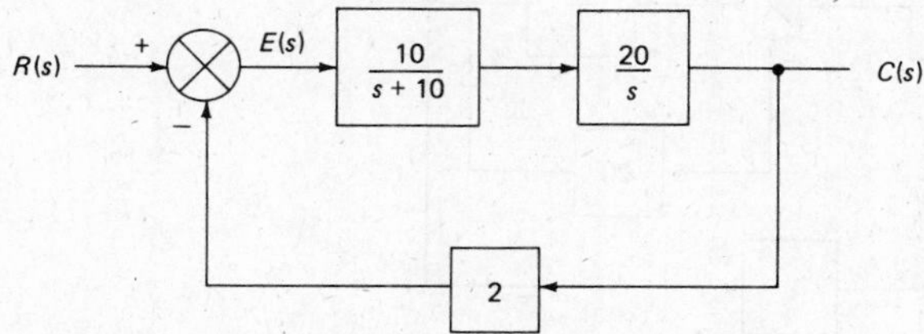

Fig. 4-11 Block diagram for Example 4-1.

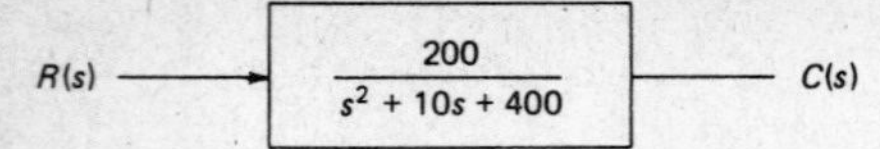

Fig. 4-12 Reduced system diagram for Example 4-1.

Characteristic equation:

$$1 + \text{loop gain} = 0$$
$$1 + HG(s) = 0$$
$$1 + \frac{400}{s(s+10)} = 0$$
$$s(s+10) + 400 = 0$$

Note that this final equation could have been obtained by setting the denominator of the closed-loop transfer function equal to zero.

Actuating-signal ratio:

$$\frac{E}{R}(s) = \frac{1}{1 + \text{loop gain}} = \frac{1}{1 + HG(s)} = \frac{1}{1 + 400/s(s+10)}$$
$$= \frac{s(s+10)}{s(s+10) + 400} = \frac{s(s+10)}{s^2 + 10s + 400}$$

Reduction of the block diagram is shown in Fig. 4-12.

Example 4-2

As another example consider the system shown in Fig. 4-13. Before solving for the various gains and transfer functions, it is necessary to simplify the diagram. To do this the transfer function between $T(s)$ and $C(s)$ is obtained

$$\frac{C}{T}(s) = \frac{G}{1 + HG} = \frac{30/s}{1 + (20)30/s} = \frac{30}{s + 600}$$

Fig. 4-13 Block diagram for Example 4-2.

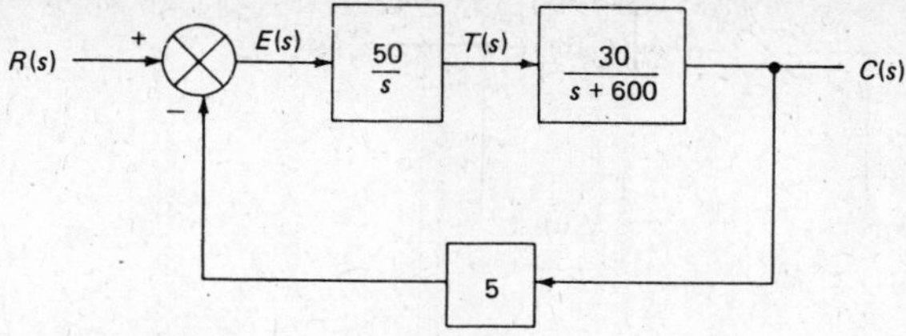

Fig. 4-14 Simplified form of Fig. 4-13.

Figure 4-13 is now redrawn in a simplified form with a single block placed between $T(s)$ and $C(s)$. It is shown in Fig. 4-14.

$$\text{DTG} = \frac{(50)(30)}{s(s+600)} = \frac{1500}{s(s+600)}$$

$$\text{loop gain} = \frac{50}{s}\,\frac{30}{s+600}\,(5) = \frac{7500}{s(s+600)}$$

$$\text{CLTF} = \frac{\text{DTG}}{1 + \text{loop gain}} = \frac{1500/s(s+600)}{1 + 7500/s(s+600)}$$

$$\frac{C}{R}(s) = \text{CLTF} = \frac{1500}{s(s+600) + 7500}$$

$$= \frac{1500}{s^2 + 600s + 7500}$$

Characteristic equation:

$$1 + \text{loop gain} = 0$$

$$1 + \frac{7500}{s(s+600)} = 0$$

$$s^2 + 600s + 7500 = 0$$

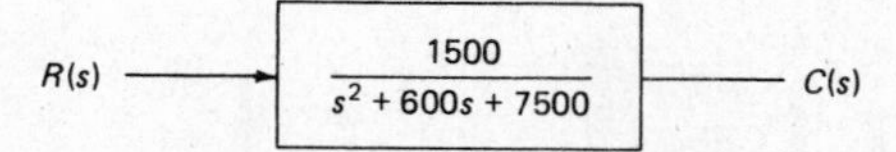

Fig. 4-15 Reduced system diagram for Example 4-2.

4-2-1 Block-Diagram Reduction

Refer to Fig. 4-16; the system will be reduced as described in the preceding section. The block diagrams in Fig. 4-17 show the reduction step by step, apply-

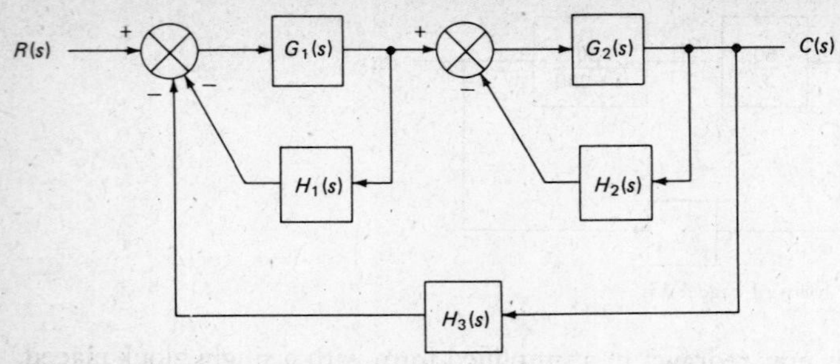

Fig. 4-16 General-system block diagram.

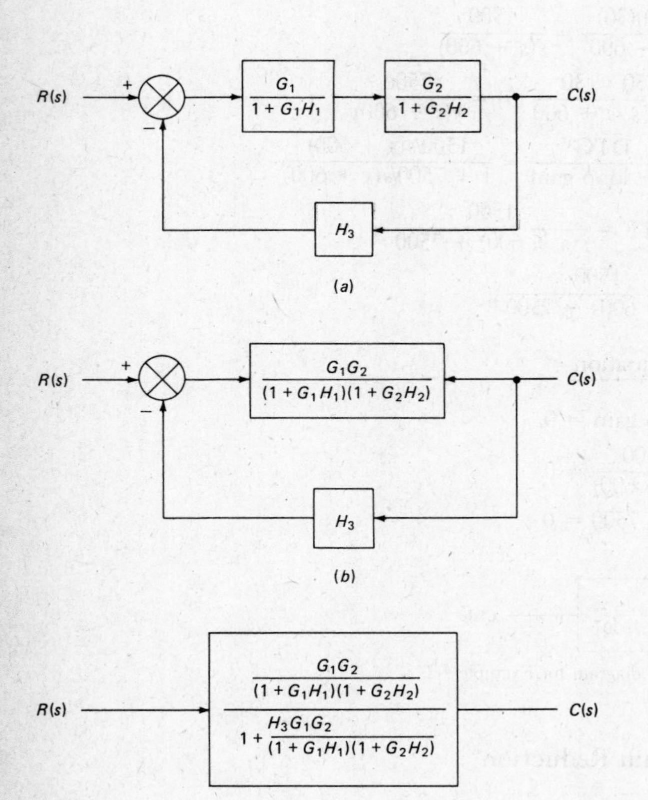

Fig. 4-17 Reduction: (*a*) step 1, (*b*) step 2, and (*c*) step 3.

ing the basic reduction formula in Fig. 4-10. The closed-loop transfer function can now be found by simplifying the expression in the box in Fig. 4-17*c*:

$$\frac{C}{R}(s) = \frac{G_1G_2}{(1 + G_1H_1)(1 + G_2H_2) + G_1G_2H_3}$$

$$= \frac{G_1G_2}{1 + G_1H_1 + G_2H_2 + G_1G_2H_1H_2 + G_1G_2H_3} \tag{4-33}$$

Equation (4-33) is the closed-loop transfer function of the system in Fig. 4-16.

As a further example consider the system described in Fig. 4-18*a*. As a first step, starting from the innermost loop, cascaded gains G_2 and G_3 are combined to form one gain block, as shown in Fig. 4-18*b*. The second step is to obtain the

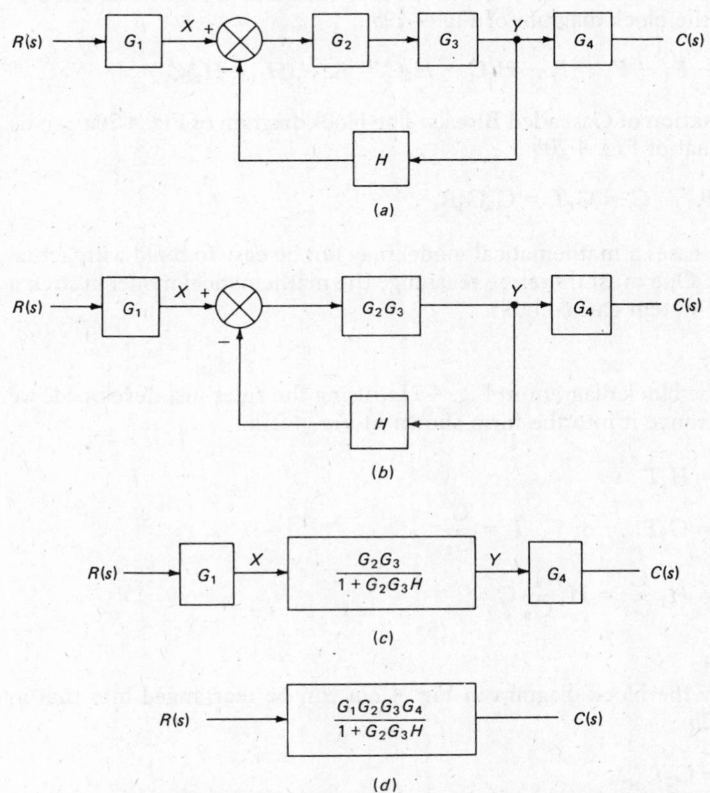

Fig. 4-18 (*a*) Block diagram of initial system. Reduction: (*b*) step 1, (*c*) step 2, (*d*) step 3.

closed-loop transfer function for the feedback loop between points X and Y, as shown in Fig. 4-18*c*. Finally, the three cascaded blocks in Fig. 4-18*c* are combined to get the final simplified block shown in Fig. 4-18*d*.

The closed-loop transfer function is the expression in the block of Fig. 4-18*d* and is given by

$$\frac{C}{R}(s) = \frac{G_1G_2G_3G_4}{1 + G_2G_3H} \tag{4-34}$$

4-2-2 Block-Diagram Algebra

In many cases, a block diagram can be reduced by some simple techniques.

Addition of Parallel Paths. Given the block diagram of Fig. 4-19*a*, one can reduce it to the block diagram of Fig. 4-19*b*:

$$E = R - F_1 - F_2 = R - H_1C - H_2C = R - (H_1 + H_2)C$$

Combination of Cascaded Blocks. The block diagram of Fig. 4-20*a* can be reduced to that of Fig. 4-20*b*:

$$T = G_1R \qquad C = G_2T = G_2G_1R$$

In some cases a mathematical model may not be easy to build with actual components. One must therefore rearrange the mathematical model in such a way that the system can be built.

Example 4-3

Given the block diagram in Fig. 4-21*a*, using the rules just developed, we can rearrange it into the form shown in Fig. 4-21*b*:

$$F_1 = H_1T$$

$$C = G_2T \qquad \text{or} \qquad T = \frac{C}{G_2}$$

$$F_1 = H_1\frac{C}{G_2} = H_1\frac{1}{G_2}C$$

Example 4-4

Similarly the block diagram in Fig. 4-22*a* can be rearranged into that in Fig. 4-22*b*:

$$C = G_2T$$

$$F_2 = H_2C = H_2G_2T$$

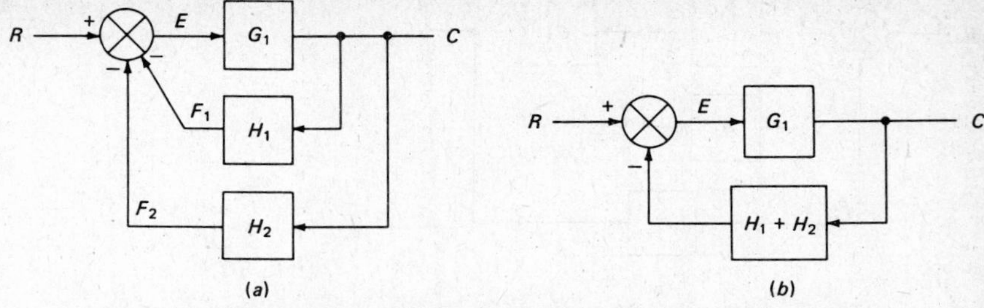

Fig. 4-19 (*a*) Simple system with parallel paths; (*b*) combination of parallel paths.

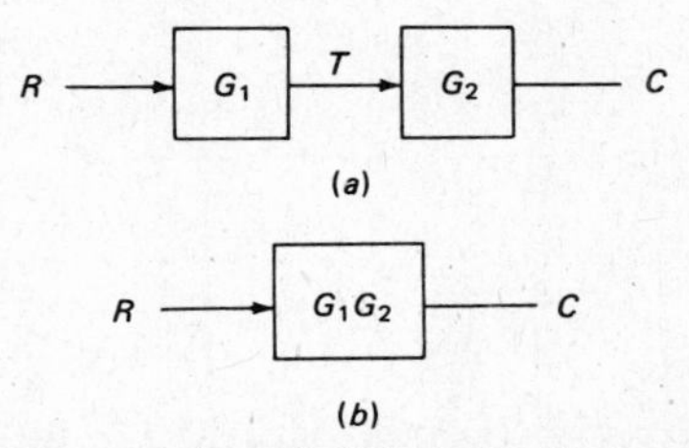

Fig. 4-20 (*a*) Cascaded blocks; (*b*) cascaded blocks combined.

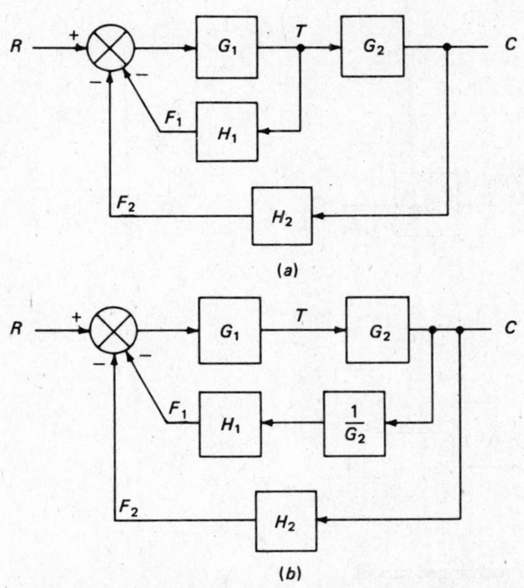

Fig. 4-21 (*a*) Mathematical model; (*b*) rearranged model.

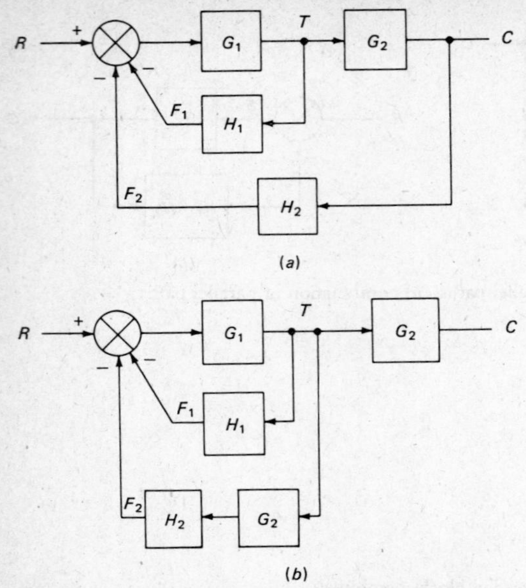

Fig. 4-22 (*a*) Mathematical model; (*b*) rearranged model.

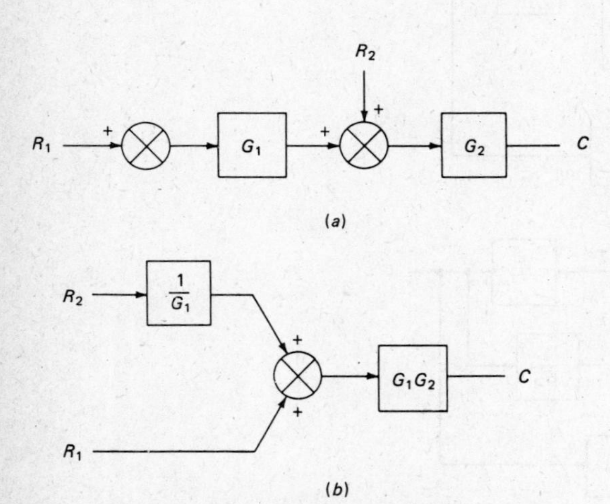

Fig. 4-23 (*a*) Mathematical model; (*b*) rearranged model.

Example 4-5

The block diagram in Fig. 4-23*a* can be rearranged into that in Fig. 4-23*b*:

$$C = G_1G_2R_1 + G_2R_2 = G_1G_2R_1 + G_2G_1\frac{1}{G_1}R_2 = G_1G_2\left(R_1 + \frac{R_2}{G_1}\right)$$

Having a better understanding of block diagrams, we can examine the dc motor in more detail. Referring to Fig. 2-28, we can begin construction of a motor block diagram. From the figure the equation for motor current can be written

$$i_a = \frac{v_m - v_b}{z_m} \tag{4-35}$$

where z_m = motor impedance
v_m = applied motor voltage
v_b = back emf voltage

Taking the transform of the above gives

$$I_a(s) = \frac{V_m(s) - V_b(s)}{Z_m(s)} = \frac{V_m(s) - V_b(s)}{r_a + sl_m} \tag{4-36}$$

The motor develops a torque from the current; thus

$$t_{dev} = K_t i_a$$

Again taking the transform of the above equation, we get

$$T_{dev}(s) = K_t I_a(s) \tag{4-37}$$

After subtracting all opposing torques from the developed torque, a net torque is left to accelerate the load. The opposing torques are made up of motor viscous friction, load viscous friction reflected to the motor, and any extraneous torques present. Thus

$$T_{ac}(s) = T_{dev}(s) - T_m(s) - aT_v(s) - T_{xm}(s) \tag{4-38}$$

where $T_m(s)$ = motor viscous friction = $f_v\dot{\Theta}_m(s)$
$aT_v(s)$ = load viscous friction reflected to motor = $aF_l\dot{\Theta}_l(s)$
$T_{xm}(s)$ = extraneous torque present at motor = $aT_l(s)$

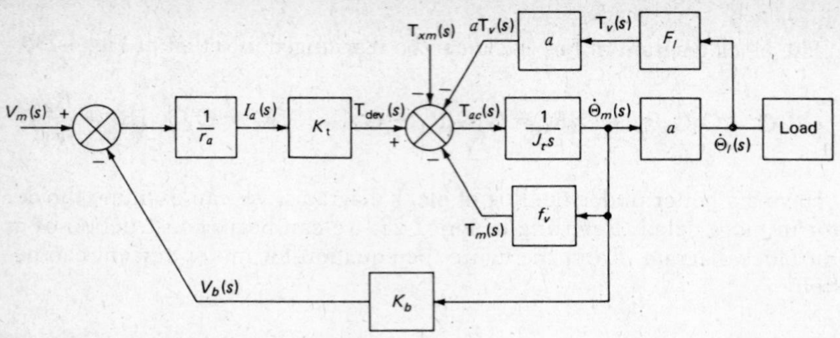

Fig. 4-24 Block diagram of a dc motor.

The motor acceleration $\ddot{\Theta}_m(s)$ can now be calculated:

$$\ddot{\Theta}_m(s) = \frac{T_{ac}(s)}{J_t} \tag{4-39}$$

$$\dot{\Theta}_m(s) = \frac{\ddot{\Theta}_m(s)}{s} = T_{ac}(s)\,\frac{1}{J_t s} \tag{4-40}$$

Combining Eqs. (4-36) to (4-40), we can construct Fig. 4-24. In Eq. (4-36) the motor impedance can be approximated by its resistance since the inductance l_m is generally very small.

It is left for the student to show (Prob. 4-24) by block-diagram reduction and manipulation that Fig. 4-24 is equivalent to Fig. 4-7.

PROBLEMS

4-1. What is the transfer function of a potentiometer with a gain K_p of 1.5 V/rad?

4-2. Using Eq. (4-8) and the final-value theorem (Appendix B), find an expression for the accelerometer output $e_o(t)$ when the acceleration being measured is a step of magnitude a.

4-3. What is the transfer function of a tachometer with a gain K_T of 40 V/(rad/s)?

4-4. Using Eq. (4-19):

(a) Derive the transfer function of the dc motor, $\dot{\Theta}(s)/V_m(s)$ when $T_l(s) = 0$.

(b) Write it in terms of K and τ where $\tau = J_m/F_m$ and $K = K_m/J_m$.

4-5. What is the motor transfer function if $K_m = 30$ in · lb/V, $J_m = 1.2$ in · s², and $F_m = 2.4$ in · lb · s?

4-6. A dc motor has the torque-speed curve shown in Fig. P4-6. Find its transfer function given that $J_m = 0.8$ in · lb · s² and $V_m = 10$ V.

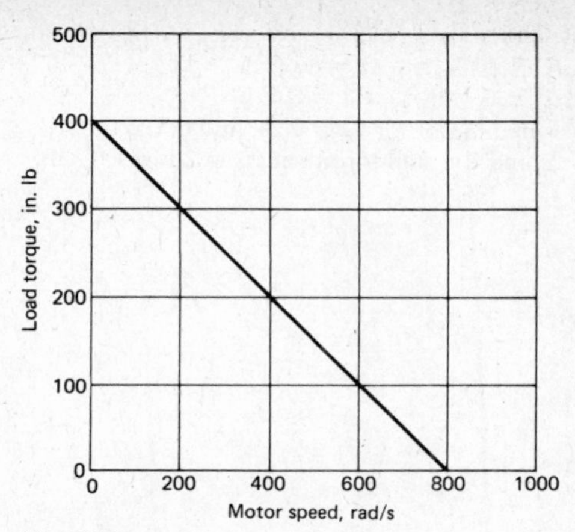

Fig. P4-6

4-7. Draw the torque-speed curves for the motor of Prob. 4-5 for motor voltages of 12, 24, and 30 V, all on one graph.

4-8. The motor of Prob. 4-6 is connected to a 100-in · lb load through a gearpass with ratio $a = 0.5$. Using the torque-speed curve given in Fig. P4-6, find:

(a) The steady-state motor speed and load speed

(b) The value of load torque that will stall the motor

4-9. The linearized characteristics of an ac servo motor are shown in Fig. P4-9. The moment of inertia is 10^{-4} slug · ft².

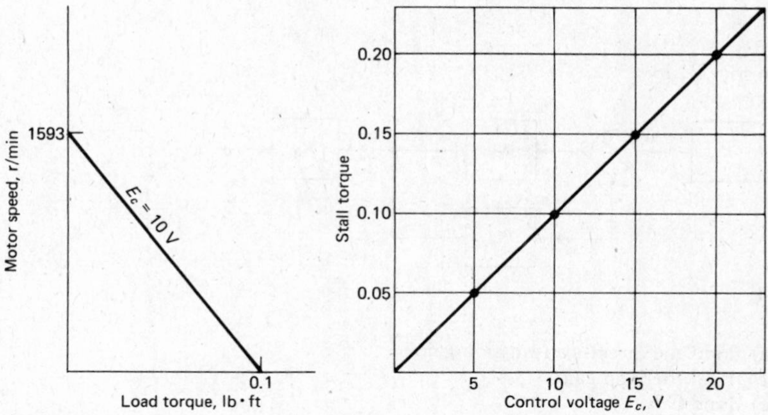

Fig. P4-9

(a) Find the motor time constant. *Hint:* $\tau = J_m/F_m$.
(b) Find the transfer function $\Omega_m(s)/E_c(s)$ where $\Omega = \dot{\Theta}$.
(c) Find the transfer function $\Theta_m(s)/E_c(s)$. *Hint:* $\Theta_m = \Omega/s$.

4-10. (a) Sketch the torque-speed curves of a motor for $v_m = 2$, 4, and 6 V if the friction coefficient is 0.04 ft · lb · s and the stall torque at $v_m = 2$ is 0.4 ft · lb.
(b) Find the motor constant K_m

4-11. For the diagram in Fig. P4-11:

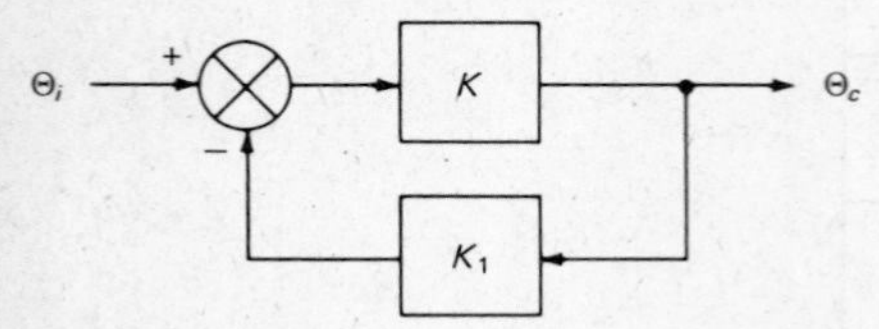

Fig. P4-11

(a) Find Θ_c/Θ_i.
(b) What is the direct-transmission gain?
(c) What is the loop gain?
(d) What is the closed-loop gain?

4-12. Find the closed-loop transfer function for Fig. P4-12.

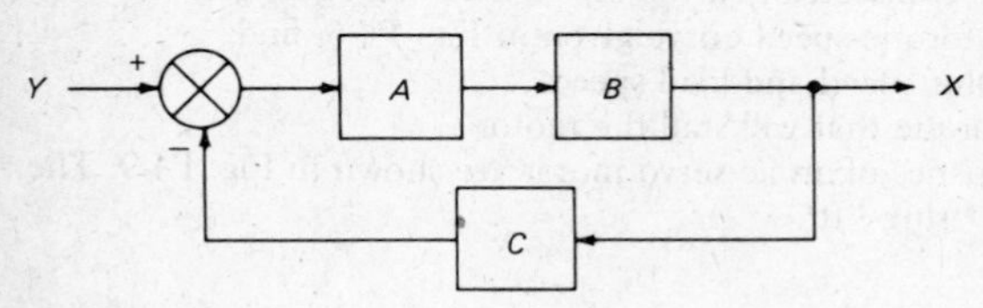

Fig. P4-12

4-13. In Fig. P4-13:

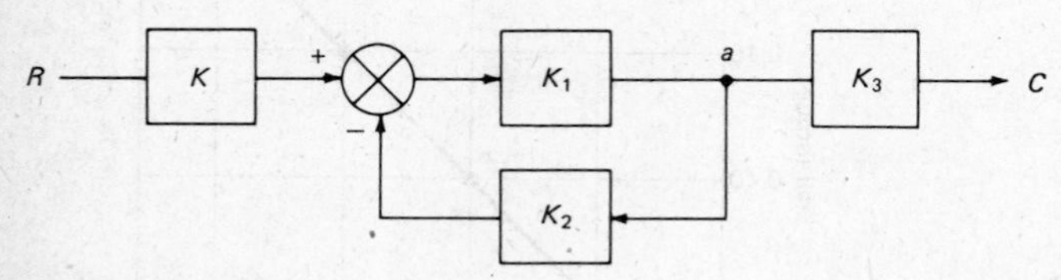

Fig. P4-13

(a) Find the direct-transmission gain.
(b) Find the loop gain.
(c) Find C/R.
(d) What is the value at point *a* in terms of C?

4-14. Find the closed-loop gain (transfer function) C/R in Fig. P4-14.

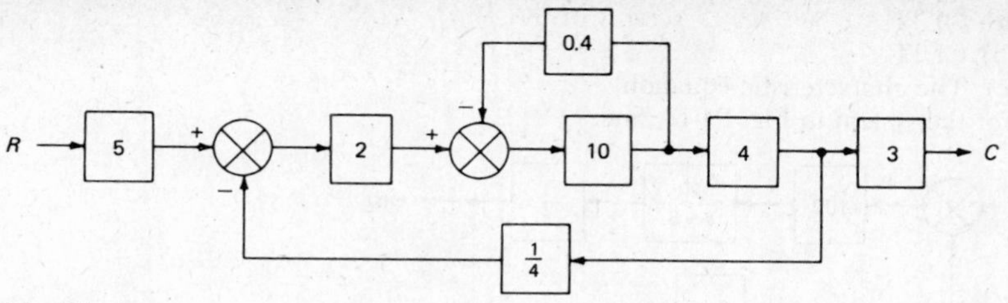

Fig. P4-14

4-15. In Fig. P4-15:

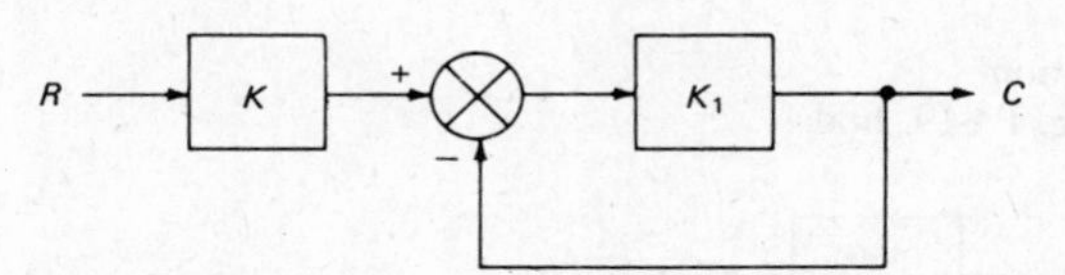

Fig. P4-15

(a) Find C/R.
(b) Why is this system called a unity-feedback system?

4-16. For the system in Fig. P4-16, find:

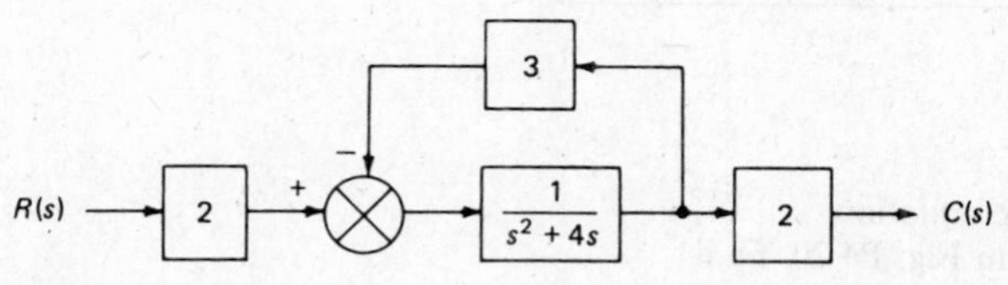

Fig. P4-16

(a) OLTF
(b) CLTF
(c) The characteristic equation

4-17. For the system in Fig. P4-17, find:

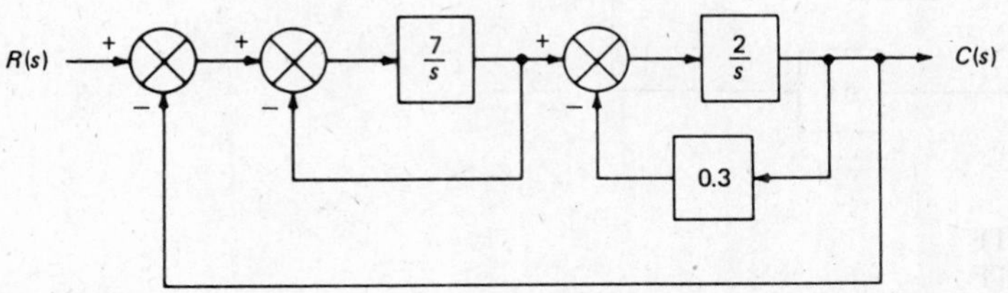

Fig. P4-17

(a) OLTF
(b) CLTF
(c) The characteristic equation

4-18. For the system in Fig. P4-18, find:

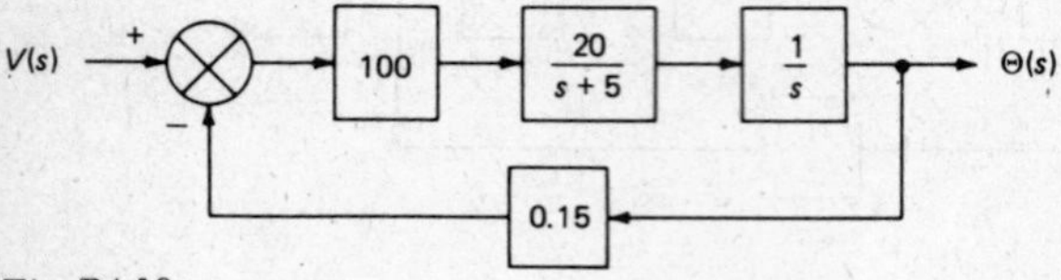

Fig. P4-18

(a) OLTF
(b) CLTF
(c) The characteristic equation

4-19. For the system shown in Fig. P4-19, find:

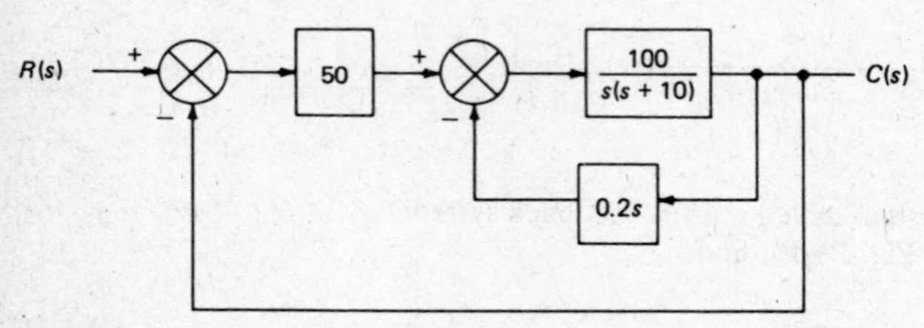

Fig. P4-19

(a) OLTF
(b) CLTF
(c) The characteristic equation

4-20. In the system shown in Fig. P4-20, find:

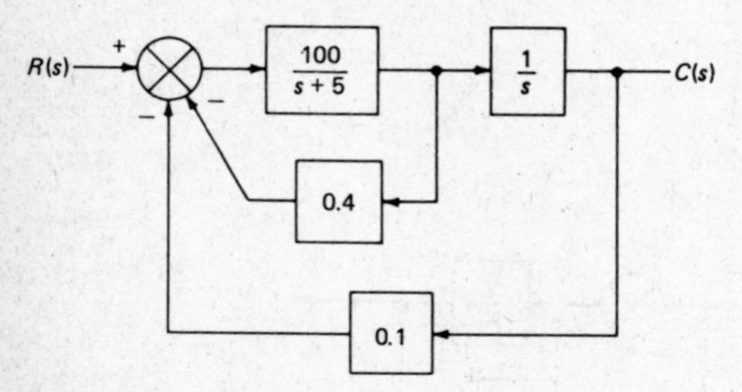

Fig. P4-20

(a) OLTF
(b) CLTF
(c) The characteristic equation

4-21. For the system shown in Fig. P4-21, find:

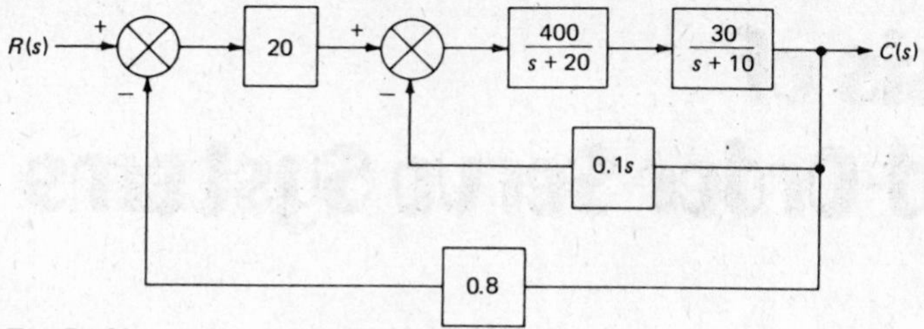

Fig. P4-21

(a) OLTF
(b) CLTF
(c) The characteristic equation

4-22. For the system shown in Fig. P4-22, find:

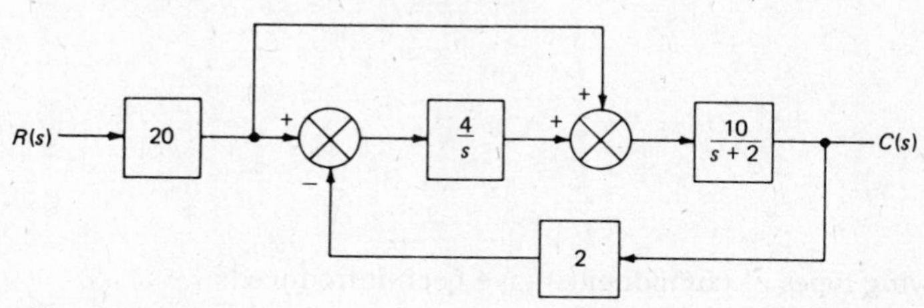

Fig. P4-22

(a) OLTF
(b) CLTF
(c) The characteristic equation

4-23. For the system in Fig. P4-23, find an expression for the output $\Theta(s)$. *Hint:* Use superposition; i.e., first find Θ/R_1 with $R_2 = 0$; then find Θ/R_2 with $R_1 = 0$, then add the two.

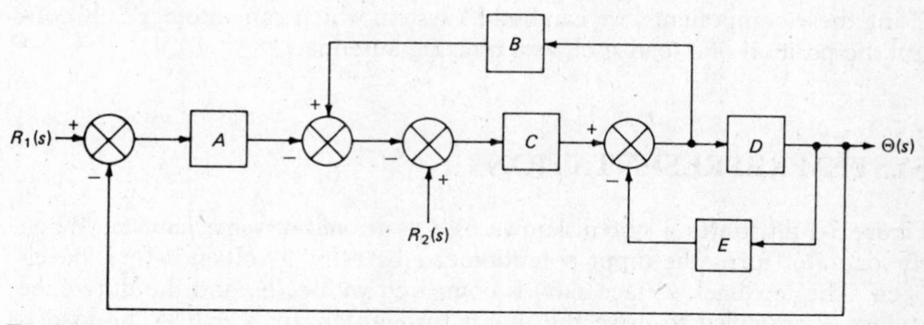

Fig. P4-23

4-24. Show that Fig. 4-7 is equivalent to Fig. 4-24.

chapter 5

Analysis of Second-Order Servo Systems

So far, the following types of components have been introduced:

1. Transducers to convert angular rotation into an electric signal, the potentiometer
2. Devices to increase signal strength and match impedance, the amplifier
3. Devices to develop torque, the motor
4. Devices to transmit torque, gears

Using these components, we can build a system which can automatically control the position of a load such as a tracking antenna.

5-1 SYSTEM REPRESENTATION

Figure 5-1 illustrates a system known as a *positional servomechanism*. When the operator turns the input potentiometer by $\theta_R(t)$, a voltage $e_R(t)$ is developed. The feedback voltage $e_F(t)$ is compared with $e_R(t)$, and the difference voltage is amplified to drive the motor through an angle $\theta_m(t)$. The load is driven by its gear through an angle $\theta_l(t)$, while the feedback potentiometer is driven through an angle $\theta_F(t)$ to develop the voltage $e_F(t)$.

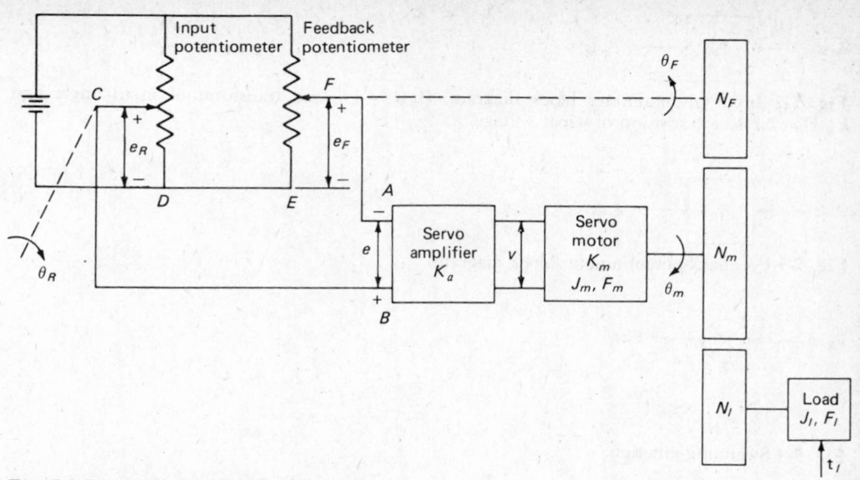

Fig. 5-1 Positional servomechanism, where
$\theta_R(t)$ = reference input angular position, rad
$\theta_m(t)$ = motor-shaft angular position, rad
$\theta_l(t)$ = load-shaft angular position, rad
$\theta_F(t)$ = feedback-gear angular position, rad
J_m = inertia of motor shaft, slug · ft²
J_l = inertia of load, slug · ft²
F_m = viscous-friction coefficient of motor, lb · ft · s/rad
F_l = viscous-friction coefficient of load, lb · ft · s/rad
K_m = motor constant, lb · ft/V
N_m = number of teeth on motor gear
N_l = number of teeth on load gear
N_F = number of teeth on feedback gear
K_a = gain of servo amplifier
t_l = additional torque acting on load, e.g., wind acting on antenna

5-1-1 Block Diagrams of System Components

In Chap. 4 we learned that each component can be represented by a mathematical model based on the Laplace transform of the equation representing the system. The resulting transfer functions can be portrayed by block diagrams as follows:

1. The input potentiometer: the transfer function is the potentiometer constant K_p (Fig. 5-2).
2. The feedback potentiometer: Assuming that the feedback potentiometer is identical to the input potentiometer, the same transfer function K_p is obtained (Fig. 5-3).

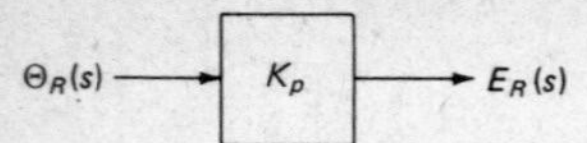

Fig. 5-2 Input-potentiometer block diagram; $\Theta_R(s)$ = Laplace transform of input angle and $E_R(s)$ = Laplace transform of wiper voltage.

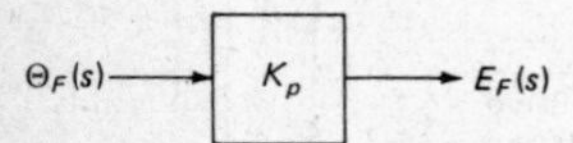

Fig. 5-3 Feedback-potentiometer block diagram.

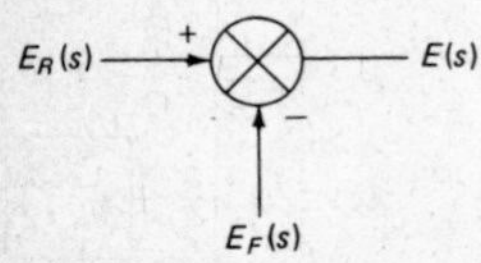

Fig. 5-4 Summing junction.

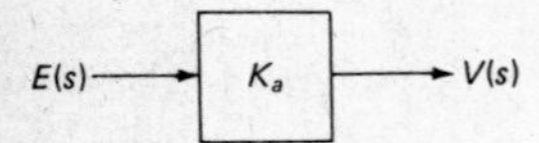

Fig. 5-5 Servo amplifier.

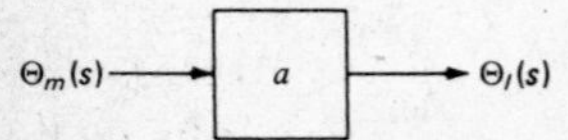

Fig. 5-6 Gear train to load; $a = N_m/N_1$.

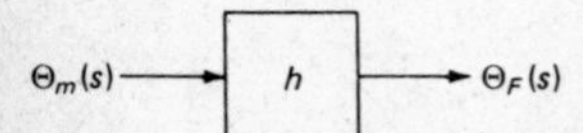

Fig. 5-7 Gear train to feedback potentiometer; $h = N_m/N_F$.

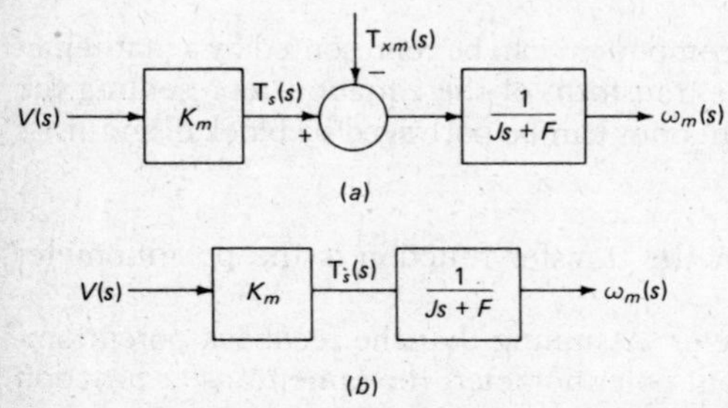

Fig. 5-8 Block diagram of (a) dc motor and (b) motor with $T_{xm} = 0$.

3. The summing junction: referring to Fig. 5-1 and writing Kirchhoff's voltage law around path $ABCDEFA$, we have

$$-e(t) + e_R(t) - e_F(t) = 0$$

or $e(t) = e_R(t) - e_F(t)$

In Laplace format

$$E(s) = E_R(s) - E_F(s) \tag{5-1}$$

Equation (5-1) can be represented by Fig. 5-4.

4. The servo amplifier: the error voltage $E(s)$ is amplified to $V(s)$ (see Fig. 5-5).
5. The gear train: motor to load (see Fig. 5-6) and motor to feedback potentiometer (see Fig. 5-7).
6. The dc motor: the voltage V from the amplifier produces the stall torque T_s in the windings of the armature. If there is an external load torque acting on the motor, T_{xm}, the net torque $T_s - T_{xm}$ is used to accelerate the total inertia and overcome the viscous friction of the motor and load; see Fig. 5-8a, in which

$$T_{xm} = aT_l \qquad J = J_m + a^2J_l \qquad F = F_m + a^2F_l$$

$$\omega_m(t) = \dot{\theta}_m(t) \qquad \text{or} \qquad \Omega_m(s) = s\Theta_m(s)$$

Figure 5-8 can be simplified if we assume that the external torque T_{xm} is zero (Fig. 5-8b) and combine the two blocks (Fig. 5-9).

Since we are dealing with a positional system, it is more convenient to change Fig. 5-9 into a form where the output is angle rather than speed. But

$$\Omega_m(s) = s\Theta_m(s) \tag{5-2}$$

or $$\frac{\Theta_m(s)}{\Omega_m(s)} = \frac{1}{s}$$

and $$\Theta_m(s) = \frac{1}{s}\,\Omega_m(s) \tag{5-3}$$

Equation (5-3) can be represented by Fig. 5-10, and Fig. 5-9 can then be represented by Fig. 5-11a, which can be simplified into Fig. 5-11b.

$$V(s) \longrightarrow \boxed{\frac{K_m}{Js + F}} \longrightarrow \Omega_m(s)$$

Fig. 5-9 Simplified block diagram of motor.

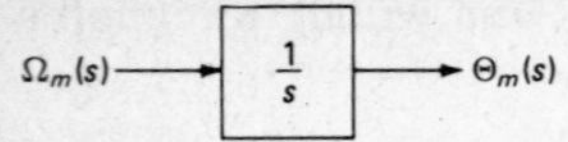

Fig. 5-10 Speed to position block diagram.

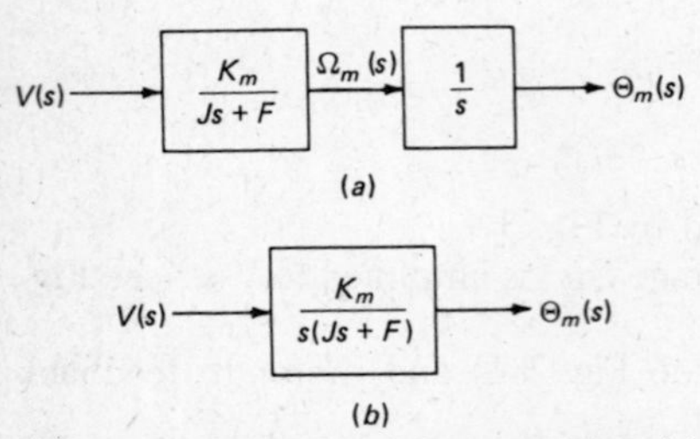

Fig. 5-11 (*a*) Motor block diagram with position output; (*b*) simplified diagram.

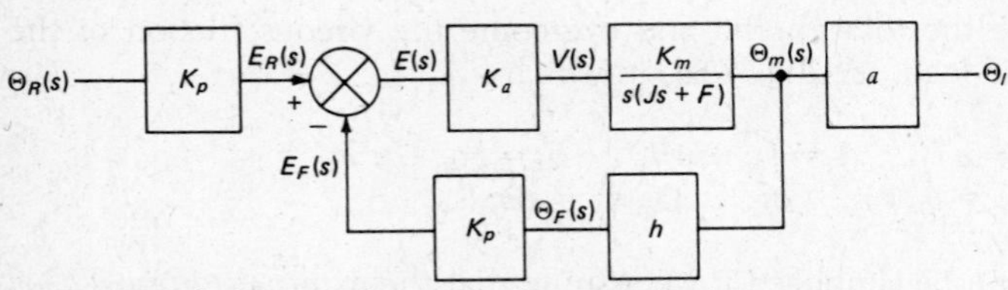

Fig. 5-12 Block diagram of a positional servomechanism.

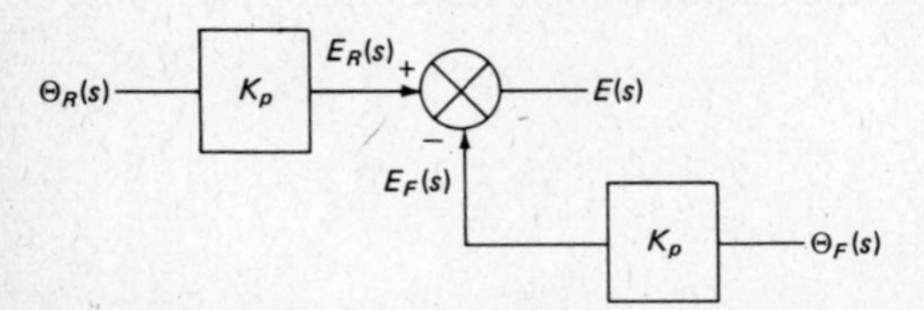

Fig. 5-13 Summing junction formed by two potentiometers.

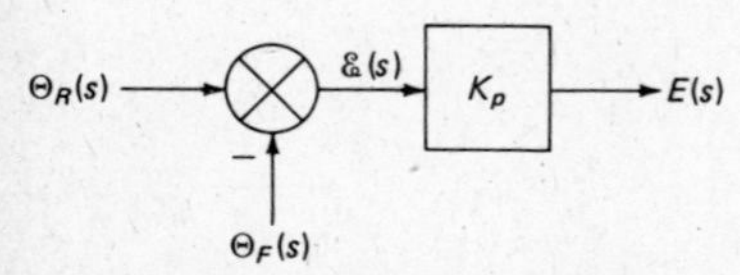

Fig. 5-14 Simplified summing junction.

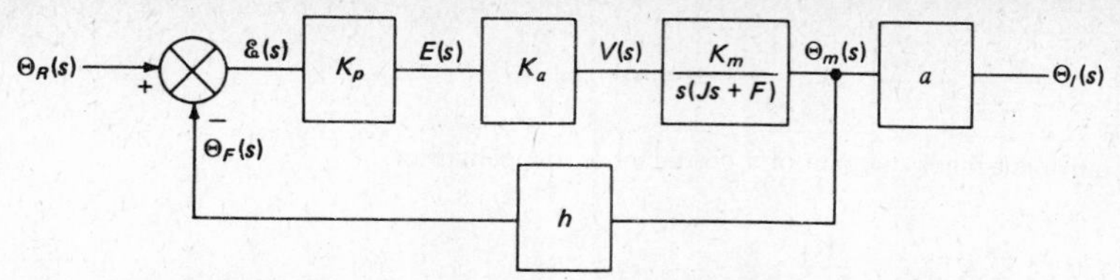

Fig. 5-15 Simplified block diagram of a positional servomechanism.

5-1-2 Overall Block Diagram

Combining the block diagrams for all the components, we see that the overall block diagram of the system of Fig. 5-1 becomes Fig. 5-12. As a further simplification, consider the summing junction formed by the two potentiometers (Fig. 5-13). Computing $E(s)$, we get

$$\begin{aligned} E(s) &= E_R(s) - E_F(s) \\ &= K_p\Theta_R(s) - K_p\Theta_F(s) \\ &= K_p[\Theta_R(s) - \Theta_F(s)] \end{aligned}$$

Let

$$\Theta_R(s) - \Theta_F(s) = \mathscr{E}(s)$$

then

$$E(s) = K_p\mathscr{E}(s)$$

and Fig. 5-13 can be redrawn as shown in Fig. 5-14.

Figure 5-14 can be interpreted as follows: even though there are two separate potentiometers, we can mathematically represent them both by one potentiometer whose input angle is the difference between the input and feedback positions. The final block-diagram version is shown in Fig. 5-15.

5-2 SYSTEM TRANSFER FUNCTION

Using block-diagram techniques, we can find the overall transfer function $\Theta_m(s)/\Theta_R(s)$ for the negative feedback system of Fig. 5-15

$$\frac{\Theta_m(s)}{\Theta_R(s)} = \frac{\text{DTG}^*}{1 + \text{LG}} \tag{5-4}$$

*DTG = direct transmission gain; LG = loop gain.

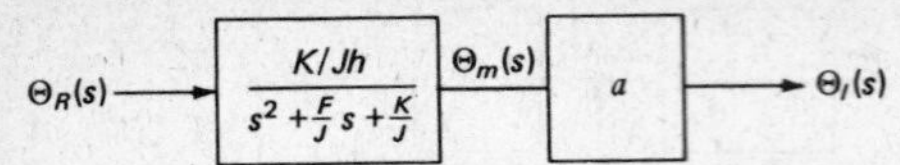

Fig. 5-16 Equivalent block diagram of a positional servomechanism.

where $$\text{DTG} = K_p K_a \frac{K_m}{s(Js + F)} \tag{5-5}$$

If we let

$$K = hK_p K_a K_m \qquad \text{or} \qquad K_p K_a K_m = \frac{K}{h} \tag{5-6}$$

Eq. (5-5) becomes

$$\text{DTG} = \frac{K/h}{s(Js + F)} = \frac{K/h}{Js^2 + Fs} = \frac{K/Jh}{s^2 + (F/J)s} \tag{5-7}$$

$$\text{and LG} = hK_p K_a \frac{K_m}{s(Js + F)} \tag{5-8}$$

$$\text{LG} = \frac{K}{Js^2 + Fs} = \frac{K/J}{s^2 + (F/J)s} \tag{5-9}$$

Equation (5-4) becomes

$$\frac{\Theta_m(s)}{\Theta_R(s)} = \frac{(K/Jh)/[s^2 + (F/J)s]}{1 + (K/J)/[s^2 + (F/J)s]} = \frac{K/Jh}{s^2 + (F/J)s + K/J} \tag{5-10}$$

Equation (5-10) indicates that mathematically the block diagram (Fig. 5-15) can be replaced by Fig. 5-16.

5-3 STEP INPUT

Consider the case where the input shaft is very quickly turned by an amount R and kept in this position. This represents a mathematical step input shown in Fig. 5-17. Then

$$\Theta_R(s) = \frac{R}{s} \tag{5-11}$$

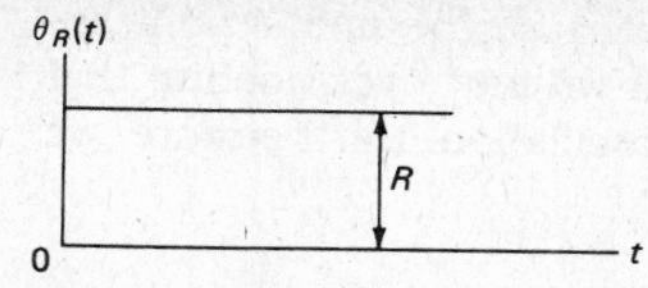

Fig. 5-17 Step input.

Rewriting Eq. (5-10) gives

$$\Theta_m(s) = \frac{K/Jh}{s^2 + (F/J)s + K/J}\,\Theta_R(s) = \frac{K/Jh}{s^2 + (F/J)s + K/J}\,\frac{R}{s}$$

$$= \frac{RK/Jh}{s[s^2 + (F/J)s + K/J]} \tag{5-12}$$

Example 5-1

Using the constants

$$R = 1 \qquad K = 6 \qquad J = 1 \qquad h = 1 \qquad F = 5$$

find $\theta_m(t)$.

Solution

Solve for $\boldsymbol{\theta_m(t)}$. Equation (5-12) becomes

$$\Theta_m(s) = \frac{6}{s(s^2 + 5s + 6)} = \frac{6}{s(s + 2)(s + 3)}$$

which can be expanded by partial fractions to

$$\Theta_m(s) = \frac{1}{s} - \frac{3}{s + 2} + \frac{2}{s + 3}$$

using the table of transform pairs gives

$$\theta_m(t) = 1 - 3e^{-2t} + 2e^{-3t}$$

which is plotted in Fig. 5-18.

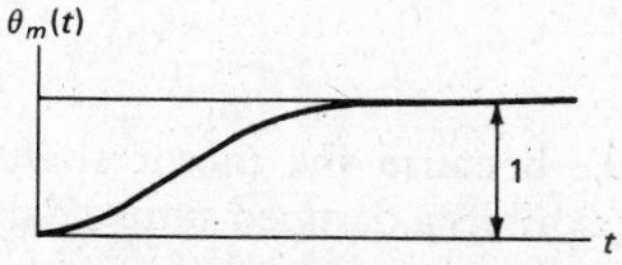

Fig. 5-18 Response for Example 5-1.

This is known as an overdamped case because the motor shaft approaches a final value (steady state) of 1 rad without overshooting that value. Notice also that there is absolutely no oscillation. Furthermore, we can divide the solution into two parts:

1. Steady state: $\theta_{m,SS} = 1$ rad, which is the same as the input, $R = 1$ rad, for this case.
2. Natural or transient response:

$$\theta_{m,N} = -3e^{-2t} + 2e^{-3t}$$

The natural response consists of two exponential terms which decay in five time constants:

e^{-2t} has a time constant of $\frac{1}{2}$ s.

e^{-3t} has a time constant of $\frac{1}{3}$ s.

The system time constant is the longer of the two, $\frac{1}{2}$ s, so that the system reaches steady state in 2.5 s.

Example 5-2

For the constants

$$R = 1 \qquad K = 2 \qquad J = 1 \qquad F = 2 \qquad h = 1$$

find $\theta_m(t)$

Solution

Solve for $\theta_m(t)$. Equation (5-12) becomes

$$\Theta_m(s) = \frac{2}{s(s^2 + 2s + 2)} = \frac{2}{s[(s + 1)^2 + 1]}$$

Using the transform pairs, we get

$$\theta_m(t) = 1 - \frac{2}{\sqrt{2}} e^{-t} \sin (t + 45°)$$

which is plotted in Fig. 5-19.

This is known as the underdamped case because the motor shaft overshoots the steady state value of 1 rad and exhibits a damped sinusoidal

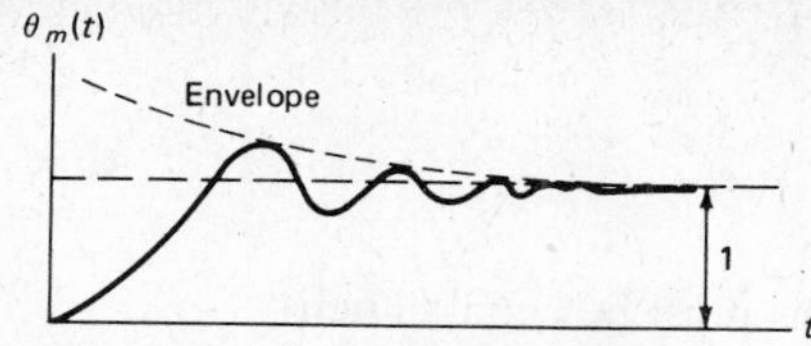

Fig. 5-19 Response for Example 5-2.

response as it approaches steady state. The time constant of the system is the time constant of the exponential envelope e^{-t}, which in this case has a value of 1 s. Therefore, the system reaches steady state in 5 s.

5-4 THE TRANSIENT RESPONSE

The form of the transient response can always be determined without solving equations. The denominator of the transfer function completely describes the transient response because the denominator is always the *characteristic equation* of the system. Referring to Eq. (5-10), we have

$$\frac{\Theta_m(s)}{\Theta_R(s)} = \frac{K/Jh}{s^2 + (F/J)s + K/J} \tag{5-10}$$

The characteristic equation is

$$s^2 + \frac{F}{J}s + \frac{K}{J} = 0 \tag{5-13}$$

but the standard form of the characteristic equation is

$$s^2 + 2\zeta\omega_n s + \omega_n^2 = 0 \tag{5-14}$$

where ω_n = natural frequency of the system, i.e., frequency of oscillation in absence of friction, rad/s
ζ = damping ratio, defined as F/F_c, where F_c = critical friction
F_c = amount of friction required to prevent oscillation in transient response
$\zeta = 1$ = critically damped case
$\zeta < 1$ = underdamped case
$\zeta > 1$ = overdamped case

Equating (5-13) and (5-14) gives

$$s^2 + \frac{F}{J}s + \frac{K}{J} = s^2 + 2\zeta\omega_n s + \omega_n^2 \qquad (5\text{-}15)$$

Since this an identity, corresponding terms must be equal. Then

$$\omega_n^2 = \frac{K}{J}$$

$$\text{or} \quad \omega_n = \sqrt{\frac{K}{J}} \qquad (5\text{-}16)$$

$$\text{and} \quad 2\zeta\omega_n = \frac{F}{J}$$

$$\text{or} \quad \zeta = \frac{F}{2J\omega_n} \qquad (5\text{-}17)$$

Substituting (5-16) into (5-17), we get

$$\zeta = \frac{F}{2J\sqrt{K/J}} = \frac{F}{2\sqrt{J^2K/J}} = \frac{F}{2\sqrt{KJ}} \qquad (5\text{-}18)$$

Equation (5-16) shows that the natural frequency does not depend on friction F. This must be true since the natural frequency can occur only if friction is not present.

Equation (5-18) shows that ζ increases as F increases or K or J decreases. In other words, friction damps the system, while gain K or inertia J can reduce damping. For a given system, F and J are normally fixed so that ζ is changed by varying K. Increasing K reduces ζ; decreasing K increases ζ.

Example 5-3

Given a second-order servomechanism with $K = 9$, $J = 1$, and $F = 0$, find the natural frequency and damping ratio.

Solution

$$\omega_n = \sqrt{\frac{K}{J}} = \sqrt{9} = 3 \text{ rad/s} \qquad \zeta = \frac{F}{2\sqrt{KJ}} = 0$$

Since $\zeta < 1$, the system is underdamped ($\zeta = 0$ is a special case known as *undamped*). Because there is no friction, the system will oscillate forever at the natural frequency of 3 rad/s, shown in Fig. 5-20.

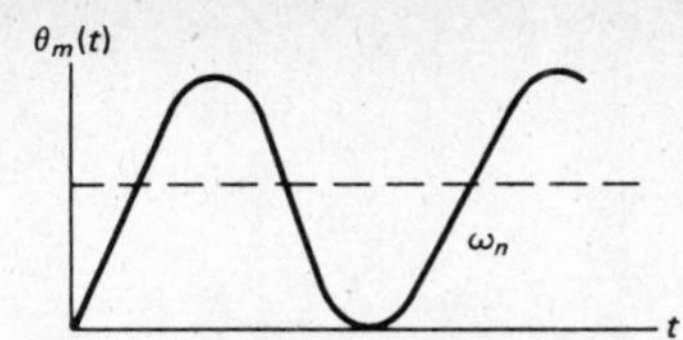

Fig. 5-20 Undamped ($\zeta = 0$) response to a step input.

Example 5-4

Given a second-order servomechanism with $K = 9, J = 1, F = 6$, find the natural frequency and damping ratio.

Solution

$$\omega_n = \sqrt{\frac{K}{J}} = 3 \text{ rad/s}$$

$$\zeta = \frac{F}{2\sqrt{KJ}} = \frac{6}{2(3)} = 1$$

Since $\zeta = 1$, the system is critically damped. The system will oscillate at 3 rad/s only if F is reduced to zero. The actual response is shown in Fig. 5-21.

Example 5-5

Given a second-order servomechanism with $K = 9, J = 1$, and $F = 9$, find the natural frequency and damping ratio.

Solution

$$\omega_n = \sqrt{\frac{K}{J}} = 3 \text{ rad/s}$$

$$\zeta = \frac{F}{2\sqrt{KJ}} = \frac{9}{2(3)} = 1.5$$

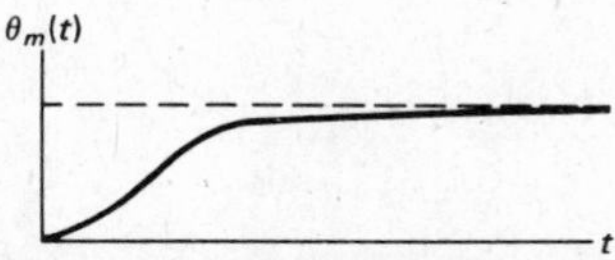

Fig. 5-21 Critically damped ($\zeta = 1$) response to a step input.

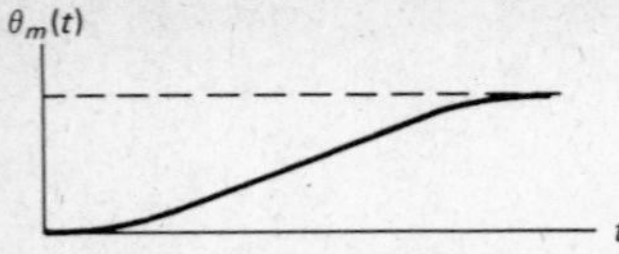

Fig. 5-22 Overdamped ($\zeta > 1$) response to a step input.

Since $\zeta > 1$, the system is overdamped. The response as shown in Fig. 5-22 is similar to the critically damped case but takes longer to reach steady state. The system will oscillate at 3 rad/s only if F is reduced to zero.

Example 5-6

Given a second-order servomechanism with $K = 9$, $J = 1$, and $F = 3$, find the natural frequency, damping ratio, damped frequency, and time constant.

Solution

$$\omega_n = \sqrt{\frac{K}{J}} = 3 \text{ rad/s}$$

$$\zeta = \frac{F}{2\sqrt{KJ}} = \frac{3}{2(3)} = 0.5$$

Since $\zeta < 1$, the system is underdamped. Overshoot occurs, and the system oscillates around the steady-state value. The oscillations are damped (get smaller with time), and eventually the system stops at the steady-state value, as shown in Fig. 5-23.

The oscillations occur at the frequency ω_d (damped frequency) where

$$\omega_d = \omega_n \sqrt{1 - \zeta^2} \tag{5-19}$$

If the system contains no friction, $F = 0$ and $\zeta = 0$, then

$$\omega_d = \omega_n \sqrt{1 - 0} = \omega_n$$

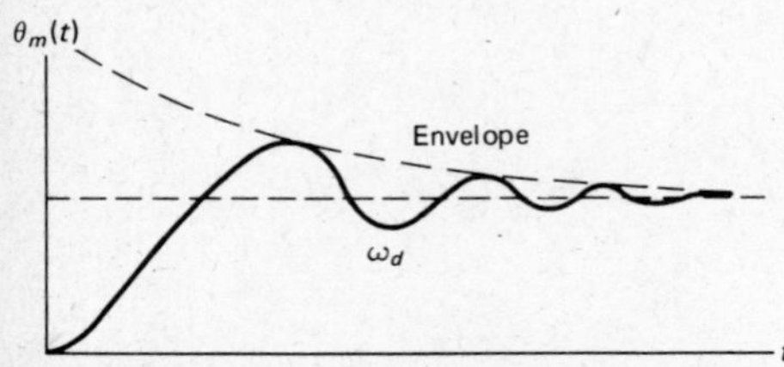

Fig. 5-23 Underdamped ($\zeta < 1$) response to a step input.

or the system will oscillate at the natural frequency when *no* damping is present (see Fig. 5-20); but for this example, $\omega_n = 3$, $\zeta = 0.5$,

$$\omega_d = \sqrt{1 - (0.5)^2} = 2.6 \text{ rad/s}$$

The envelope is a decaying exponential of the form

$$e^{-\zeta\omega_n t}$$

The time constant of the envelope is

$$\tau = \frac{1}{\zeta\omega_n} \tag{5-20}$$

For this example

$$\tau = \frac{1}{0.5(3)} = 0.667 \text{ s}$$

the system settles at the steady-state value in 5 time constants, or

$$5(0.667) = 3.34 \text{ s}$$

5-5 THE STEADY STATE

For a step input of amplitude R, Eq. (5-12) showed that the position of the motor shaft is given by

$$\Theta_m(s) = \frac{RK/Jh}{s[s^2 + (F/J)s + K/J]} \tag{5-12}$$

Practically, the steady state is reached after 5 time constants of the transient response. But in precise mathematical terms, the steady state is reached when the time approaches infinity, or

$$\theta_{m,SS} = \lim_{t\to\infty} \theta_m(t) \tag{5-21}$$

Using the final-value theorem shown in Appendix B, we get

$$\lim_{t\to\infty} \theta_m(t) = \lim_{s\to 0} s\Theta_m(s)$$

$$\text{or} \quad \theta_{m,SS} = \lim_{s\to 0} s\Theta_m(s) = \lim_{s\to 0} s \frac{RK/Jh}{s[s^2 + (F/J)s + K/J]}$$

$$= \lim_{s\to 0} \frac{RK/Jh}{s^2 + (F/J)s + K/J}$$

$$= \frac{RK/Jh}{K/J}$$

$$= \frac{R}{h} \tag{5-22}$$

Equation (5-22) shows that the final shaft position of the motor is given by the input potentiometer rotation divided by the feedback gear ratio, but for the special case of unity feedback

$$\theta_{m,SS} = R \qquad \text{for } h = 1 \tag{5-23}$$

In Fig. 5-14, we defined

$$\mathscr{E} = \theta_R - \theta_F \tag{5-24}$$

$$\text{or} \quad \mathscr{E} = \theta_R - h\theta_m \tag{5-25}$$

For a step input, $\theta_R = R$; since $\theta_{m,SS} = R/h$, Eq. (5-25) becomes

$$\mathscr{E}_{SS} = \theta_R - h\theta_{m,SS}$$

$$= R - h\frac{R}{h}$$

$$= R - R$$

$$= 0 \tag{5-26}$$

But we expected this result all along. If the motor reaches a steady-state value and stops, the motor torque must be zero. This means that the voltage applied to the motor must be zero or the error signal $\mathscr{E}$ must be zero.

5-6 GENERAL CONSIDERATIONS FOR A SECOND-ORDER SYSTEM

From our knowledge of the steady-state value for a step input, as well as the form of the transient depending on ζ, we can sketch the family of curves shown

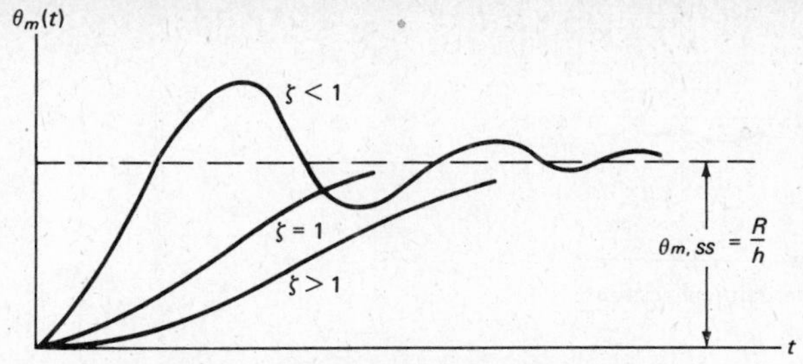

Fig. 5-24 Second-order responses for a step input R.

in Fig. 5-24. Instead of using Fig. 5-24, most of the literature employs a normalized second-order response curve, in which the ordinate is

$$y = \frac{\theta_m(t)}{\theta_{m,ss}}$$

thus $y_{SS} = 1$

and the abscissa is

$$x = \omega_n t$$

The normalized curve is shown in Fig. 5-25, but in practice it is difficult to use. For example, suppose an underdamped system yielded the result shown in

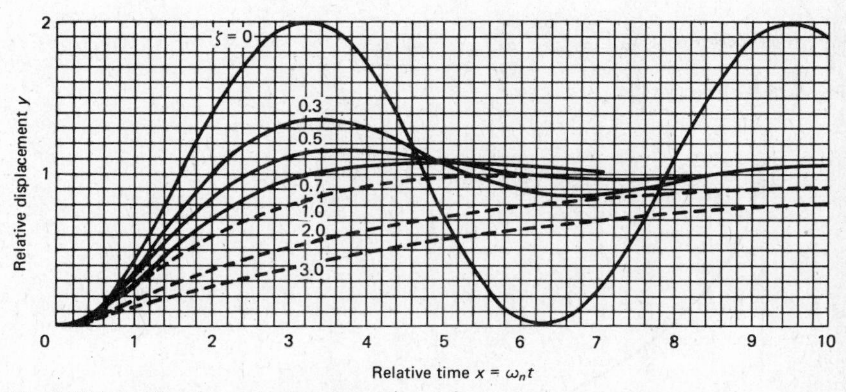

Fig. 5-25 Normalized second-order response.

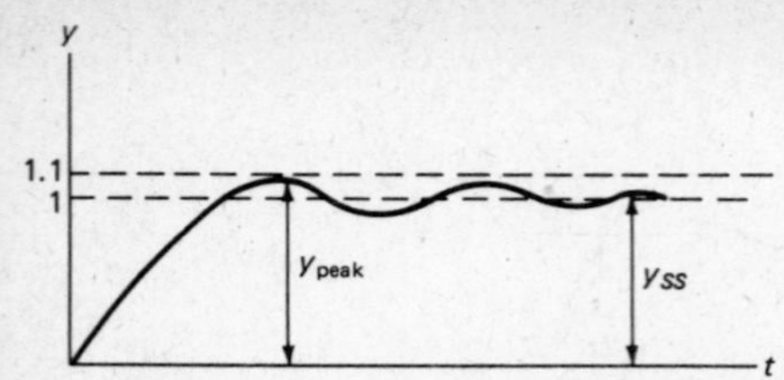

Fig. 5-26 Example of an underdamped response.

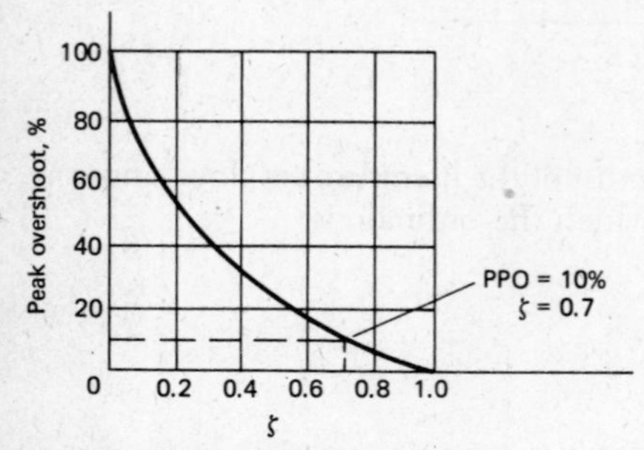

Fig. 5-27 Peak-overshoot curve.

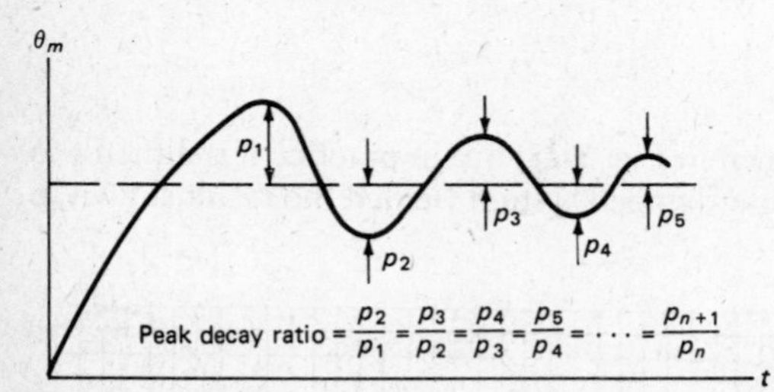

Fig. 5-28 Peak decay ratio.

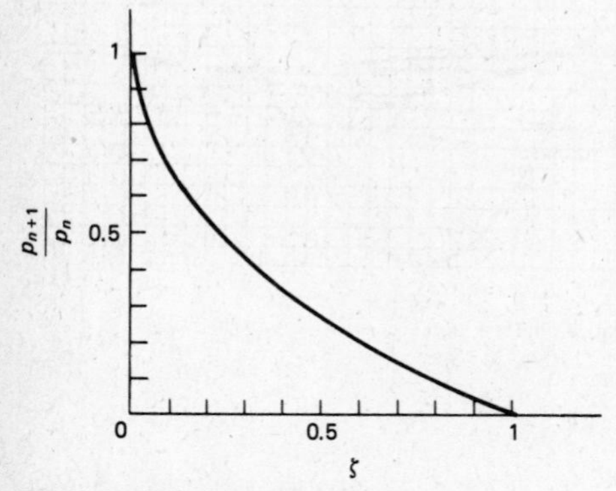

Fig. 5-29 Peak-decay-ratio curve.

Fig. 5-26 on a recorder. If we define peak percent overshoot PPO as

$$\text{PPO} = \frac{y_{\text{peak}} - y_{SS}}{y_{SS}} \times 100$$

then for the response shown above

$$\text{PPO} = \frac{1.1 - 1}{1} \times 100 = 10\%$$

It would be very helpful if we had a way of converting PPO directly into ζ. Figure 5-27 provides us with the conversion. For our example above, where PPO = 10 percent, $\zeta = 0.7$.

Another curve that proves useful is the peak decay ratio vs. ζ. Starting with any peak overshoot (peak value minus steady-state value), determine the ratio of the next peak overshoot to the preceding one. Refer to Fig. 5-28. Using the value of peak decay ratio, ζ can be found from the curve shown in Fig. 5-29.

5-7 SYSTEM DESIGN

When a servomechanism is designed, the speed of response is very important. For example, a pilot flying at 700 mi/h and wanting to turn the aircraft simply cannot afford too many seconds to pass for the maneuver to take place. On the other hand, excessive overshoot and resulting lengthy decaying oscillations are not practical either. In most cases, for proper design, ζ is set between 0.3 and 0.8.

The speed of response is defined in terms of rise time and settling time. The *rise time* is the time required for the system to move from 10 to 90 percent of the steady-state value (before the first overshoot). Highly oscillatory systems (low values of ζ) have a very short rise time; i.e., they respond very quickly. Therefore, we have to pay for fast response with more overshoot and oscillation. This low degree of stability is not very desirable (see Fig. 5-30).

The *settling time* of the system is defined as the time required to reach within 2 percent or less of the steady-state value and stay confined within this tolerance. The settling time is simply 5 time constants. For the underdamped case, the value is $5/\zeta\omega_n$. For the overdamped case it depends on the longer of the 2 time constants of the transient.

From an examination of the settling and rise times, it is obvious that the overdamped case is not practical in design. Underdamped cases where ζ is very small are not desirable because in spite of a fast rise time they result in excessive overshoot and lengthy settling times. The range $0.3 < \zeta < 0.8$ generally yields acceptable responses.

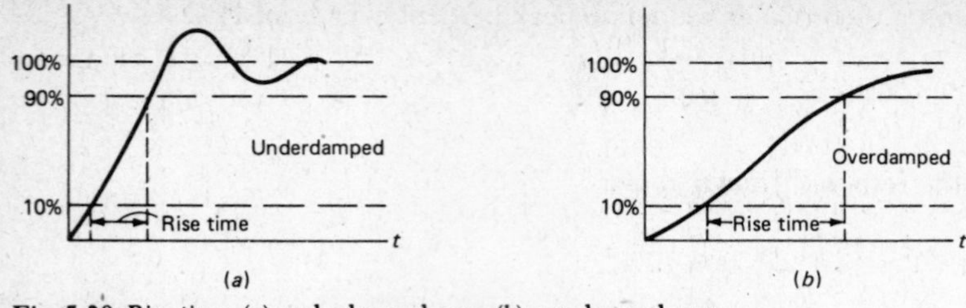

Fig. 5-30 Rise time: (*a*) underdamped case; (*b*) overdamped case.

Example 5-7

A positional servomechanism has the following parameters:

$$h = 1 \qquad K_a = 10 \qquad K_p = 1 \qquad K_m = 1 \qquad J = 0.1 \qquad F = 0.38$$

The system is tested by applying a step input $R = 0.1$ rad. The result (motor position vs. time) obtained on a recorder is shown in Fig. 5-31. Determine the rise time, settling time (approximate), frequency of response ω_d, ζ, and the steady-state error $\mathscr{E}$ from the recording. Determine the theoretical values of ζ, ω_d, settling time, and steady-state error and compare them with the measured values.

Solution

$$\text{Rise time} = 0.2 \text{ s} \qquad \text{Settling time} = 2.4 \text{ s} \qquad \text{Period } T = 0.6 \text{ s}$$

$$\omega_d = \frac{2\pi}{T} = \frac{2\pi}{0.6} = 10.05 \text{ rad/s}$$

$$\text{PPO} = \frac{0.15 - 0.095}{0.095} \times 100 = 58\%$$

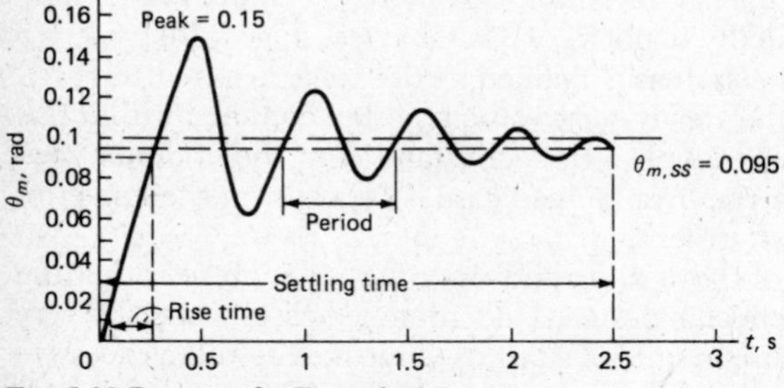

Fig. 5-31 Response for Example 5-7.

from the PPO curve

$$\zeta = 0.18$$

and $\mathscr{E}_{SS} = R - h\theta_{m,ss} = 0.1 - 0.095 = 0.005$

Using the parameters

$$K = hK_aK_pK_m = 10 \quad \text{and} \quad \omega_n = \sqrt{\frac{K}{J}} = 10$$

we get

$$\zeta = \frac{F}{2\sqrt{KJ}} = \frac{0.38}{2\sqrt{10(0.1)}} = 0.19$$

but $\omega_d = \omega_n \sqrt{1 - \zeta^2}$

$$= 10\sqrt{1 - 0.19^2} \approx 10$$

and $\tau = \dfrac{1}{\zeta\omega_n} = \dfrac{1}{0.19 \times 10} = 0.527$

The settling time is $5\tau = 2.64$ s and $\mathscr{E}_{SS} = 0$.
The fact that the numbers are not in exact agreement is due to

1. Measurement errors
2. Inexact knowledge of the parameters
3. Inexact model (the positional servo is not a pure *linear* second-order system)

5-8 RESPONSE OF A SERVO TO DISTURBANCES

A major advantage of a feedback control system over an open-loop system is the ability to maintain the response fairly steady while the environment changes. In a home heating system the thermostat is set for a desired temperature and even though such external conditions as wind velocity and outside temperature change, the inside temperature remains fairly constant. In an antenna rotator system, a dial is turned to cause the antenna to point toward a particular station. Should the wind velocity change and attempt to turn the antenna, the change in feedback signal alerts the motor to turn the antenna back to the proper position (assuming that the velocity increase does not exceed the capacity of the motor).

Mathematically, the wind on the antenna creates a constant disturbance torque, T_{xm}, or

$$T_{xm}(s) = \frac{T_{xm}}{s}$$

When we use the model for a motor with a disturbance acting on the output (see Fig. 5-8*a*) and combine it with Fig. 5-15, the block diagram of a positional servomechanism becomes Fig. 5-32.

Assuming that no input is applied, $\theta_R = 0$, the motor shaft θ_m should also be at zero position. If a constant disturbance occurs at the motor shaft, how much does the shaft turn? Initially, there will be a transient response with the same value of ζ and ω_n determined previously for a step input. In the steady state, however, the shaft will not return to the original position; i.e., there will be a steady-state error. The error must exist to provide voltage to the motor to maintain a constant countertorque that will balance out the disturbing torque. If this situation did not occur, the resulting torque would cause the antenna (or any load) to keep turning.

To determine Θ_m find the transfer function between $T_{xm}(s)$ and Θ_m,

$$\frac{\Theta_m(s)}{T_{xm}(s)} = \frac{\text{DTG}}{1 + \text{LG}} = \frac{(-1)[1/s(Js + F)]}{1 + hK_pK_aK_m/s(Js + F)}$$

$$= \frac{-1}{Js^2 + Fs + K} \qquad \text{where } K = hK_aK_pK_m$$

or
$$\frac{\Theta_m(s)}{T_{xm}(s)} = \frac{-1/J}{s^2 + (F/J)s + K/J} \tag{5-27}$$

Notice that the denominator of the transfer function is the usual characteristic equation.

Since the disturbance is assumed to be a constant

$$T_{xm}(s) = \frac{T_{xm}}{s}$$

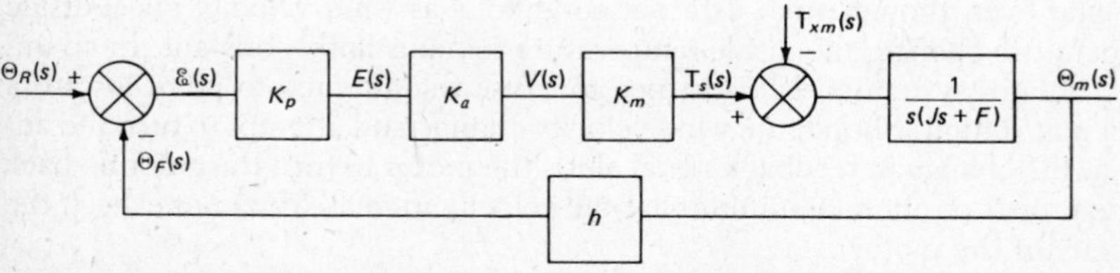

Fig. 5-32 Block diagram of a positional servomechanism showing a disturbance T_{xm}.

solving (5-27) for $\Theta_m(s)$ gives

$$\Theta_m(s) = \frac{-\mathsf{T}_{xm}/J}{s[s^2 + (F/J)s + K/J]} \tag{5-28}$$

The steady state value of θ_m is determined by the final-value theorem

$$\begin{aligned}\theta_{m,SS} &= \lim_{t\to\infty} \theta_m(t) = \lim_{s\to 0} s\Theta_m(s) \\ &= \lim_{s\to 0} s \frac{-\mathsf{T}_{xm}/J}{s[s^2 + (F/J)s + K/J]}\end{aligned}$$

$$\text{or} \quad \theta_{m,SS} = \frac{-\mathsf{T}_{xm}/J}{K/J} = \frac{-\mathsf{T}_{xm}}{K} \tag{5-29}$$

$$\text{and} \quad \mathscr{E}_{SS} = \theta_R - h\theta_{m,ss}$$

But we assumed $\theta_R = 0$, so that

$$\mathscr{E}_{SS} = 0 - h\frac{-\mathsf{T}_{xm}}{K} = \frac{h\mathsf{T}_{xm}}{K} \tag{5-30}$$

An important question to be answered is what happens when the disturbance disappears (the wind subsides)? After 5 time constants, the system will return to the original steady-state position.

Example 5-8

An antenna-positioning servo has $h = 1$, $K = 25$, $J = 1$, and $F = 4$. The input potentiometer is rapidly turned from 0 to 0.1 rad, and 1 h later the wind creates a constant disturbance of 1 ft · lb; then 1 h later the wind disappears. Completely describe the movement of the motor shaft over the 2-h period.

Solution

$$\omega_n = \sqrt{\frac{K}{J}} = 5$$

$$\zeta = \frac{F}{2\sqrt{KJ}} = \frac{4}{10} = 0.4$$

The system is underdamped.

$$\text{Settling time} = \frac{5}{\zeta\omega_n} = \frac{5}{2} = 2.5 \text{ s}$$

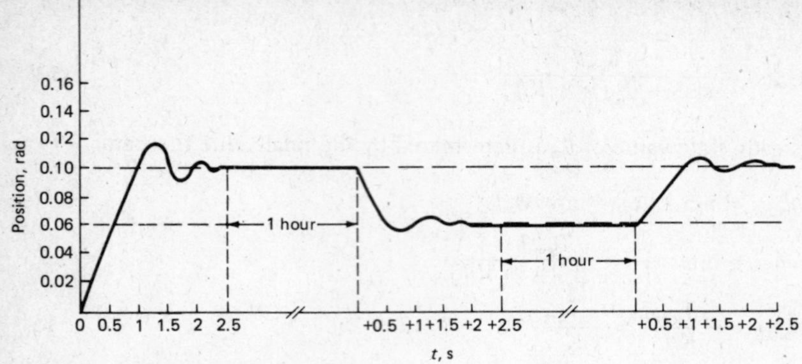

Fig. 5-33 Positioning antenna servo.

(a) A step input of 0.1 is applied. After 2.5 s the system will reach a steady-state value of

$$\theta_{m,SS} = \frac{R}{h} = \frac{0.1}{1} = 0.1 \text{ rad}$$

(b) Then 1 h later, a step disturbance of $T_{xm} = 1$ ft · lb occurs; 2.5 s after that, the system will move by a steady-state amount

$$\theta_{m,SS} = -\frac{T_{xm}}{K} = -\frac{1}{25} = -0.04 \text{ rad}$$

This means that the new shaft position is

$$\theta_m = 0.1 - 0.04 = 0.06 \text{ rad}$$

(c) Then 1 h later the disturbance disappears; 2.5 s later, the shaft returns to the original position of 0.1 rad. The overall response is shown in Fig. 5-33.

It should be noted that the PPO when taken as a percentage of the *change* in steady-state value remains constant for each transient condition.

5-9 ERROR COEFFICIENT

The fact that steady-state errors do not always go to zero leads us to derive a general equation which can be used in all linear systems to solve for the error.

Equation (4-31) relates the error to the input

$$\frac{E}{R}(s) = \frac{1}{1 + HG(s)} \tag{4-31}$$

Solving for the error, we get

$$E(s) = \frac{R(s)}{1 + HG(s)} \tag{5-31}$$

and applying the final-value theorem gives

$$\text{Steady-state error} = e(\infty) = \lim_{t\to\infty} e(t) = \lim_{s\to 0} sE(s) = \lim_{s\to 0} \frac{sR(s)}{1 + HG(s)} \tag{5-32}$$

We thus obtain a general equation which can be used for any linear system.

Example 5-9

Consider a system whose loop gain is

$$HG(s) = \frac{60}{(s + 2)(s + 10)}$$

A step input of 5° is applied

$$r(t) = 5 \qquad R(s) = \frac{5}{s}$$

Find the steady-state error.

Solution

Using (5-32), we have

$$e(\infty) = \lim_{s\to 0} \frac{s(5/s)}{1 + 60/[(s + 2)(s + 10)]} = \lim_{s\to 0} \frac{5}{1 + \{60/[(s + 2)(s + 10)]\}}$$
$$= \frac{5}{1 + [60/(2)(10)]} = \frac{5}{1 + 3} = 1.25°$$

This example illustrates a second-order servo which has a finite steady-state error for step inputs.

Example 5-10

For a system with loop gain

$$HG(s) = \frac{50}{s(s+10)} \qquad R(s) = \frac{5}{s}$$

find the steady-state error.

Solution

Again we have a second-order system with a step input of magnitude 5°. However, when Eq. (5-32) is applied,

$$\begin{aligned} e(\infty) &= \lim_{s\to 0} \frac{s(5/s)}{1 + 50/[s(s+10)]} = \lim_{s\to 0} \frac{5}{1 + 50/[s(s+10)]} \\ &= \frac{5}{1 + 50/0} \\ &= \frac{5}{1+\infty} = \frac{5}{\infty} = 0 \end{aligned}$$

we see that the steady-state error is equal to zero.

Both examples had step inputs. The only difference was the form of $HG(s)$, the loop gain. In the case where zero error was obtained, $HG(s)$ had a pure s in the denominator, whereas in the case where a finite error was obtained s did not appear by itself.

$$HG(s) = \begin{cases} \dfrac{50}{s(s+10)} & \text{zero error} \\ \dfrac{60}{(s+2)(s+10)} & \text{finite error} \end{cases}$$

A simpler equation will now be derived to calculate the error in any system for step inputs.

Suppose a step of magnitude r_0 is applied to any system. Then, the Laplace transform of the input is r_0/s. Applying Eq. (5-32), we get

$$e(\infty) = \lim_{s\to 0} \frac{s(r_0/s)}{1 + HG(s)}$$

or, more simply,

$$e(\infty) = \lim_{s\to 0} \frac{r_0}{1 + HG(s)}$$

If $HG(s)$ as s goes to zero is denoted by a new variable K_e, we obtain for the error

$$e(\infty) = \frac{r_0}{1 + K_e} \tag{5-33}$$

$$\text{where} \quad K_e = \lim_{s \to 0} HG(s) \tag{5-34}$$

Thus it can be easily seen that the steady-state error is inversely proportional to the loop gain. In other words, as the loop gain is increased, the error will decrease.

Example 5-11

A position servo has the loop gain and input given below. Determine the steady-state error.

Solution

$$HG(s) = \frac{2000}{(s + 2)(s + 10)}$$

$$\text{Input} = 10 \text{ rad, step}$$

Therefore $r_0 = 10$

First solve for the error coefficient using Eq. (5-34):

$$K_e = \lim_{s \to 0} HG(s) = \lim_{s \to 0} \frac{2000}{(s + 2)(s + 10)} = \frac{2000}{20}$$

$$= 100$$

Now solve for the steady-state error using Eq. (5-33):

$$e_{SS} = e(\infty) = \frac{10}{1 + 100} = \frac{10}{101} = 0.099 \approx 0.1 \text{ rad}$$

PROBLEMS

5-1. Draw the block diagram of a second-order position servo with the following specifications:

Potentiometer const = 0.9 V/rad
Motor const = 0.01 lb · ft/V
F_{tot} = 0.1 lb · ft · s
N_l = 100 teeth
Amplifier gain = 10
J_{tot} = 0.01 slug · ft²
N_m = 50 teeth
N_F = 50 teeth

5-2. Find the steady-state error of the system in Fig. P5-2.

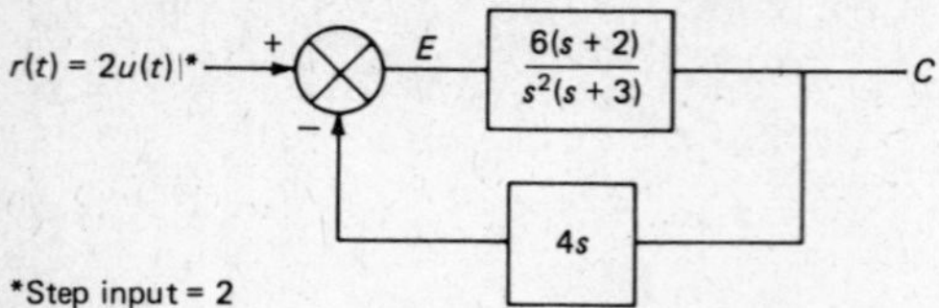

Fig. P5-2

5-3. A step-input signal of 1 rad is applied to a second-order system with the following constants:

$$K_p = 0.1 \quad K_a = 10 \quad K_m = 6 \quad h = 1 \quad J = 1 \quad F = 5$$

(a) Find the total response (steady state and transient) for the motor shaft position θ_m.
(b) Sketch the result θ_m as a function of time.
(c) How long does it take to reach steady state?
(d) Is the system overdamped or underdamped?

5-4. Repeat Prob. 5-3, if

$$K_p = 0.1 \quad K_a = 10 \quad K_m = 2 \quad h = 1 \quad J = 1 \quad F = 2$$

5-5. The characteristic equation of a second-order system is

$$s^2 + 0.6s + 9 = 0$$

Find:
(a) The natural frequency ω_n
(b) The damping ratio ζ
(c) The damped natural frequency ω_d
(d) The time constant (envelope time constant)
(e) The percentage peak overshoot

5-6. An input step $\theta_R = 0.1$ rad is applied to a second-order servo with the constants

$$K = 50 \quad h = 1 \quad J = 2 \quad F = 10$$

(a) Find the steady-state motor-shaft position θ_m.
(b) Determine ζ and ω_n.
(c) Is the system overdamped or underdamped?
(d) Find the new value of ζ if the amplifier gain is doubled.

5-7. A second-order system has the constants

$$K_p = 0.1 \quad K_a = 10 \quad h = 1, \frac{K_m}{J} = 4 \quad \frac{F}{J} = 2$$

Determine the values of ζ and ω_n.

5-8. A disturbance $T_{xm} = 10$ ft · lb is applied to a second-order servo whose constants are

$$K = 20 \qquad h = 0.2 \qquad F = 5 \qquad J = 1.25$$

Determine:
(a) The steady-state error in radians
(b) The steady-state shaft position θ_m in radians

5-9. An input $\theta_R = 0.1$ rad is applied to a servo whose constants are

$$h = 1 \qquad K_a = 100 \qquad K_m = 1 \qquad K_p = 1$$

If a disturbance $T_{xm} = 1$ ft · lb occurs at the shaft, find:
(a) The steady-state error in radians
(b) The steady-state shaft position θ_m in radians

5-10. The response of a second-order servo to a step input is shown in Fig. P5-10. Find the values of ζ, τ (time constant), ω_n, and ω_d.

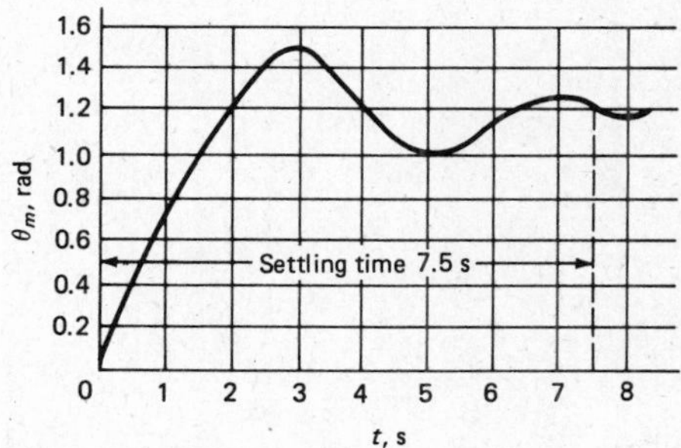

Fig. P5-10

5-11. The parameters of a second-order servo are

$K_p = 57$ V/rad $\quad K_m = 1.82 \times 10^{-4}$ lb · ft/V $\quad F_m = 4 \times 10^{-5}$ lb · ft · s
$F_l = 0 \quad J_m = 4 \times 10^{-6}$ slug · ft² $\quad J_l = 8.64 \times 10^{-4}$ slug · ft²

$$h = \text{feedback gear ratio} = \frac{N_m}{N_f} = \frac{1}{12}$$

$$a = \text{gear ratio from motor to load} = \frac{N_m}{N_l} = \frac{1}{12}$$

T_l = extraneous load torque = 20.88

The input command is $\theta_R = 0.1$ rad.
(a) Draw the block diagram of the system.
(b) Find the value of the amplifier gain K_a required to yield a steady-state error E of 0.174 rad.
(c) Determine ω_n and ζ.

5-12. From Fig. P5-12 find:

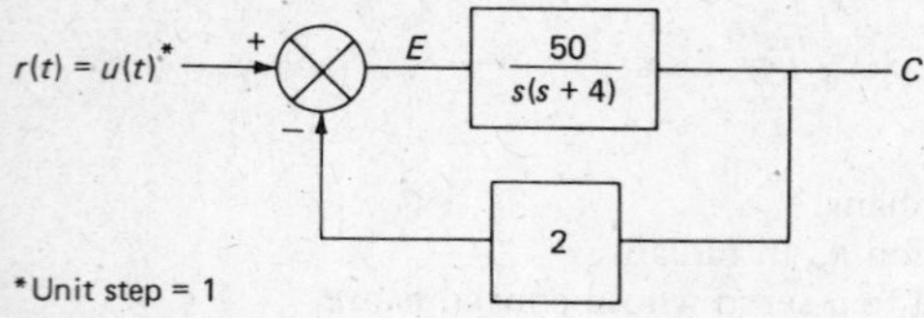

Fig. P5-12

(a) The damping factor ζ
(b) The natural frequency ω_n
(c) The damped frequency ω_d
(d) The percentage overshoot at output

5-13. From Fig. P5-13 find:

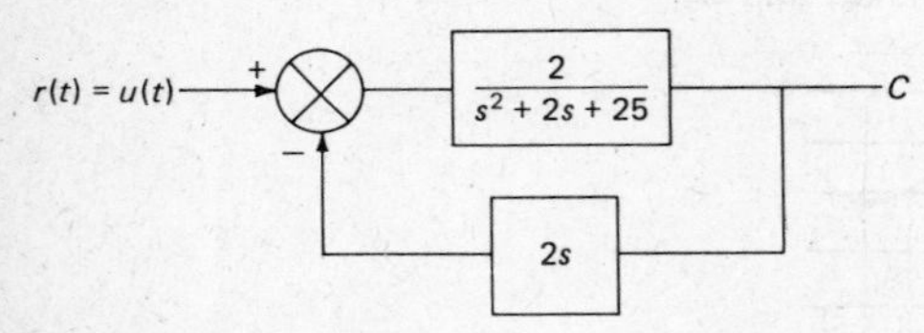

Fig. P5-13

(a) The natural frequency ω_n
(b) The damping factor ζ
(c) The damped frequency ω_d
(d) The percentage overshoot at output

5-14. In Fig. P5-14:

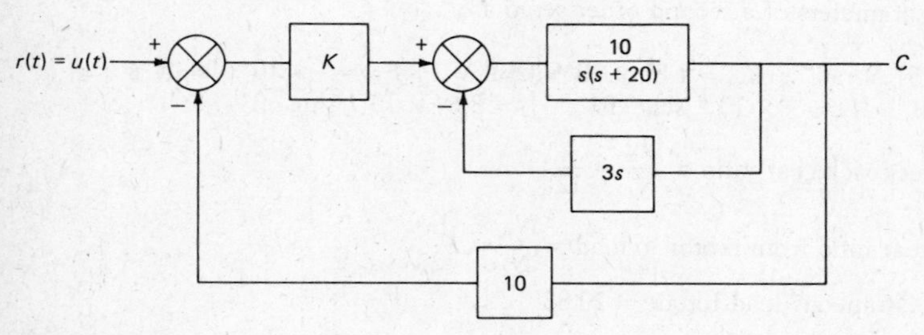

Fig. P5-14

(a) If $K = 25$, find the percentage overshoot.
(b) Find K for critical damping.
(c) Find the steady-state error for each value of K above.

5-15. Explain the difference between rise time and the time to reach steady state (settling time) in a second-order system.

5-16. In feedback control systems the designer often fixes the value of ζ, the damping ratio of second-order systems, so that $0.3 < \zeta < 0.8$. Explain the reasons for selecting this range of values in terms of overshoot, rise time, and time to reach steady state (settling time). Remember that the time necessary to reach steady state varies inversely with ζ. For the limits of ζ given, is the system oscillatory (underdamped) or nonoscillatory (overdamped)?

5-17. A position servo with an input step of 1 rad has following loop gain

$$HG(s) = \frac{100}{s(s + 5)}$$

Determine the steady-state error using the error-coefficient technique.

5-18. For the block diagram in Fig. P5-18:

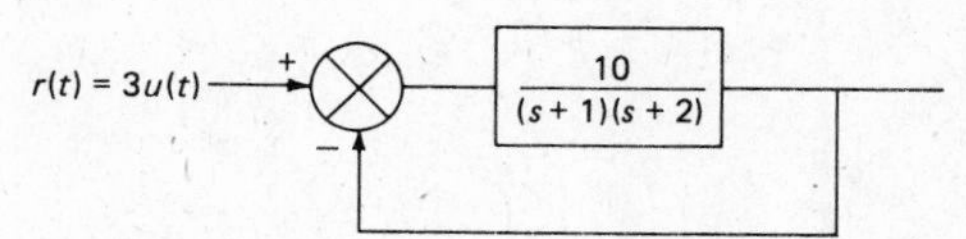

Fig. P5-18

(a) Find the error coefficient K_e.
(b) Find the steady-state error.

chapter 6

Frequency-Response Analysis

Frequency-response analysis is a method for predicting and adjusting system performance without actually building hardware or solving differential equations. The method requiring the solution of differential equations is known as *time-domain analysis*.

6-1 INTRODUCTION

Frequency-response analysis is merely a method for expressing the system transfer function in the sinusoidal steady state. The overall system is characterized by a group of *blocks*, e.g., amplifiers, motors, or transducers. Each block is described by its own transfer function, which can be expressed analytically or determined experimentally by making measurements of input and output magnitude and phase at various frequencies. Approximate expressions for the transfer functions can be obtained by curve-fitting the experimental data.

Frequency-response or transfer-function analysis ignores initial conditions, but it is valid since initial conditions have no effect on the form of the steady-state and natural responses. Transfer functions of networks are easily obtained by writing the impedances in the form R, sL, and $1/sC$. If a differential equation is given, the ratio of output to input is determined after replacing d/dt by s, d^2/dt^2 by s^2, etc. However, Laplace transforms are generally used because they normally provide a faster solution than differential equations.

In order to determine system behavior, various test signals have to be applied. Fourier analysis shows that all signals can be broken down into the sum of many sinusoidal signals. Therefore, it is only important to determine the response as a function of frequency. This is done mathematically by replacing s with $j\omega$ in the transfer function.

If experimental data are required, simple techniques are available to curve-fit an approximate transfer function. The usual method depends on the logarithmic gain and phase-vs.-frequency plot (*Bode plot*), which describes the transfer function in terms of connected straight lines of different slope.

Example 6-1

Find $V_o(s)/V_i(s)$ using Fig. 6-1.

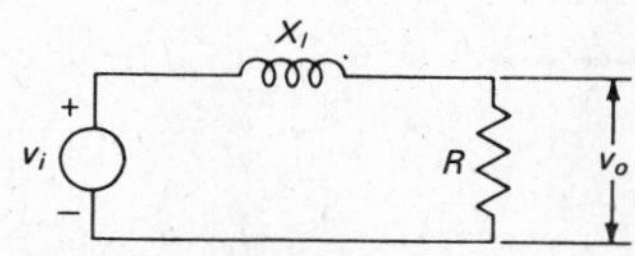

Fig. 6-1 Example 6-1.

Solution

Since the impedance of the inductor is $Z_L = sL$, by the voltage-divider rule

$$T(s) = \frac{V_o(s)}{V_i(s)} = \frac{R}{R + sL} = \frac{R/L}{s + R/L}$$

Example 6-2

Given

$$\frac{A\,d^2i_o}{dt^2} + \frac{B\,di_o}{dt} + Ci_o = i_i$$

find $I_o(s)/I_i(s)$.

Solution

Assuming zero initial conditions and taking Laplace transforms of both sides of the differential equation, we have

$$As^2I_o(s) + BsI_o(s) + CI_o(s) = I_i(s)$$

or $\quad (As^2 + Bs + C)I_o(s) = I_i(s)$

Then $\quad T(s) = \dfrac{I_o(s)}{I_i(s)} = \dfrac{1}{As^2 + Bs + C}$

Example 6-3

Find $|V_o/V_i|$ and ϕ if $V_i = A \sin \omega t$ and $V_o = B \cos \omega t$.

Solution

$$T(s) = \left|\frac{V_o}{V_i}\right| = \frac{B}{A} \qquad \phi = 90°$$

Example 6-4

Find $T(s) = V_o(s)/V_i(s)$ using Fig. 6-2.

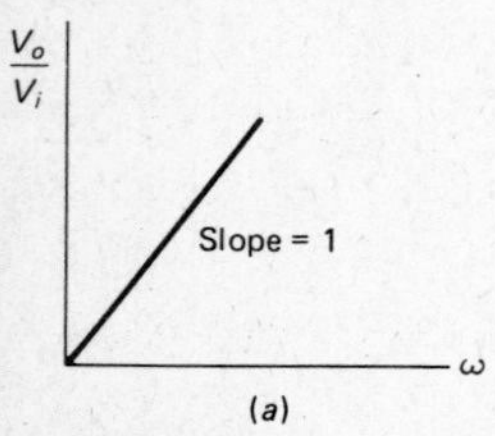

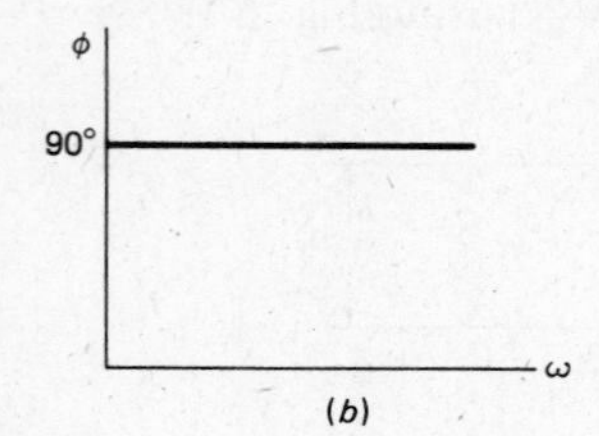

Fig. 6-2 Example 6-4.

Solution

From Fig. 6-2*a*

$$\left|\frac{V_o}{V_i}\right| = \omega$$

Then $T(j\omega) = \dfrac{V_o}{V_i} = j\omega$

Checking the phase, we have

$$\phi = \tan^{-1}\frac{\omega}{0} = 90°$$

Thus $T(s) = s$

6-2 STABILITY

Stability is a measure of the ability of a system to avoid sustained oscillations. According to the usual definition of stability, ***a stable system is one in which a bounded input produces a bounded output***. A bounded signal is one that never exceeds a finite value. A ramp input to a position servo is not bounded because as time increases to infinity, the signal approaches infinity.

Consider the block diagram shown in Fig. 6-3. $G(s)$ represents all the elements in the forward path lumped into one block, that is, $G(s) = G_1(s)\,G_2(s)$, and so forth. Suppose the input R is a sinusoid, $R = R_m \sin \omega t$; then let $s = j\omega$ and

$$G(s) = G(j\omega) \qquad H(s) = H(j\omega)$$

In Chap. 4 the transfer function for a negative-feedback system was shown to be

$$T(s) = \frac{G(s)}{1 + G(s)H(s)} \tag{6-1}$$

Letting $s = j\omega$, we have

$$T(j\omega) = \frac{G(j\omega)}{1 + G(j\omega)H(j\omega)} \tag{6-2}$$

and, assuming unity feedback,

$$H(j\omega) = 1 \tag{6-3}$$

$$\text{or} \quad T(j\omega) = \frac{G(j\omega)}{1 + G(j\omega)} \tag{6-4}$$

If there is one frequency ω_1 at which

$$G(j\omega) = -1 \qquad \text{or} \qquad G = 1 \qquad \text{and} \qquad \phi = 180° \tag{6-5}$$

$$\text{then} \quad T(j\omega_1) = \frac{-1}{1 - 1} = \infty \tag{6-6}$$

Equation (6-6) states that if a sinusoid of frequency ω_1 is applied to the closed-loop system, the output is ∞ (unbounded output) and so the system is un-

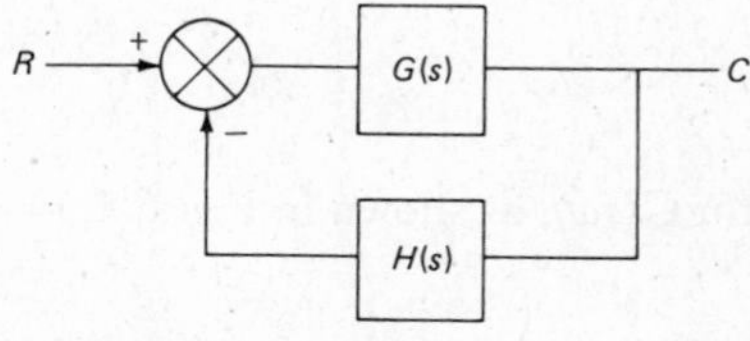

Fig. 6-3 Block diagram.

stable. One might argue that instability can be avoided by merely never applying the frequency ω_1 to the system. But the typical signal applied to the servo is a step function or some arbitrary waveshape. These signals are made up of all frequencies (Fourier series). The instant the system senses the component at frequency ω_1 it "takes off." Even if the input signal did not contain ω_1, noise or any other disturbance is very likely to contain ω_1. The servo designer's job is to ensure that $G(j\omega)$ is never equal to -1. This can be done by adding compensation, which is covered in Chap. 7, or selecting new components for the system.

By analogy to other physical systems, there is an expression known as the *resonant mode*. For example, bridges have been known to go into violent oscillations and collapse because wind disturbances have occurred at the resonant frequencies. The resonant modes are the natural frequencies determined by the characteristic equation of the closed-loop system.

Remember that sinusoidal inputs are merely used to characterize the transfer function of the system in a manner which lends itself to easy analysis, but all signals can be represented by sinusoids. Frequency analysis proceeds by finding, without the use of equations, the frequencies at which

$$G(j\omega) = -1 \tag{6-7}$$

This is much more practical than mathematically solving the characteristic equation

$$1 + G(s) = 0 \tag{6-8}$$

6-3 FREQUENCY-RESPONSE PLOTS

All frequency-response plots are graphical portrayals of gain and phase as a function of frequency. Since $G(j\omega)$ is a complex number, it can be written in rectangular form

$$G(j\omega) = A(\omega) + jB(\omega)$$

The gain is defined as the magnitude, or absolute value, of the complex number or

$$|G(j\omega)| = \sqrt{[A(\omega)]^2 + [B(\omega)]^2}$$

The phase is defined as the angle of the vector $G(j\omega)$, as shown in Fig. 6-4, or

$$\phi(j\omega) = \tan^{-1} \frac{B(\omega)}{A(\omega)}$$

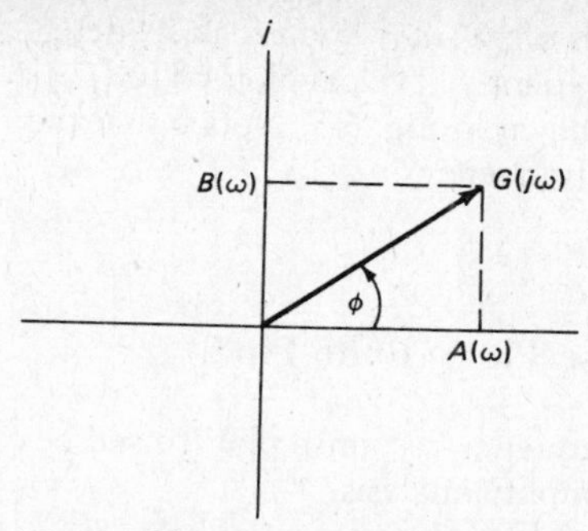

Fig. 6-4 G(jω) in complex form.

6-3-1 Rectangular Frequency-Response Characteristics

The rectangular frequency-response characteristics are simply graphs of gain and phase as a function of frequency. The standard second-order system is illustrated below:

$$G(s) = \frac{1}{s^2 + 2\zeta\omega_n s + \omega_n^2}$$

Then $G(j\omega) = \dfrac{1}{(j\omega)^2 + 2\zeta\omega_n(j\omega) + \omega_n^2} = \dfrac{1}{(\omega_n^2 - \omega^2) + 2\zeta\omega_n\omega j}$

and $|G(j\omega)| = \dfrac{1}{\sqrt{(\omega_n^2 - \omega^2)^2 + (2\zeta\omega_n\omega)^2}}$

$$\phi = -\tan^{-1}\frac{2\zeta\omega_n}{\omega_n^2 - \omega^2}$$

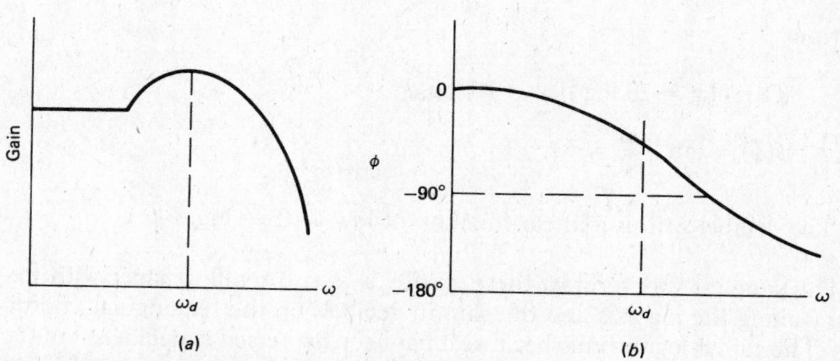

Fig. 6-5 Rectangular frequency response characteristics.

For specified values of ζ and ω_n, the values of $|G(j\omega)|$ and ϕ can be determined by substituting values of ω from zero to infinity. Typical plots of $|G(j\omega)|$ and ϕ for the underdamped case, $\zeta < 1$, are shown in Fig. 6-5. Notice that the magnitude plot peaks at the damped natural frequency

$$\omega_d = \omega_n \sqrt{1 - \zeta^2}$$

6-3-2 Logarithmic Magnitude and Phase Plots (Bode Plots)

A logarithmic rather than the rectangular frequency-response plot is used because approximations can be made which simplify analysis.

The gain is usually expressed in decibels (dB), where

$$G_{dB} = |G(j\omega)|_{dB} = 20 \log |G(j\omega)| \tag{6-9}$$

Another advantage of logarithmic representation is that multiplication and division become addition and subtraction; i.e.,

$$\log |G_1(j\omega)| \cdot |G_2(j\omega)| = \log |G_1(j\omega)| + \log |G_2(j\omega)|$$

$$\log \frac{|G_1(j\omega)|}{|G_2(j\omega)|} = \log |G_1(j\omega)| - \log |G_2(j\omega)|$$

$$\log \frac{|G_1(j\omega)| \cdot |G_2(j\omega)|}{|G_3(j\omega)|} = \log |G_1(j\omega)| + \log |G_2(j\omega)| - \log |G_3(j\omega)|$$

Example 6-5

Consider the function

$$G(s) = s \qquad \text{or} \qquad G(j\omega) = j\omega$$

Solution

$$|G(j\omega)|_{dB} = 20 \log |j\omega| = 20 \log \omega$$

$$\phi(\omega) = \tan^{-1} \frac{\omega}{0} = 90°$$

The decibel gain is a linear function of low ω. (See Fig. 6-6.)

The simplest way to draw these graphs is to use semilog paper with frequency along the log axis and the gain in decibels on the rectangular coordinates. The gain is logarithmic because it has been expressed in decibels, but the semilog paper eliminates the need to find the log of the frequency.

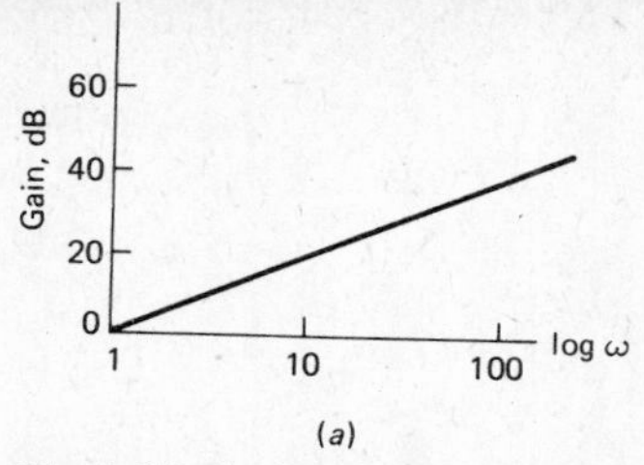

Fig. 6-6 Bode plots for Example 6-5.

Notice that decibel gain is 0 when $\omega = 1$ because $\log 1 = 0$, but at $\omega = 10$

$$\text{dB} = 20 \log 10 = 20(1) = 20$$

By definition, when two frequencies have a ratio of 10:1 they are separated by a *decade*. Thus, the slope of the decibel gain is 20 dB/decade. Each time frequency goes up 1 decade the gain goes up 20 dB. Notice that at $\omega = 100$, the gain is 40 dB.

In general, if the decibel gain is $20 \log \omega$ and at ω_1 dB $= 20 \log \omega_1$, then if $\omega_2 = 10\omega_1$,

$$\begin{aligned} \text{dB} &= 20 \log \omega_2 = 20 \log 10\omega_1 \\ &= 20 \log 10 + 20 \log \omega_1 = 20 + 20 \log \omega_1 \end{aligned}$$

Similarly, if $\omega_2 = 0.1\omega_1$,

$$\begin{aligned} \text{dB} &= 20 \log \omega_2 = 20 \log 0.1\omega_1 \\ &= 20 \log 0.1 + 20 \log \omega_1 = 20(-\log 10) + 20 \log \omega_1 \\ &= -20 + 20 \log \omega_1 = 20 \log \omega_1 - 20 \end{aligned}$$

We have just defined the decade as a frequency ratio of 10:1, but another common term is the *octave*. Two frequencies are separated by an octave if their ratio is 2:1. Then if $\omega_2 = 2\omega_1$,

$$\begin{aligned} \text{dB} &= 20 \log \omega_2 = 20 \log 2\omega_1 = 20 \log 2 + 20 \log \omega_1 \\ &= 20(0.3) + 20 \log \omega_1 = 6 + 20 \log \omega_1 \end{aligned}$$

If the frequency changes by 1 octave, the gain changes by 6 dB and the slope of the decibel-gain plot is 6 dB/octave. But since the slopes represent the same function,

$$G(j\omega) = j\omega$$

the slopes must be equivalent and

$$6 \text{ dB/octave} = 20 \text{ dB/decade} \tag{6-10}$$

Example 6-6
Consider a more general function

$$G(s) = 10s \qquad \text{or} \qquad G(j\omega) = 10j\omega$$

Solution

$$\begin{aligned} G_{dB} &= 20 \log |10j\omega| = 20 \log 10\omega = 20 \log 10 + 20 \log \omega \\ &= 20 + 20 \log \omega \end{aligned}$$

When $\omega = 0.01$, $G_{dB} = 20 - 40 = -20$. Note that

$$\log 0.01 = -\log \frac{1}{0.01} = -\log 100 = -2$$

$$G_{dB} = \begin{cases} 20 - 20 = 0 & \text{when } \omega = 0.1 \\ 20 + 0 = 20 & \text{when } \omega = 1 \end{cases}$$

The magnitude plot is shown in Fig. 6-7.

Observe that the line in Fig. 6-7 crosses the 0-dB axis at $\omega = 0.1$. In general, if $G(s) = Ks$, the line will cross at $\omega = 1/K$. Therefore, Ks can be plotted by simply drawing a line whose slope is 20 dB/decade (or 6 dB/octave) and which crosses at $\omega = 1/K$.

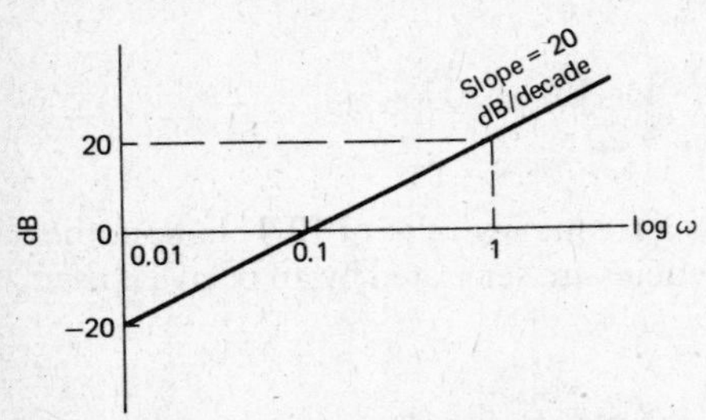

Fig. 6-7 Magnitude Bode plot for Example 6-6.

Example 6-7
Consider the function

$$G(s) = \frac{10}{s} \qquad \text{or} \qquad G(j\omega) = \frac{10}{j\omega}$$

Solution

$$G_{dB} = 20 \log \frac{10}{\omega} = 20 \log 10 - 20 \log \omega = 20 - 20 \log \omega$$

$$\text{Then} \quad G_{dB} = \begin{cases} 20 - 0 = 20 & \text{when } \omega = 1 \\ 20 - 20 = 0 & \text{when } \omega = 10 \\ 20 - 40 = -20 & \text{when } \omega = 100 \end{cases}$$

The magnitude Bode plot is shown in Fig. 6-8.

In Example 6-7 notice that the slope of the magnitude plot is -20 dB/decade (or -6 dB/octave). In other words, an s term in the numerator results in a positive slope, but the same term in the denominator results in a negative slope. Also, observe that the line crosses the 0-dB axis at $\omega = 10$. In other words, if $G(s) = K/s$, the Bode plot will cross the axis at $\omega = K$. In order to plot K/s, simply sketch a line whose slope is -20 dB/decade and which crosses the axis at $\omega = K$.

In order to facilitate sketching of the Bode plot, always factor to obtain terms of the form $s/a + 1$. For example,

$$s + 10 = 10\left(\frac{s}{10} + 1\right)$$

$$\frac{1}{s + 10} = \frac{1}{10(s/10 + 1)} = \frac{0.1}{s/10 + 1}$$

$$(s + 5)(s + 10) = 50\left(\frac{s}{5} + 1\right)\left(\frac{s}{10} + 1\right)$$

$$\frac{10s(s + 8)}{s + 4} = \frac{10(8s)(s/8 + 1)}{4(s/4 + 1)} = \frac{20s(s/8 + 1)}{s/4 + 1}$$

$$(s + 10)^2 = 10(10)\left(\frac{s}{10} + 1\right)^2 = 100\left(\frac{s}{10} + 1\right)^2$$

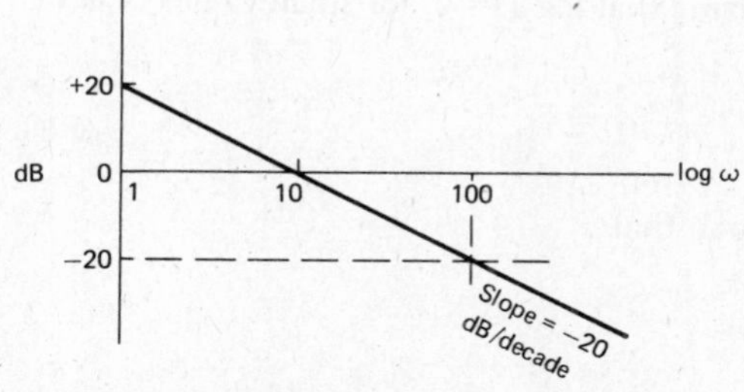

Fig. 6-8 Magnitude Bode plot for Example 6-7.

Continuing with examples of Bode plots, consider the expression

$$G(s) = \frac{1}{s + a} \tag{6-11}$$

where a is any positive number. Then

$$G(s) = \frac{1}{a(s/a + 1)} = \frac{1/a}{s/a + 1}$$

and $$G(j\omega) = \frac{1/a}{j\omega/a + 1}$$

or $$|G(j\omega)| = \frac{1/a}{\sqrt{1 + (\omega/a)^2}} \tag{6-12}$$

and
$$\begin{aligned} |G(j\omega)|_{dB} &= 20 \log \frac{1/a}{\sqrt{1 + (\omega/a)^2}} \\ &= 20 \log \frac{1}{a} - 20 \log \sqrt{1 + \left(\frac{\omega}{a}\right)^2} \\ &= 20 \log \frac{1}{a} - 10 \log \left[1 + \left(\frac{\omega}{a}\right)^2\right] \end{aligned} \tag{6-13}$$

The first term in equation (6-13), 20 log $(1/a)$, is just a constant and does not depend on frequency. Consider the second term for small values of frequency such that

$$\frac{\omega}{a} \ll 1$$

Then $$G_{dB} = |G(j\omega)|_{dB} = 20 \log \frac{1}{a} - 10 \log 1$$

Defining $K = 20 \log 1/a$, and realizing that $\log 1 = 0$, for small values of ω we have

$$G_{dB} = K \tag{6-14}$$

Now consider large values of ω such that

$$\frac{\omega}{a} \gg 1$$

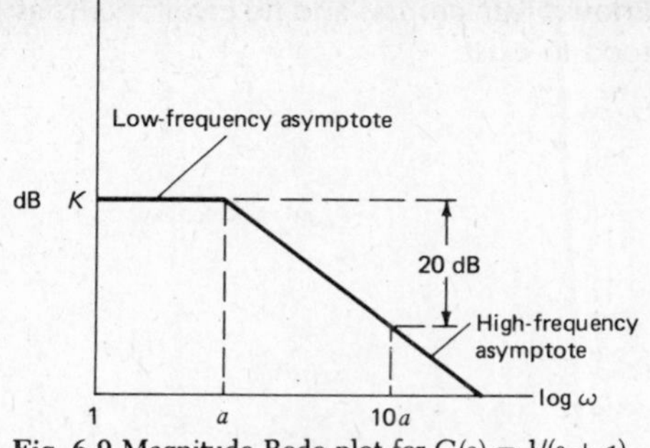

Fig. 6-9 Magnitude Bode plot for $G(s) = 1/(s + a)$.

Thus $1 + \left(\frac{\omega}{a}\right)^2 \approx \left(\frac{\omega}{a}\right)^2$

$$G_{dB} = K - 10 \log \left(\frac{\omega}{a}\right)^2$$

$$= K - 20 \log \frac{\omega}{a} \qquad \text{for large values of } \omega \tag{6-15}$$

Let $\omega = a$ in Eq. (6-15); then $20 \log (\omega/a) = 20 \log 1 = 0$ and

$$G_{dB} = K \qquad \text{for } \omega = a \tag{6-16}$$

The high- and low-frequency approximations result in the same gain at $\omega = a$. This point is called a *break frequency* (or *corner frequency*).

If we neglect the fact that these approximations only hold at very high and very low frequencies, the log magnitude plot becomes two straight lines which intersect at the point $\omega = a$. For $\omega \leq a$, the low-frequency asymptote* is

$$G_{dB} = K \tag{6-17}$$

and for $\omega \geq a$ the high-frequency asymptote is

$$G_{dB} = K - 20 \log \frac{\omega}{a} \tag{6-18}$$

which is a line with a slope of −20 dB/decade. Both lines are plotted in Fig. 6-9.

In actuality, the plot should not be these two straight lines but a smooth curve. However, in practice, the true curve can be approximated by a series of

* An asymptote is a line that is approached but never reached.

straight lines which intersect at the various *break points*, and no error occurs as long as the approximation is understood to exist.

Example 6-8

Given the transfer function

$$G(s) = \frac{100s}{s + 10}$$

determine the gain Bode plot.

Solution

$$G(j\omega) = \frac{100j\omega}{j\omega + 10} = \frac{10j\omega}{1 + j\omega/10}$$

$$|G(j\omega)| = \frac{10\omega}{\sqrt{1 + (\omega/10)^2}}$$

$$G_{dB} = 20 \log 10\omega - 10 \log \left[1 + \left(\frac{\omega}{10}\right)^2\right]$$

$$= 20 + 20 \log \omega - 10 \log \left[1 + \left(\frac{\omega}{10}\right)^2\right]$$

Then for $\omega < 10$

$$G_{dB} = 20 + 20 \log \omega$$

$$\text{and} \quad G_{dB} = \begin{cases} 20 & \text{for } \omega = 1 \\ 20 + 20 \log 10 = 40 & \text{for } \omega = 10 \end{cases}$$

For $\omega < 10$, the plot is a straight line with a slope of +20 dB/decade; for $\omega > 10$,

$$G_{dB} = 20 + 20 \log \omega - 20 \log \frac{\omega}{10}$$

$$= 20 + 20 \log \omega - (20 \log \omega - 20 \log 10)$$
$$= 20 + 20 \log \omega - 20 \log \omega + 20 = 40$$

The gain Bode plot is shown in Fig. 6-10.

In Example 6-8, the break point is $\omega = 10$. For $\omega < 10$ it was shown that

$$G_{dB} = 20 + 20 \log \omega$$

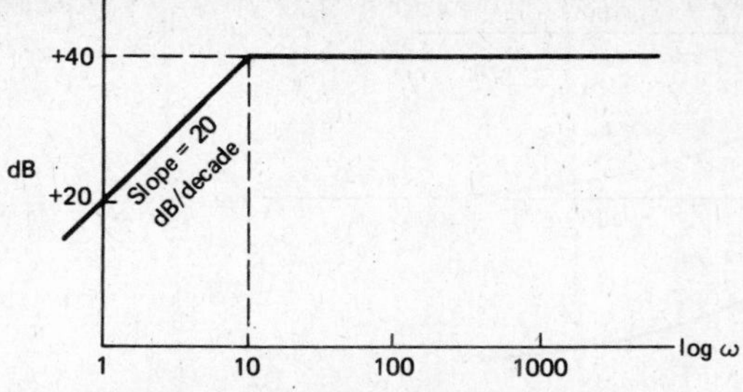

Fig. 6-10 Magnitude Bode plot for $G(s) = 100s/(s + 10)$.

This is a straight line of the form $y = mx + b$. The y intercept is $y = b$ at $x = 0$; on logarithmic plots $x = \log \omega = 0$ when $\omega = 1$. Beyond $\omega = 10$, the term $-20 \log (\omega/10)$ appears and causes the plot to level off at 40 dB.

At this juncture, a way to simplify the sketching of the plots emerges: treat each part of the transfer function separately and then add up the results:

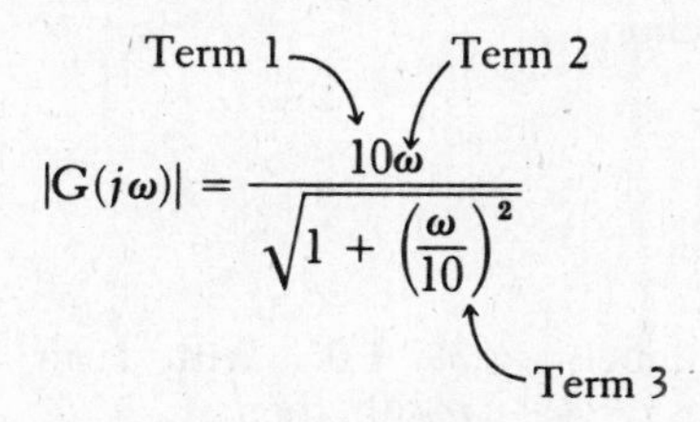

$$|G(j\omega)| = \frac{10\omega}{\sqrt{1 + \left(\frac{\omega}{10}\right)^2}}$$

Figure 6-11 illustrates this procedure for Example 6-8.

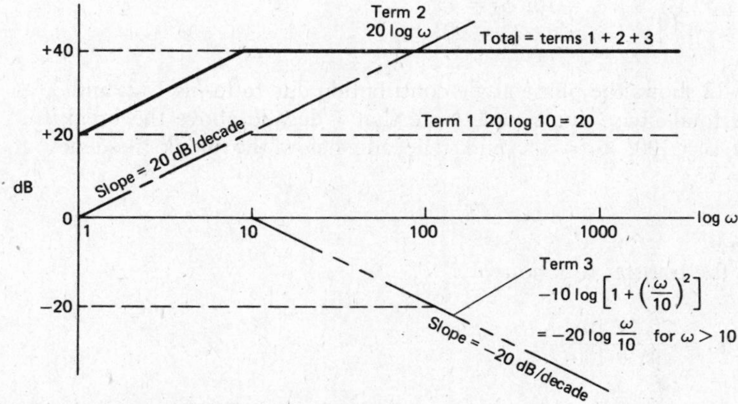

Fig. 6-11 Constructing a Bode plot.

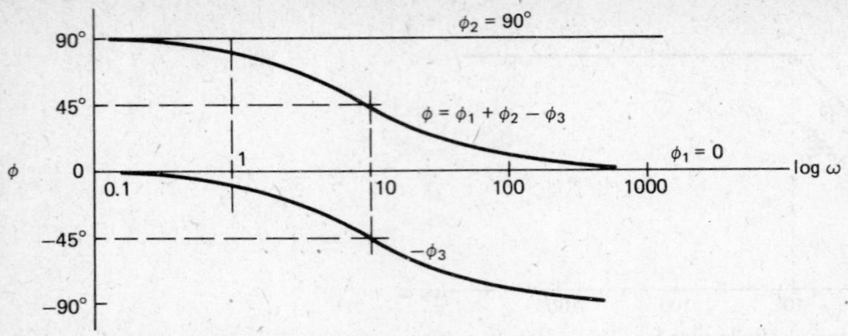

Fig. 6-12 Phase Bode plot for $G(s) = 100s/(s + 10)$.

Now consider the phase ϕ; from a knowledge of complex numbers, it can be shown that the total phase angle is

$$\phi = \text{phase angle of numerator} - \text{phase angle of denominator} \tag{6-19}$$

Referring to the original transfer function,

$$G(j\omega) = \frac{10\,j\omega}{1 + \dfrac{j\omega}{10}}$$

let term 1 = 10; this is a real number, so $\phi_1 = 0°$. With term 2 = $j\omega$, $\phi_2 = \tan^{-1}\omega/0 = 90°$. Finally, let term 3 = $1 + (j\omega/10)$; then

$$\phi_3 = \tan^{-1}\frac{\omega}{10} \quad \begin{cases} \omega = 0, \ \phi_3 = 0° \\ \omega = 10, \ \phi_3 = 45° \\ \omega = \infty, \ \phi_3 = 90° \end{cases}$$

Figure 6-12 shows the phase angle contribution due to terms 1, 2, and 3, as well as the total phase angle plot. Note that 1 decade above the break frequency, at $\omega = 100$, $\phi_3 = 84°$, and 1 decade below the break frequency at $\omega = 1$, $\phi_3 = 6°$.

Example 6-9

Given the transfer function

$$G(s) = \frac{s + 10}{s(s + 100)}$$

determine the magnitude and phase Bode plots.

Solution

$$G(s) = \frac{10(s/10 + 1)}{100s(s/100 + 1)} = \frac{0.1(s/10 + 1)}{s(s/100 + 1)}$$

The break frequencies are $\omega_1 = 10$ rad/s and $\omega_2 = 100$ rad/s.

Term 1: $20 \log 0.1 = -20$

Term 2: $20 \log \left| \frac{s}{10} + 1 \right|$

Term 3: $-20 \log |s|$

Term 4: $-20 \log \left| \frac{s}{100} + 1 \right|$

We can now take many shortcuts: (1) don't bother writing $s = j\omega$, just make a mental note of it; (2) don't take magnitudes (instead put magnitude signs in the expressions). Then note all the break points. The resulting Bode plot is shown in Fig 6-13.

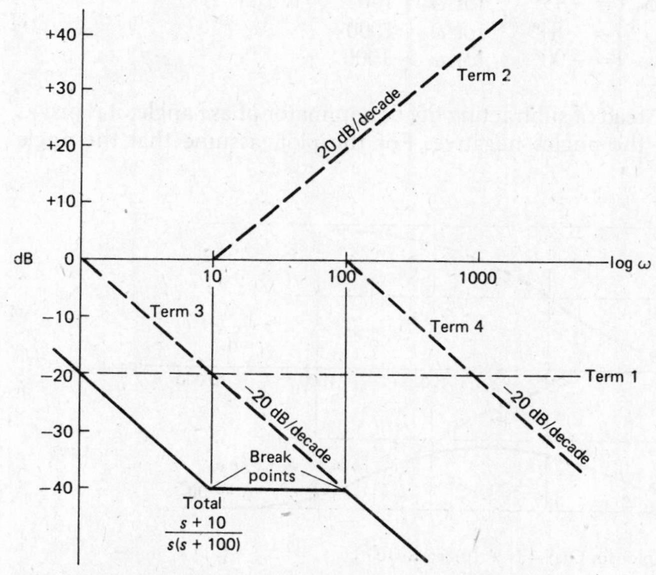

Fig. 6-13 Magnitude Bode plot for $G(s) = s + 10/[s(s + 100)]$.

To determine the phase plot, the numerator is

Term 1: 0.1 = real number $\quad \phi_1 = 0$

Term 2: $\dfrac{s}{10} + 1 \qquad \phi_2 = \tan^{-1}\dfrac{\omega}{10}$

or

$$\phi_2 \begin{cases} = 0 & \text{for } \omega \ll 10 \\ \approx 6^\circ & \text{for } \omega = 1 \\ = 45^\circ & \text{for } \omega = 10 \\ \approx 84^\circ & \text{for } \omega = 100 \\ = 90^\circ & \text{for } \omega \gg 10 \end{cases}$$

and the denominator is

Term 3: $s \qquad \phi_3 = -\tan^{-1}\dfrac{\omega}{0} = -90^\circ$

Term 4: $\dfrac{s}{100} + 1 \qquad \phi_4 = -\tan^{-1}\dfrac{\omega}{100}$

or

$$\phi_4 \begin{cases} = 0 & \text{for } \omega \ll 100 \\ \approx -6^\circ & \text{for } \omega = 10 \\ = -45^\circ & \text{for } \omega = 100 \\ \approx -84^\circ & \text{for } \omega = 1000 \\ = -90^\circ & \text{for } \omega \gg 1000 \end{cases}$$

Notice that instead of subtracting the denominator phase angle, it is just as easy to make the angles negative. For the plot, assume that the angle

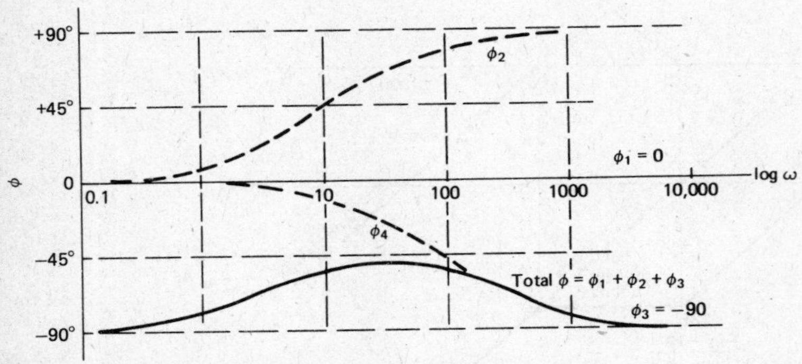

Fig. 6-14 Phase Bode plot for $G(s) = (s + 10)/[s(s + 100)]$.

reaches its asymptote 2 decades away from the break frequency. The phase plot is shown in Fig. 6-14.

Example 6-10

Given the transfer function

$$G(j\omega) = \frac{2}{(j\omega/2 + 1)(j\omega/8 + 1)}$$

determine the magnitude and phase Bode plots.

Solution

Term 1: $20 \log 2 = 6$

Term 2: $-20 \log \left| j\frac{\omega}{2} + 1 \right|$

Term 3: $-20 \log \left| j\frac{\omega}{8} + 1 \right|$

The magnitude plot is shown in Fig. 6-15. At the second break point ($\omega = 8$), merely add the two slopes algebraically $(-6 + -6 = -12)$ to obtain the slope of the total log magnitude plot.

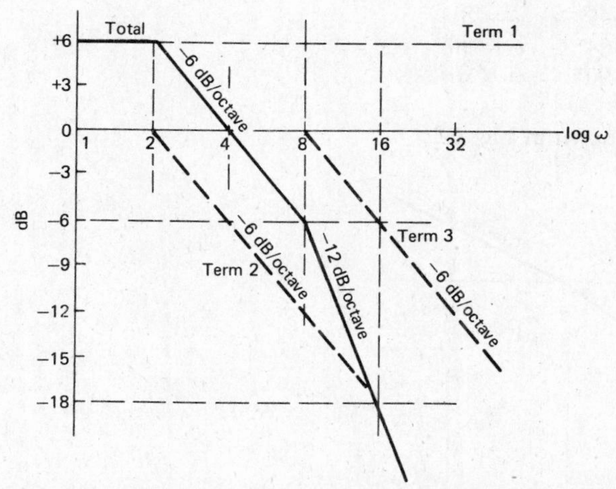

Fig. 6-15 Magnitude Bode plot for $G(j\omega) = 2/[(j\omega/2 + 1)(j\omega/8 + 1)]$.

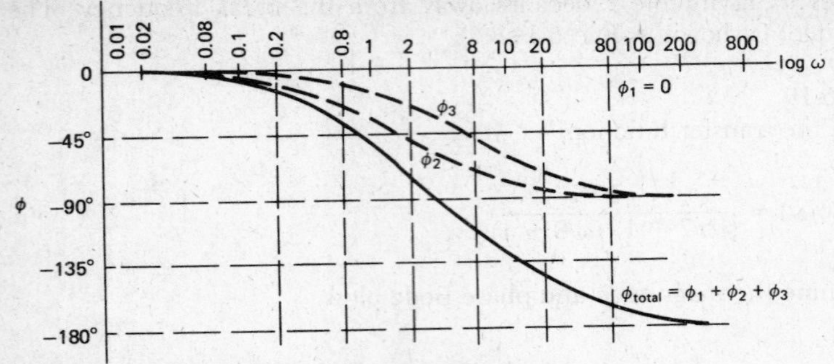

Fig. 6-16 Phase Bode plot for $G(j\omega) = 2/[(j\omega/2 + 1)(j\omega/8 + 1)]$.

Using the rules for the phase plot, we get

Term 1: $\phi_1 = 0$

$$\text{Term 2:}\quad \phi_2 \begin{cases} = 0 & \omega \ll 2 \\ \approx -6^\circ & \omega = 0.2 \\ = -45^\circ & \omega = 2 \\ \approx -84^\circ & \omega = 20 \\ \approx -90^\circ & \omega \gg 2 \end{cases}$$

$$\text{Term 3:}\quad \phi_3 \begin{cases} = 0 & \omega \ll 8 \\ \approx -6^\circ & \omega = 0.8 \\ = -45^\circ & \omega = 8 \\ \approx -84^\circ & \omega = 80 \\ = -90^\circ & \omega \ll 8 \end{cases}$$

The phase plot is shown in Fig. 6-16.

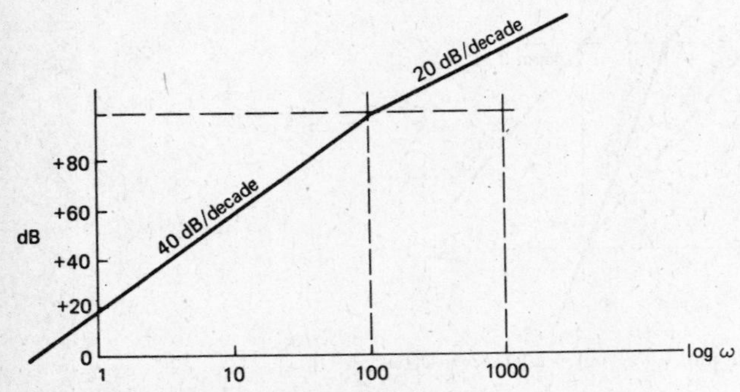

Fig. 6-17 Bode plot for $G(s) = 1000s^2/(s + 100)$.

Example 6-11

Determine the magnitude Bode plot for

$$G(s) = \frac{1000s^2}{s + 100}$$

Solution

$$G(s) = \frac{10s^2}{s/100 + 1}$$

$$G_{dB} = 20 \log 10 + 20 \log |s^2| - 20 \log \left| \frac{s}{100} + 1 \right|$$

$$= 20 + 40 \log |s| - 20 \log \left| \frac{s}{100} + 1 \right|$$

The magnitude plot is shown in Fig 6-17.

6-3-3 Bode Plots with Square Terms

Consider the transfer function

$$G(s) = \frac{100}{s^2}$$

or $G_{dB} = 20 \log |G(j\omega)| = 20 \log \dfrac{100}{\omega^2}$

$$= 20 \log 100 - 20 \log \omega^2 = 20 \log 100 - 40 \log \omega$$
$$= 40 - 40 \log \omega$$

Substituting various values of ω in the equation gives

$$G_{dB} = \begin{cases} 40 & \text{when } \omega = 1 \\ 0 & \text{when } \omega = 10 \\ -40 & \text{when } \omega = 100 \end{cases}$$

The resulting Bode plot is shown in Fig. 6-18. Referring to Example 6-11, observe that the Bode plot is a straight line whose slope is -40 dB/decade (which can be shown to be -12 dB/octave) and which crosses the horizontal axis at $\omega = 10$.

To determine the phase

$$G(j\omega) = \frac{100}{(j\omega)^2} = \frac{100}{-\omega^2} = \frac{-100}{\omega^2}$$

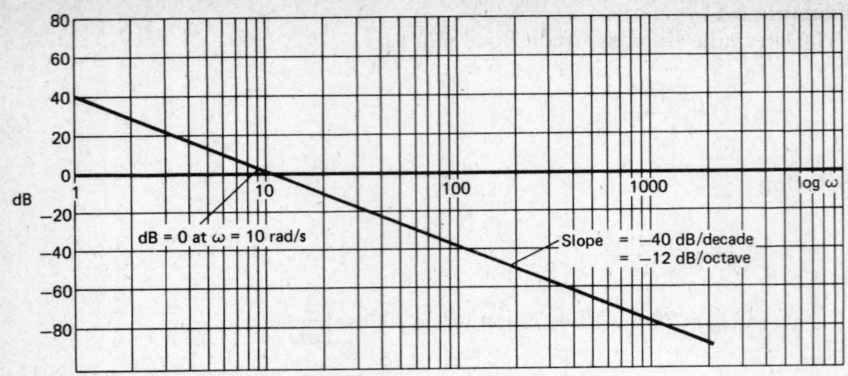

Fig. 6-18 Bode plot for $G(s) = 100/s^2$.

Because the real term is negative and the imaginary term is zero, the phase is

$$\phi = \tan^{-1}\frac{0}{-100/\omega^2} = \pm 180°$$

It can be shown that for the general expression

$$G(s) = \frac{K}{s^2}$$

the phase angle is a constant ±180°, and the magnitude Bode plot is a straight line whose slope is −40 dB/decade (−12 dB/octave) and which crosses the horizontal axis at $\omega = \sqrt{K}$, as shown in Fig. 6-19.

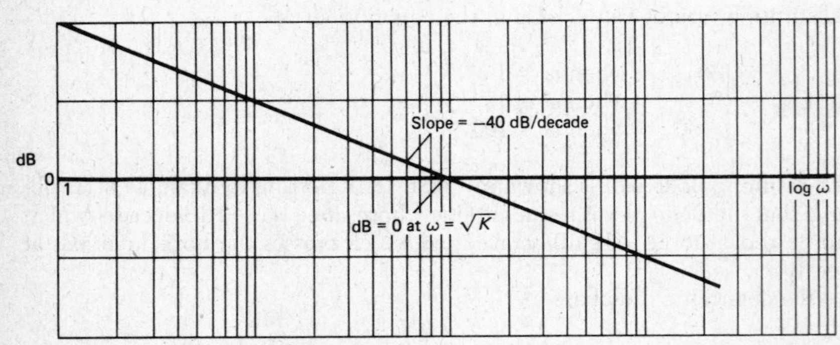

Fig. 6-19 Bode plot for $G(s) = K/s^2$.

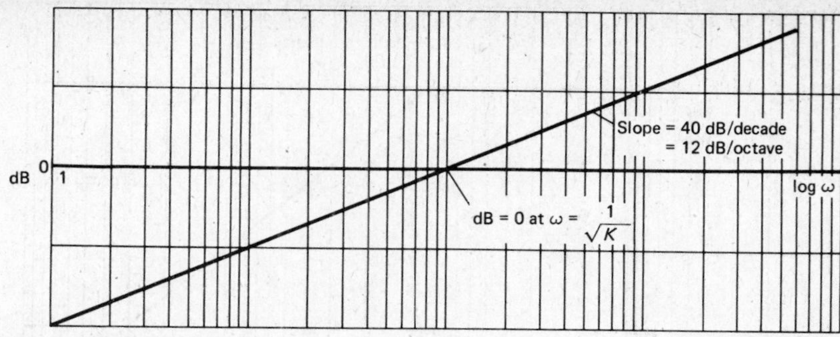

Fig. 6-20 Bode plot for $G(s) = Ks^2$.

It can also be demonstrated that if

$$G(s) = Ks^2$$

the phase is a constant $\pm 180°$ and the magnitude Bode plot is a straight line whose slope is 40 dB/decade (12 dB/octave) and which crosses the horizontal axis at $\omega = \sqrt{1/K}$ (Fig. 6-20).

For

$$G(s) = \frac{1}{(s/a + 1)^2}$$

it can be shown that the Bode plot is a horizontal line at 0 dB until the break frequency $\omega = a$. After the break frequency, the magnitude is a straight line whose slope is -40 dB/decade, as shown in Fig. 6-21.

To determine the phase

$$G(s) = \frac{1}{s/a + 1} \frac{1}{s/a + 1}$$

$$\text{or} \quad G(j\omega) = \frac{1}{j\omega/a + 1} \frac{1}{j\omega/a + 1}$$

Since this expression is the product of two identical complex numbers, the phase angle ϕ can be shown to be

$$\phi = 2\left(-\tan \frac{\omega}{a}\right)$$

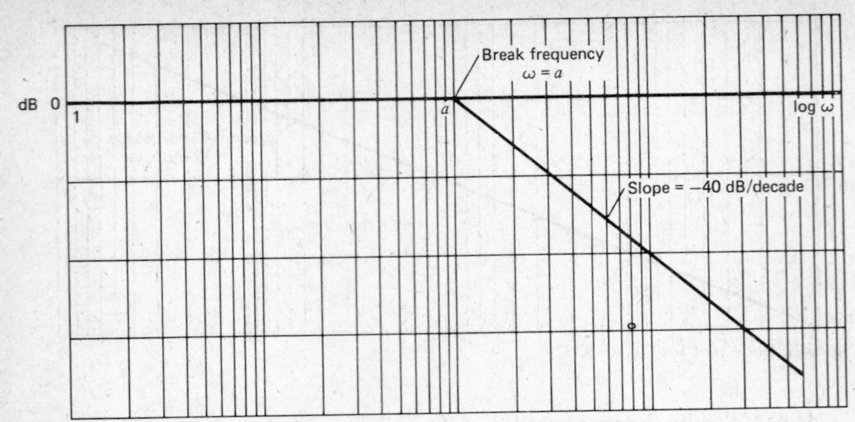

Fig. 6-21 Bode plot for $G(s) = 1/(s/a + 1)^2$.

Substituting various values of ω in this expression, we get

$$\phi = \begin{cases} 0 & \text{for } \omega = 0 \\ 2(-45^\circ) = -90^\circ & \text{for } \omega = a \\ 2(-90^\circ) = -180^\circ & \text{for } \omega \gg a \end{cases}$$

These angles are double the phase angles for the expression

$$G(s) = \frac{1}{s/a + 1}$$

The phase plot (Fig. 6-22) shows that $\phi = 90^\circ$ at the break point $\omega = a$ and asymptotically approaches 0° for low frequencies and −180° for high frequencies.

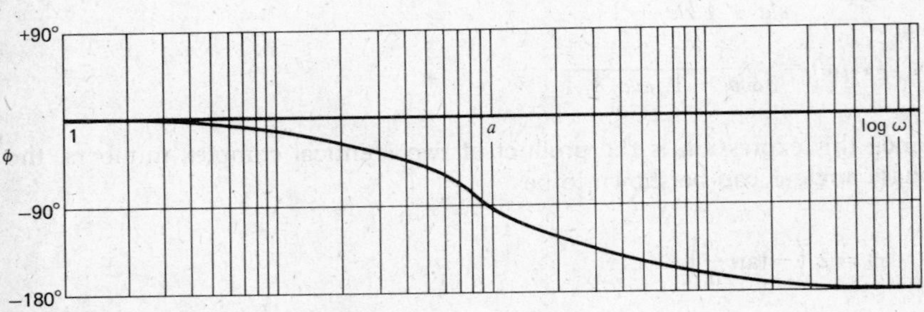

Fig. 6-22 Phase plot for $G(s) = 1/(s/a + 1)^2$.

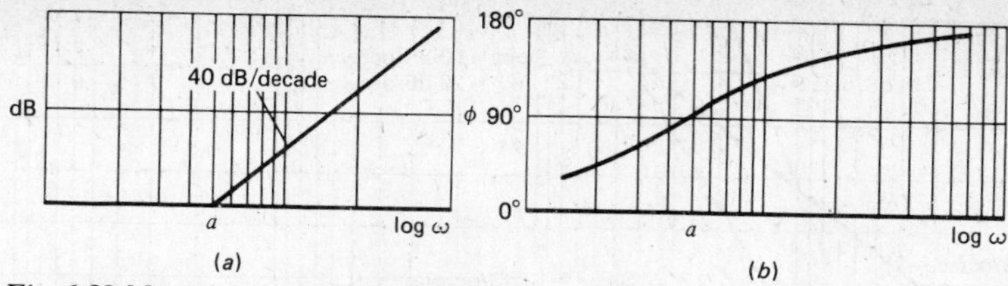

Fig. 6-23 Magnitude and phase plots for $G(s) = (s/a + 1)^2$.

The Bode plot for

$$G(s) = \left(\frac{s}{a} + 1\right)^2$$

is the mirror image of the plot for

$$G(s) = \frac{1}{(s/a + 1)^2}$$

as seen in Fig. 6-23.

Example 6-12

Given the transfer function

$$G(s) = \frac{s}{200}(s + 10)^2$$

determine the magnitude and phase Bode plots.

Solution

$$G(s) = \frac{(10)(10)s}{200}\left(\frac{s}{10} + 1\right)^2 = \frac{s}{2}\left(\frac{s}{10} + 1\right)^2$$

The magnitude and gain plots are shown in Fig. 6-24.

Example 6-13

Given the transfer function

$$G(s) = \frac{s^2}{(s + 10)^2}$$

determine the magnitude and phase Bode plots.

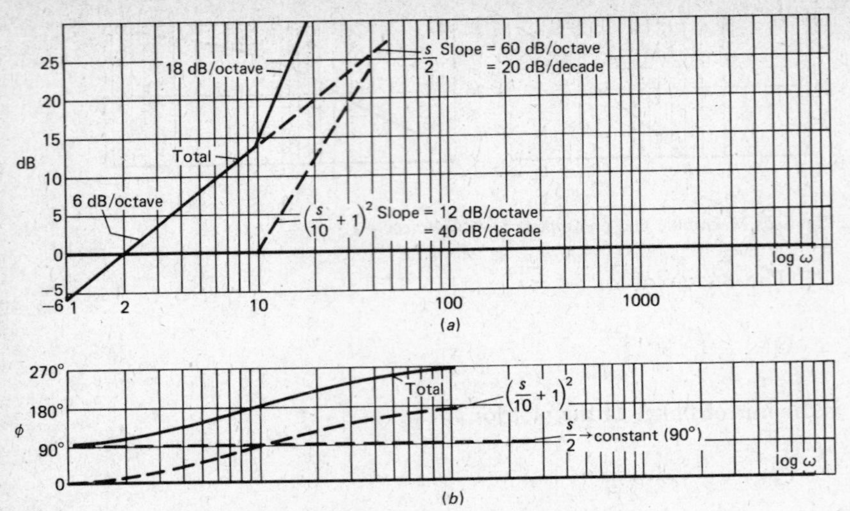

Fig. 6-24 Magnitude and phase plots for $G(s) = (s/2)[(s/10 + 1)^2]$.

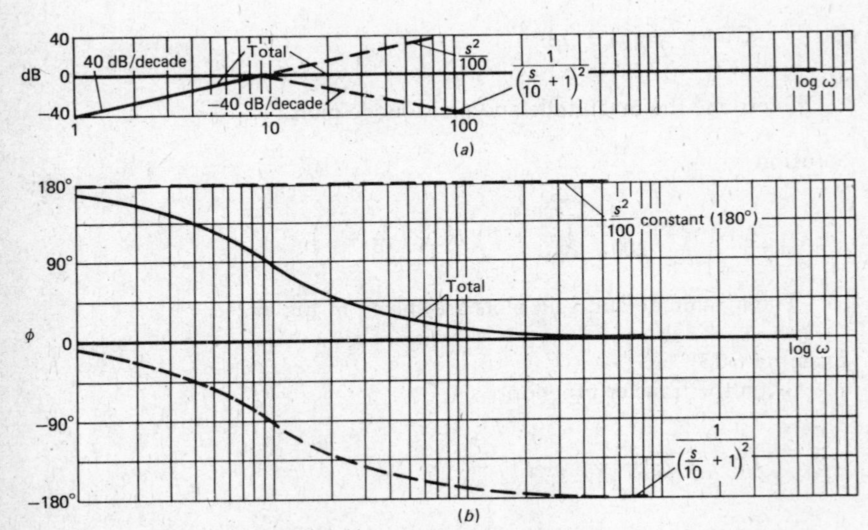

Fig. 6-25 Magnitude and phase plots for $G(s) = s^2/[100(s/10 + 1)^2]$.

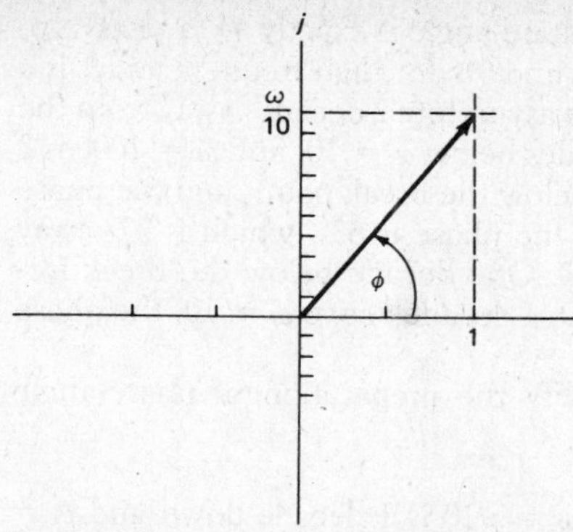

Fig. 6-26 Phase angle for $G(j\omega) = j\omega/10 + 1$.

Solution

$$G(s) = \frac{s^2}{100(s/10 + 1)^2}$$

and the resulting plots are shown in Fig. 6-25.

6-3-4 Shortcut Techniques for Phase Plots

Consider the expression

$$G(s) = \frac{s}{10} + 1 \quad \text{or} \quad G(j\omega) = \frac{j\omega}{10} + 1$$

Since $\phi = \tan^{-1} \omega/10$ (see Fig. 6-26), Table 6-1 can be prepared.

TABLE 6-1
Phase vs. Frequency for a First-Order Break

ω	ϕ
0.1	$0.573° \approx 0$
1	$5.7° \approx 6°$
5	$26.6° \approx 27°$
10	$45°$
20	$63.4° \approx 63°$
100	$84.3° \approx 84°$
1000	$89.4° \approx 90°$

At the break frequency, $\omega = 10$, the phase angle is exactly 45° and asymptotically approaches 0° for low frequencies and 90° for high frequencies. It is a good assumption that the angle reaches the asymptote 2 decades away from the break point. In this case, $\omega = 0.1$ is 2 decades below $\omega = 10$, and $\omega = 1000$ is 2 decades above $\omega = 10$. $\omega = 5$ is 1 octave below the break point, and the phase is 27°. At $\omega = 20$, 1 octave above $\omega = 10$ the phase is 63°, which is 27° away from the high-frequency asymptote of 90°. One decade below the break frequency, $\omega = 1$, the phase is 6°. At $\omega = 100$, 1 decade above $\omega = 10$, the phase angle is 84°, or 6° away from 90°.

To prepare a general chart to simplify the preparation of phase-angle plots, note the following observations:

1. For a break frequency of $\omega = a$, $\omega = a/10$ is 1 decade down and $\omega = 10a$ is 1 decade up, $\omega = a/100$ is 2 decades down and $\omega = 100a$ is 2 decades up, $\omega = a/2$ is 1 octave down and $\omega = 2a$ is 1 octave up.
2. The phase angles for the expression $G(s) = 1/(s/a + 1)$ are the negatives of the angles for $G(s) = s/a + 1$.
3. The phase angles for the expression $G(s) = (s/a + 1)^2$ are twice* the angles for $G(s) = (s/a + 1)$.
4. Terms of the form s, $1/s$, s^2, $1/s^2$ provide constant phase.

The chart is shown in Table 6-2.

6-3-5 Correction Factors for Straight-Line Bode Plots

Up to now, all magnitude Bode plots have been prepared assuming that all segments are straight lines. However, the straight lines are really asymptotes, and the true Bode plots should be smooth curves which approach the asymptotes. The following technique employs a series of correction factors to change the straight-line approximation into the actual curve.

Consider the expressions

$$G(s) = \frac{s}{10} + 1 \qquad \text{or} \qquad |G(j\omega)| = \sqrt{1 + \left(\frac{\omega}{10}\right)^2}$$

$$\text{and} \quad G_{dB} = 20 \log \sqrt{1 + \left(\frac{\omega}{10}\right)^2} = 10 \log \left[1 + \left(\frac{\omega}{10}\right)^2\right]$$

Substituting values for ω leads to Table 6-3. Plotting the actual values of gain and the asymptotes on the same axis gives Fig. 6-27. At 1 decade above and below the break frequency, the actual curve coincides with the asymptotes. At the break frequency, there is a 3-dB difference, and 1 octave up and 1 octave down the difference is 1 dB.

* If $G(s) = (s/a + 1)^3$, the angles are tripled, etc.

TABLE 6-2
Phase vs. Frequency for Some Common Terms, degrees

ω	G(s)							
	s	$\frac{1}{s}$	s^2	$\frac{1}{s^2}$	$\frac{s}{a}+1$	$\frac{1}{s/a+1}$	$\left(\frac{s}{a}+1\right)^2$	$\frac{1}{(s/a+1)^2}$
$\frac{a}{100}$	90	−90	180	−180	0	0	0	0
$\frac{a}{10}$	90	−90	180	−180	6	−6	12	−12
$\frac{a}{2}$	90	−90	180	−180	27	−27	54	−54
a	90	−90	180	−180	45	−45	90	−90
$2a$	90	−90	180	−180	63	−63	126	−126
$10a$	90	−90	180	−180	84	−84	168	−168
$100a$	90	−90	180	−180	90	−90	180	−180

TABLE 6-3
Decibels vs. Frequency for the First-Order Break

ω	G_{dB}
1	0
5	$0.97 \approx 1$
10	3
20	7
100	20

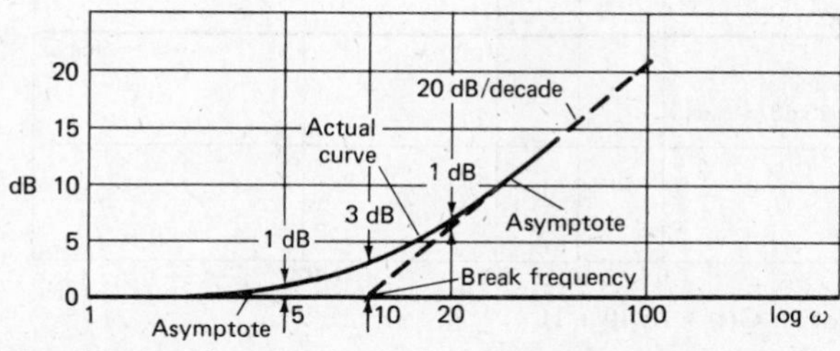

Fig. 6-27 Actual magnitude plot for a first-order break $G(s) = s/10 + 1$.

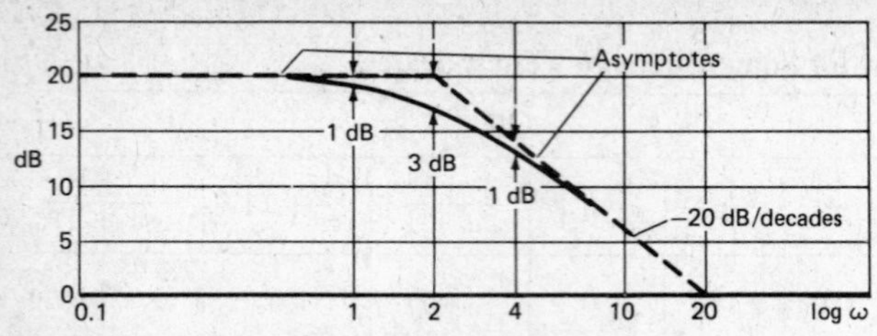

Fig. 6-28 Actual magnitude plot for $G(s) = 10/(s/2 + 1)$.

The Bode plots in Figs. 6-28 and 6-29 illustrate the use of the correction factors of 3 dB at the break point and 1 dB an octave away. A word of caution: for square terms $(s/a + 1)^2$ the corrections must be doubled.

6-4 DETERMINING $G(s)$ FROM THE BODE PLOTS

Suppose the log magnitude plot is given and it is required to determine $G(s)$. This example deals with the problem and attempts to show the curve-fitting method used when experimental data are available.

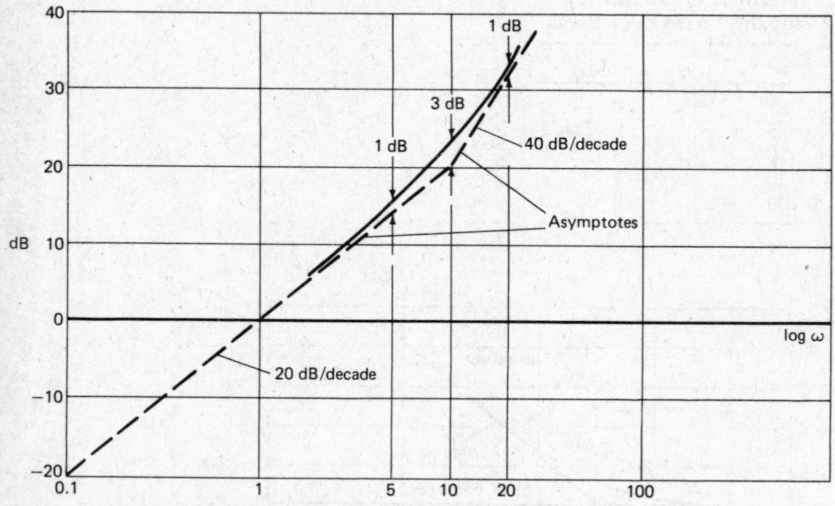

Fig. 6-29 Actual magnitude plot for $G(s) = s(s/10 + 1)$.

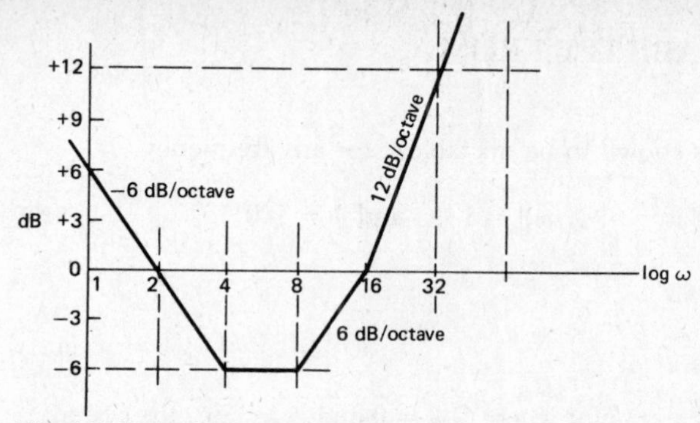

Fig. 6-30 Bode plot for Example 6-14.

Example 6-14
See Fig. 6-30.

Solution
The break frequencies are $\omega = 4$, 8, and 16 rad/s. Then

$$G(s) = \frac{K(s/4 + 1)(s/8 + 1)(s/16 + 1)}{s}$$

For frequencies less than $\omega = 4$, the only term that contributes to G_{dB} is $K/s = K/j\omega$. But $G_{dB} = 6$ when $\omega = 1$ or

$$G_{dB} = 20 \log \left|\frac{K}{j\omega}\right| = 20 \log \frac{K}{\omega} = 20 \log \frac{K}{1}$$

or $20 \log K = 6$ and $\log K = \frac{6}{20} = 0.3$, so that

$$K = \text{antilog } 0.3 = 2$$

Therefore

$$G(s) = \frac{2(s/4 + 1)(s/8 + 1)(s/16 + 1)}{s}$$

$$= \frac{2}{4(8)(16)} \frac{(s + 4)(s + 8)(s + 16)}{s} = \frac{(s + 4)(s + 8)(s + 16)}{256s}$$

6-5 DETERMINING STABILITY FROM THE BODE PLOTS

In Sec. 6-2, a system was shown to be unstable if for any frequency

$$G(j\omega) = -1 \qquad \text{that is} \qquad |G(j\omega)| = 1 \qquad \text{and } \phi = 180° \tag{6-20}$$

If $|G(j\omega)| = 1$,

$$20 \log |G(j\omega)| = 0 \tag{6-21}$$

If the Bode plot contains any point where $G_{dB} = 0$ and $\phi = 180°$, the system is unstable.

Although many systems may not exhibit such a point, they may be made to go unstable as a result of slight changes in the components due to aging or changes in environment. It is then desirable to indicate the degree of stability of a system as the amount of change in gain and phase required to cause instability. The degree of stability is defined in terms of gain and phase margin.

From Fig. 6-31, it can be seen that (1) phase margin ϕ_{PM} is the change in phase required to produce a phase of $\pm 180°$ when the gain is 0 dB and (2) gain margin (GM) is the number of decibels which must be algebraically added to the gain to produce a gain of 0 dB when the phase is $\pm 180°$. Therefore, if the

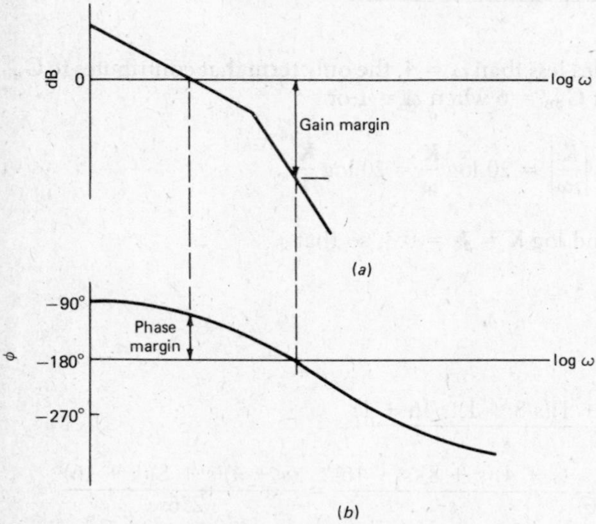

Fig. 6-31 (a) Sample magnitude plot; (b) sample phase plot.

gain increases by an amount equal to the gain margin or the phase changes by an amount equal to the phase margin, the system becomes unstable. It is the designer's responsibility to ensure that this does not occur by making the margin sufficiently large, perhaps by adding compensation networks or redesigning the system.

For stability, the gain should be less than 1 (negative decibels) when the phase is 180°. In other words, the gain margin should be a positive number. The minimum recommended gain margin is approximately 5 dB.

For stability, the phase should be less than −180° when the gain is 0 dB. In other words, −60° is acceptable but −230° is not. Therefore, for stable systems, the phase margin turns out to be a negative number. The phase margin should be in the range of −40 to −60°. The following example illustrates the determination of gain and phase margin from the Bode plots.

Example 6-15

Given the transfer function

$$G(s) = \frac{1}{s(s/3 + 1)^2}$$

determine the stability.

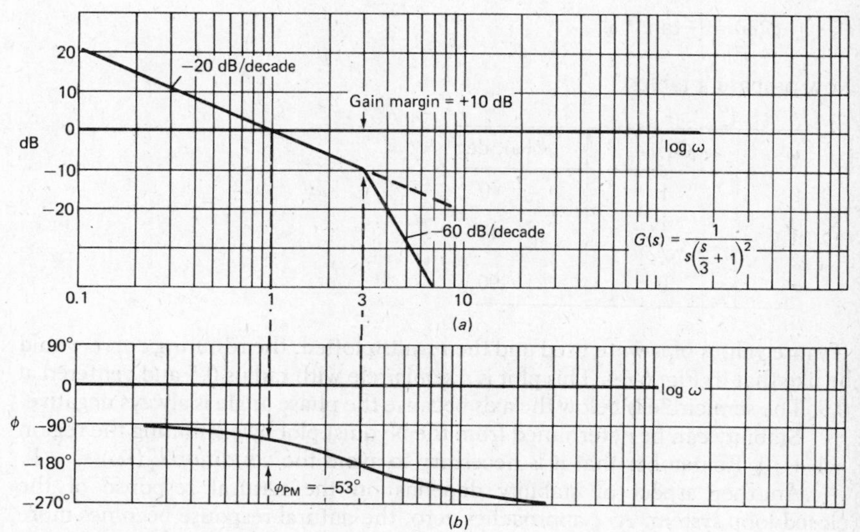

Fig. 6-32 Bode plots for Example 6-15.

Solution

See Fig. 6-32. This system meets the general requirements for stability since the gain margin is positive and greater than 5 dB while the phase margin is negative and between 40 and 60°.

When systems are outside the acceptable limits for stability, the techniques for compensation discussed in Chap. 7 are used.

6-6 NYQUIST PLOTS

The Nyquist plot is a polar plot which depicts magnitude and phase simultaneously. Given a function

$$G(j\omega) = A(\omega) + jB(\omega)$$

the polar plot relates $A(\omega)$ to $B(\omega)$ as a function of ω. Suppose $G(j\omega) = 5 + j\omega$; the Nyquist plot is shown in Fig. 6-33. Consider

$$G(j\omega) = \frac{1}{1 + j\omega}$$

$$|G(j\omega)| = \frac{1}{\sqrt{1 + \omega^2}}$$

$$\phi(\omega) = -\tan^{-1} \omega$$

Now prepare a table:

ω	$\lvert G(j\omega)\rvert$	$\phi(\omega)$, deg
0	1	0
1	$\frac{1}{\sqrt{2}}$	−45
∞	0	−90

If more values of ω were used and the results plotted, the resulting curve would be like that in Fig. 6-34. This plot is a semicircle with radius 0.5 and centered at 0.5. The semicircle is below the axis because the phase angle is always negative.

Stability can be determined from the Nyquist plot by examining the region $-1 + j0$. Remember that it is necessary to avoid the condition $G(j\omega) = -1$.

Another aspect of stability depends on the natural response of the closed-loop system. As ζ approaches zero, the natural response becomes more

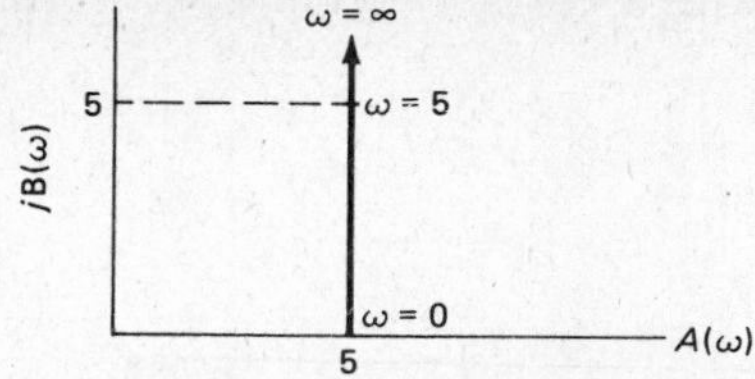

Fig. 6-33 Nyquist plot for $G(j\omega) = 5 + j\omega$.

oscillatory, which is not a desirable situation. Even though the open-loop transfer function $G(s)$ may have high enough damping, the closed-loop transfer function $T(s) = G(s)/[1 + G(s)]$ may be highly oscillatory. The natural response depends on the characteristic equation $1 + G(s) = 0$. The roots of this equation are a function of the gain of $G(s)$. By examining the Nyquist plot, the gain margin can be determined, and it can be shown that ζ is a function of the gain margin.

Consider the second-order system

$$G(s) = \frac{K}{(1 + T_1 s)(1 + T_2 s)}$$

The gain K is assumed to be a variable which can be adjusted to any positive value. The Nyquist plot is shown in Fig. 6-35. This plot indicates that as K is increased, the phase margin is decreased. The phase margin ϕ_{PM} is defined as the additional phase shift required to cause $\pm$ 180° phase shift when the magnitude of $G(j\omega)$ is 1. Note that ϕ_{PM} is determined on the Nyquist plot by drawing a circle of radius 1 with the origin as the center and noting the point at which the circle and the plot intersect. A line is then drawn between the origin and the intersection, and ϕ_{PM} is the angle between this line and the negative real axis. The phase margin for $K = 3$ is ϕ_{PM3} and for $K = 4$ is ϕ_{PM4}. It should be apparent that $\phi_{PM4} < \phi_{PM3}$. If the gain becomes too large, ϕ_{PM} approaches zero and the system is definitely unstable.

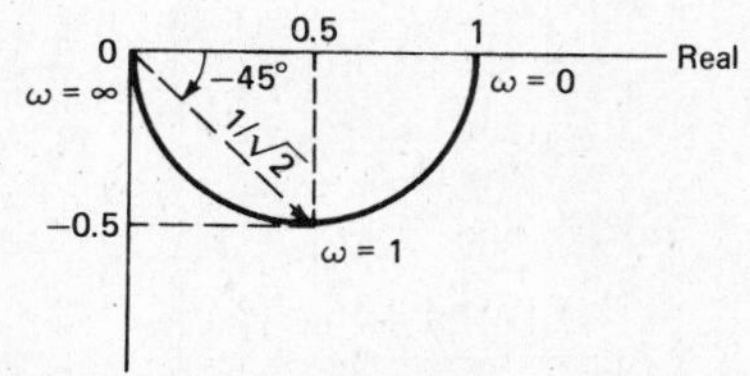

Fig. 6-34 Nyquist plot for $G(j\omega) = 1/(1 + j\omega)$.

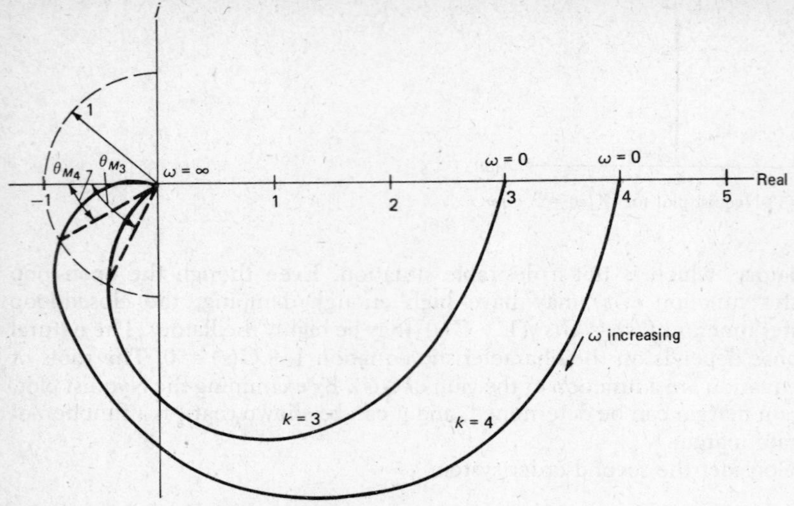

Fig. 6-35 Nyquist plot for $G(s) = 1/[(1 + T_1s)(1 + T_2s)]$.

6-7 RELATIONSHIP BETWEEN THE PHASE MARGIN AND DAMPING RATIO

An approximate relationship between ϕ_{PM} and ζ is graphed in Fig. 6-36. For a linear second-order system, there is a definite relationship between the natural response and the steady-state sinusoidal response. Not only does the phase margin give an indication of how close we are to instability for a given input frequency, it also shows the amount of oscillation to expect from the natural response. But remember that the form of the natural response is independent of the input signal.

Suppose it is found that at $\omega = 10$ rad/s

$$G(j\omega) = -1$$

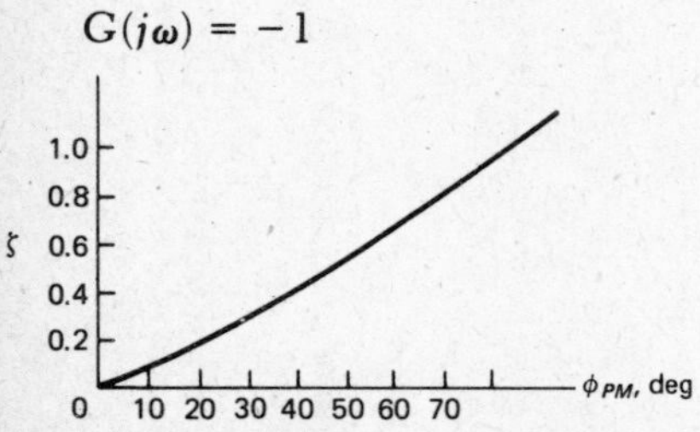

Fig. 6-36 Damping ratio ζ vs. phase margin.

Then if an input sinusoid whose frequency is 10 rad/s is applied,

$$T(j10) = \frac{G(j10)}{1 + G(j10)} = \frac{-1}{1 - 1} = \infty$$

But if this frequency is never applied, the transfer function will not blow up. Does this mean the system will remain stable? No, because the phase margin is zero, so $\zeta = 0$. Thus, the natural response never dies out.

PROBLEMS

6-1. On one 5-cycle semilog paper, plot the following transfer functions in decibels versus log ω. Label the ω scale from 1 to 100,000 rad/s and place the 0-dB axis halfway up the paper.

(a) $\frac{3}{s}$ (b) $\frac{1600}{s^2}$ (c) $\frac{2000}{s}$ (d) $\frac{8}{s^3}$ (e) $\frac{20{,}000}{s}$

6-2. Sketch the phase angle of each of the transfer functions of Prob. 6-1 on the lower half of the same semilog paper as the decibel (magnitude) plots.

6-3. On one semilog (5-cycle paper) sheet plot the following transfer functions in decibel versus log ω. Label the ω scale as in Problem 6-1.

(a) $\frac{s}{4}$ (b) $\frac{s^2}{900}$ (c) $\frac{s}{3000}$ (d) $\frac{s^3}{216}$ (e) $\frac{s}{50{,}000}$

6-4. Sketch phase angle versus log ω on the bottom half of the sheet of Prob. 6-3.

6-5. Synthesize the decibel vs. log ω plots in Fig. P6-5a and b by giving the transfer function in terms of s.

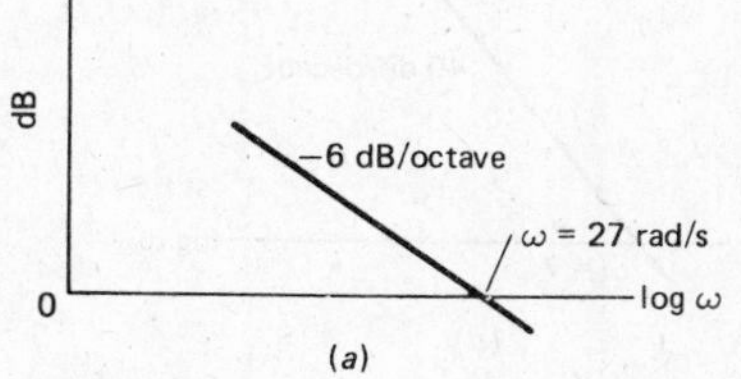

dB
−20 dB/decade
ω = 5 rad/s
log ω
(b)

Fig. P6-5

6-6. Repeat Prob. 6-5 for the plots in Fig. P6-6a and b.

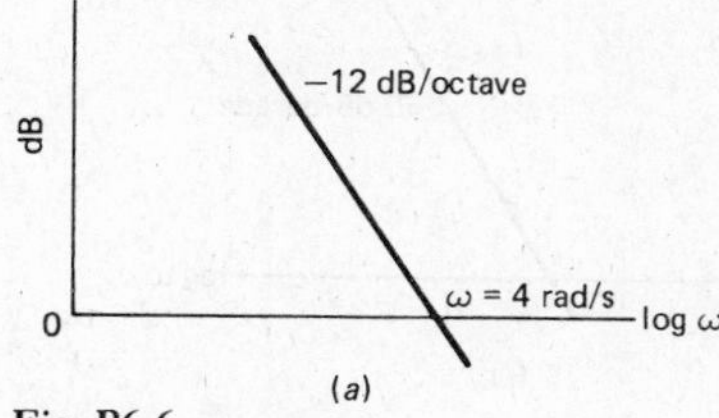

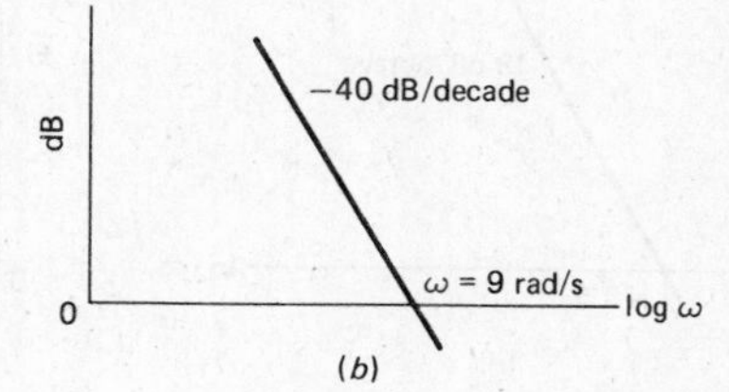

Fig. P6-6

6-7. Repeat Prob. 6-5 for the plots in Fig. 6-7*a* and *b*.

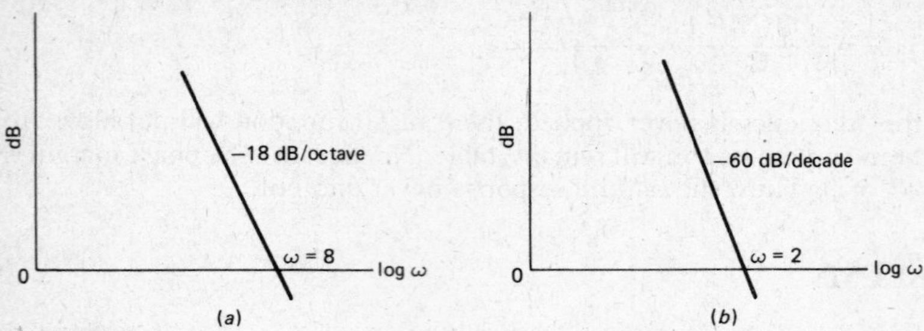

Fig. P6-7

6-8. Repeat Prob. 6-5 for the plots in Fig. 6-8*a* to *f*.

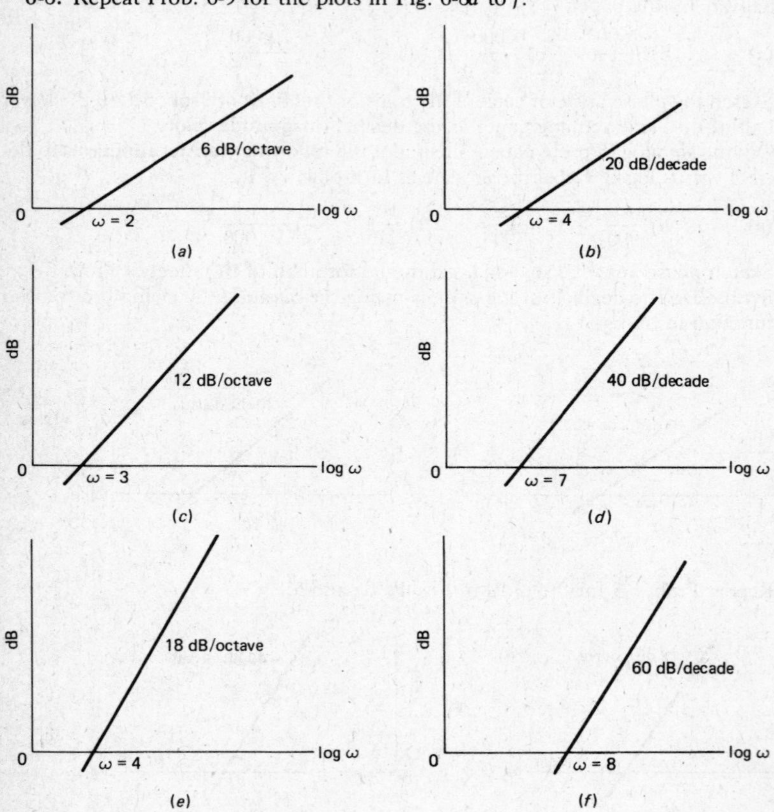

Fig. P6-8

6-9. On one semilog sheet draw the decibel and phase angle vs. log ω for the transfer function

$$G(s) = \frac{100}{0.1s + 10}$$

Use-straight line approximations for the magnitude plot and label ω from 1 to 100,000.

6-10. Repeat Prob. 6-9 for $G(s) = 100(0.2s + 1)$.

6-11. Prepare the decibel plot only for

$$G(s) = \frac{100(0.1s + 1)}{(0.01s + 1)(0.001s + 1)}$$

6-12. Repeat Prob. 6-9 for

$$G(s) = \frac{10(0.002s + 1)^2}{(0.1s + 1)(0.0001s + 1)}$$

6-13. Plot decibel versus log ω for

(a) $G(s) = \dfrac{10}{5(0.01s + 1)}$ (b) $G(s) = \dfrac{s}{100(0.001s + 1)}$

(c) After the straight-line Bode plots are prepared, add the correction factors to obtain the true curves.

6-14. Obtain the transfer function (in terms of s) for the plot in Fig. P6-14.

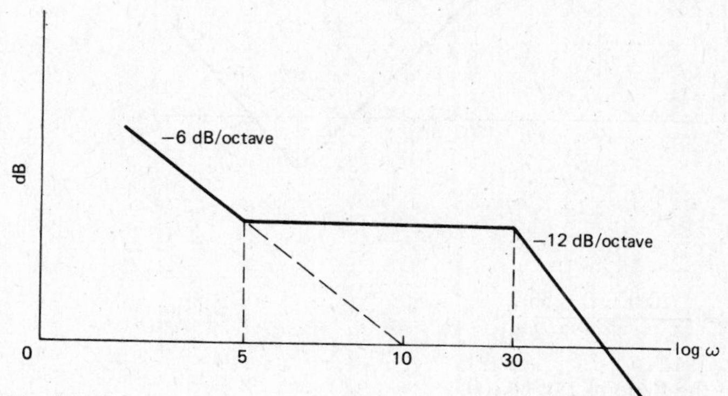

Fig. P6-14

6-15. Repeat Prob. 6-14 for the plot shown in Fig. P6-15.

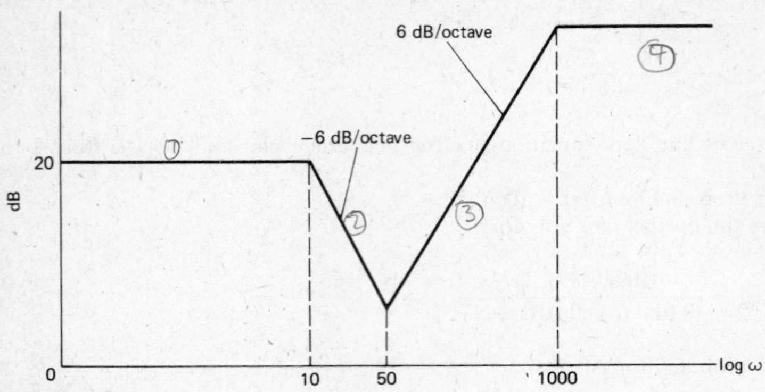

Fig. P6-15

6-16. Repeat Prob. 6-14 for the plot shown in Fig. P6-16.

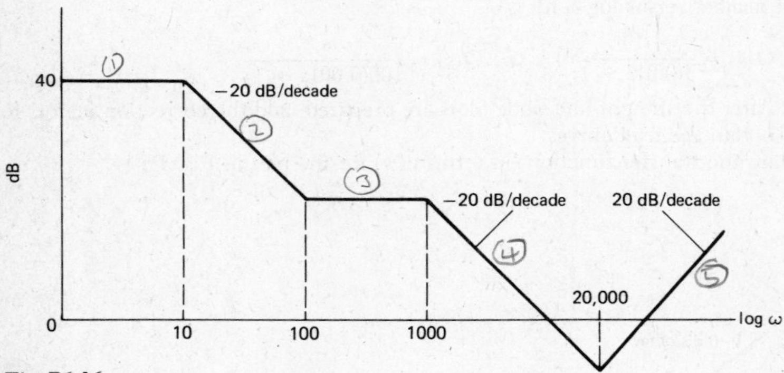

Fig. P6-16

6-17. For

$$HG(s) = \frac{20{,}000(s + 80)}{s(s + 40)(s + 200)}$$

(a) Obtain the Bode plot of $HG(s)$.
(b) Calculate the phase margin and gain margin.

6-18. (a) Obtain the Bode plot for

$$HG(s) = \frac{120}{s(s + 1)(s + 12)}$$

(b) Find the gain margin and phase margin.
(c) Is the system stable? Why?

6-19. (a) Obtain the Bode plot for

$$HG(s) = \frac{20}{s(1 + s/2)(1 + s/6)}$$

(b) Calculate the gain and phase margin.
(c) Is the system stable? Why?

6-20. Given

$$HG(s) = \frac{K}{(s + 1)^3}$$

Find the maximum value of K before the system will oscillate continuously, that is, 0-dB gain margin. *Hint:* Use semilog paper and start by determining the $-180°$ phase point.

6-21. For the system shown in Fig. P6-21, it is desired to have a steady-state error less than 0.2 and a gain margin greater than 6 dB. Determine the range of gain K permissible.

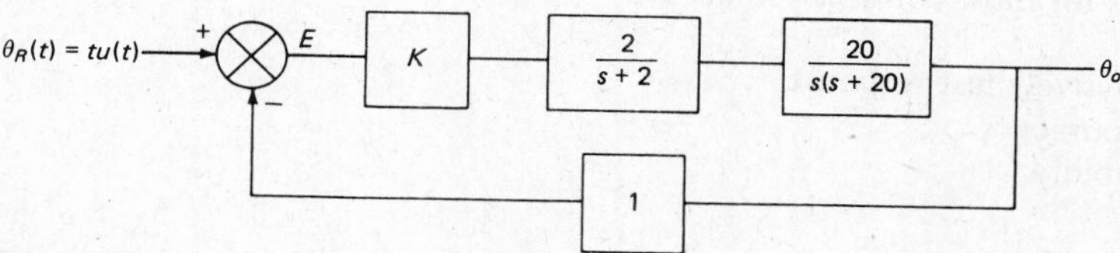

Fig. P6-21

chapter 7

Compensation

The design of a control system often involves conflicting requirements. After the initial design stage, in which a basic system configuration is established, gains must be selected to meet performance requirements. The general requirements for most control systems are

1. Relatively fast response
2. Accuracy
3. Stability

7-1 EFFECT OF GAIN ON SYSTEM PERFORMANCE

A fast and accurate response is achieved with sufficiently high gain. In Chap. 5 it was shown that the damping ratio ζ can be found from

$$\zeta = \frac{F}{2\sqrt{KJ}}$$

Increasing K reduces the value of ζ, and systems with low values of ζ exhibit short rise times, i.e., respond quickly. Furthermore, it was shown that the steady-state error due to a disturbance torque T_{xm} is

$$\mathscr{E}_{SS} = \frac{h\mathsf{T}_{xm}}{K}$$

In other words, high gain results in a small error.

It was also demonstrated that a low value of ζ means a long settling time marked by excessive oscillation. If ζ is low enough, the system can become unstable.

This chapter is concerned with compensation, a method for altering the system so that high gain can be used without loss of stability. The methods for compensation covered here include (1) rate or velocity feedback and (2) compensation networks (lead, lag, and lead-lag).

7-2 RATE FEEDBACK

The simplest and most common way of improving system stability without simply lowering gains is to provide output rate feedback. In a position control system the damping factor can be improved by the introduction of a feedback signal proportional to the velocity or rate of the output signal. Figure 4-24 shows that the dc motor has rate feedback inherently present. The motor's back-emf gradient K_b is rate feedback; i.e., a voltage is fed back proportional to the motor's speed. In Chap. 2 the motor equations were derived and the viscous coefficient f_m was defined

$$f_m = \frac{K_b K_t}{r_a}$$

The motor's total viscous coefficient F_m was also defined

$$F_m = f_m + f_v \approx f_m \qquad \text{since } f_m \gg f_v$$

Thus, it can be seen that increasing K_b will increase f_m, thereby increasing F_m.

Figure 7-1 represents a position control system with unity feedback. The characteristic equation is

$$1 + HG(s) = 0$$

$$1 + \frac{K_m/J}{s(s + F_m/J)} = 0$$

$$s^2 + \frac{F_m}{J}s + \frac{K_m}{J} = 0 \tag{7-1}$$

When the following values for the motor constants are assumed

$$F_m = 6 \text{ in} \cdot \text{lb} \cdot \text{s} \qquad K_m = 100 \text{ in} \cdot \text{lb/V} \qquad J = 1 \text{ in} \cdot \text{lb} \cdot \text{s}^2$$

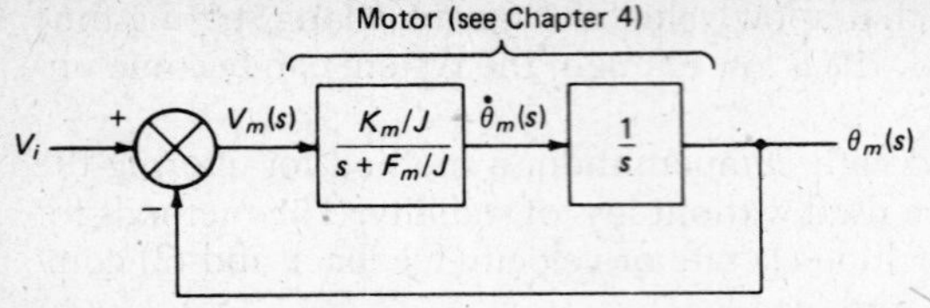

Fig. 7-1 Position control system.

Eq. (7-1) becomes

$$s^2 + 6s + 100 = 0 \tag{7-2}$$

which can be solved to give

$$\begin{aligned} \omega_n^2 &= 100 \\ \omega_n &= 10 \text{ rad/s} \\ 2\zeta\omega_n &= 6 \\ \zeta &= \frac{6}{2(10)} = 0.3 \end{aligned}$$

The damping factor of 0.3 is relatively low, and the motor's response is underdamped. If the motor's back emf were doubled, thus approximately doubling F_m without changing K_m or J, Eq. (7-1) would become

$$s^2 + 12s + 100 = 0 \tag{7-3}$$

$$\begin{aligned} \omega_n &= \sqrt{100} = 10 \text{ rad/s} \\ 2\zeta\omega_n &= 12 \\ \zeta &= 0.6 \end{aligned}$$

Notice that increasing the motor's back emf, which essentially is increasing the rate feedback already present in the motor, improves the system damping factor.

7-3 TACHOMETER FEEDBACK

Although increasing the motor's back emf in the previous section improved the damping factor, it is not a practical solution to the problem. Redesign of the motor will change other constants and result in new problems. A practical solution is the addition of a component in the system of Fig. 7-1 to provide rate feedback to the motor. The tachometer is precisely the component necessary

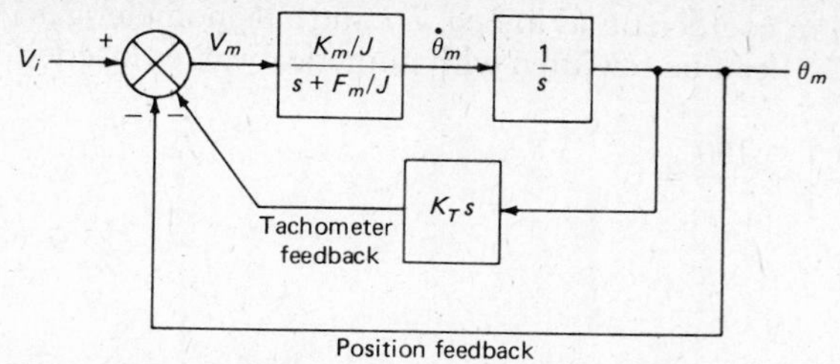

Fig. 7-2 System of Fig. 7-1 with tachometer feedback added.

to do the job. Recalling from Chap. 2 that a tachometer produces a voltage proportional to its shaft speed, if it were connected to the motor shaft, it would produce a voltage proportional to the motor's speed. This is equivalent to supplying additional back emf.

Figure 7-1 is redrawn as Fig. 7-2, the only change being the addition of a tachometer with gain K_T V/(rad · s). Note that the transfer function of the tachometer is $K_T s$, as derived in Chap. 4, and the output of the tachometer is fed to the input summing junction. Figure 7-2 can be reduced to Fig. 7-3 simply by combining the two parallel paths feeding back.

The characteristic equation for Fig. 7-3 is derived with $H = 1 + K_T s$ and is

$$1 + HG(s) = 0$$

$$1 + \frac{(1 + K_T s)K_m/J}{s(s + F_m/J)} = 0$$

$$s^2 + \frac{sF_m}{J} + \frac{K_m}{J} + \frac{sK_T K_m}{J} = 0$$

$$s^2 + s\left(\frac{F_m}{J} + \frac{K_T K_m}{J}\right) + \frac{K_m}{J} = 0 \tag{7-4}$$

Notice that the system natural frequency is still given by $\sqrt{K_m/J}$; however, the damping factor will be different.

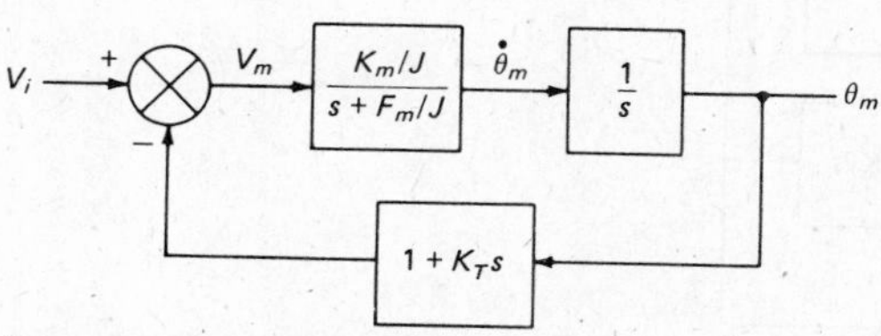

Fig. 7-3 Reduction of Fig. 7-2.

If we take the same motor coefficients as in Sec. 7-2 and a tachometer gain $K_T = 0.06$ V/(rad · s), Eq. (7-4) can be rewritten with numerical values inserted

$$s^2 + s\left[\frac{6}{1} + \frac{(0.06)(100)}{1}\right] + \frac{100}{1} = 0$$

$$s^2 + 12s + 100 = 0 \tag{7-5}$$

$$\omega_n = \sqrt{100} = 10$$

$$2\zeta\omega_n = 12$$

$$\zeta = \frac{12}{2\omega_n} = \frac{12}{20} = 0.6$$

$\omega_n = 10$ and $\zeta = 0.6$ are the same results obtained in the previous section when the motor's back emf was doubled.

If a further increase in damping factor is desired, all that need be done is to increase the tachometer gain. This can be accomplished by placing an amplifier with an adjustable gain directly after the tachometer, as shown in Fig. 7-4. In this manner the rate feedback can be increased or decreased to obtain the desired response. It should be noted here that too much rate feedback will increase the damping factor to a point where the system is sluggish and has an extremely slow response. An extremely slow response can be less desirable than an underdamped one.

Figure 7-5 shows the system response for three different damping factors. In many cases the most desirable response is the one which comes within a certain percentage of the steady-state value in the shortest time. As can be seen from the figure, the best response would be b. It has a small overshoot and falls within the percentage band at time t_1. The interesting point is that response a, even though it is poorly damped, falls within the percentage band at t_2 well before response c, which is overdamped. For this reason, a would be chosen over c. The rate feedback gain can be adjusted to obtain response a, b, c, or any other desired response. The response selected depends on the system specifications.

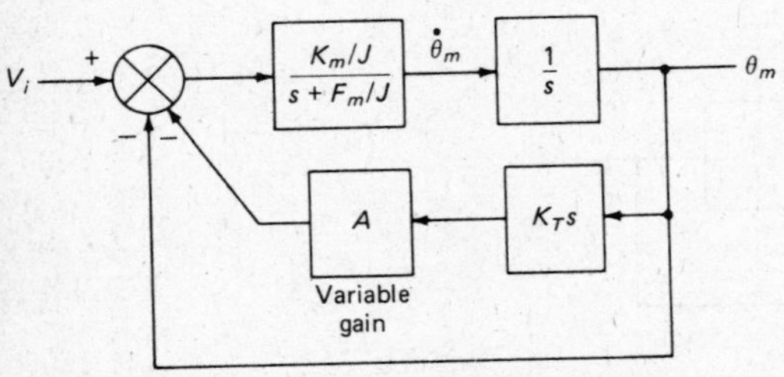

Fig. 7-4 Inclusion of variable-rate-gain feedback.

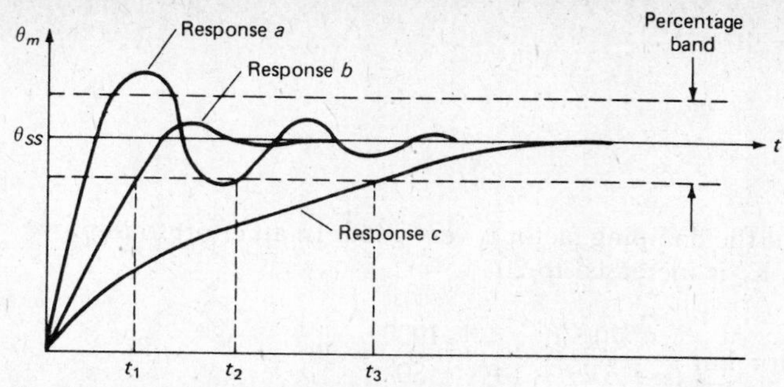

Fig. 7-5 Three different system responses.

7-4 STABILIZATION USING RATE FEEDBACK

Consider the unity feedback system of Fig. 7-6, where the input is a 5° step. If $K_a = 1$, the steady-state error is given by (refer to Chap. 5),

$$e(\infty) = \frac{r_0}{1 + K_e} = \frac{5}{1 + K_e}$$

$$\text{where} \quad K_e = \lim_{s \to 0} HG(s) = \lim_{s \to 0} \left[K_a \frac{50}{(s + 5)(s + 10)} \right]$$

$$= \lim_{s \to 0} \frac{1(50)}{(s + 5)(s + 10)} = \frac{50}{50} = 1$$

$$\text{and} \quad e(\infty) = \frac{5}{1 + 1} = 2.5°$$

We obtain the characteristic equation as follows:

$$1 + HG(s) = 0$$

$$1 + \frac{50}{(s + 5)(s + 10)} = 0 \qquad \text{or} \qquad (s + 5)(s + 10) + 50 = 0$$

$$\text{and} \quad s^2 + 15s + 50 + 50 = 0 \qquad \text{or} \qquad s^2 + 15s + 100 = 0$$

Fig. 7-6 Simple position servomechanism.

Then $\omega_n = \sqrt{100} = 10$

$$2\zeta\omega_n = 15$$

$$\zeta = \frac{15}{2(10)} = 0.75$$

The error is high, but the damping factor is very good. In an effort to improve (decrease) the error, K_a is increased to 20

$$K_e = \lim_{s\to 0} HG(s) = \lim_{s\to 0} \frac{20(50)}{(s+5)(s+10)} = \frac{1000}{50} = 20$$

and $$e(\infty) = \frac{r_0}{1+K_e} = \frac{5}{1+20} = \frac{5}{21} \approx 0.24°$$

Note that the error has been greatly reduced. What has become of the damping ratio?

$$1 + HG(s) = 0$$

$$1 + \frac{20(50)}{(s+5)(s+10)} = 0$$

$$s^2 + 15s + 1050 = 0$$

$$\omega_n = \sqrt{1050} \approx 32$$

$$2\zeta\omega_n = 15$$

$$\zeta = \frac{15}{2(32)} \approx 0.25$$

The damping ratio is quite small now and the system would be more oscillatory.

In an effort to decrease the error, the system has become quite underdamped. The problem can be solved by the introduction of rate feedback, as shown in Fig. 7-7. For this system

$$HG(s) = \frac{K_a[50/(s+5)(s+10)]}{1 + K_T s(50)/[(s+5)(s+10)]} = \frac{50K_a}{(s+5)(s+10) + 50K_T s}$$

$$= \frac{50K_a}{s^2 + (50K_T + 15)s + 50}$$

with $K_a = 20$ as before

$$K_e = \lim_{s\to 0} \frac{50(20)}{s^2 + (50K_T + 15)s + 50} = \frac{50(20)}{50} = 20$$

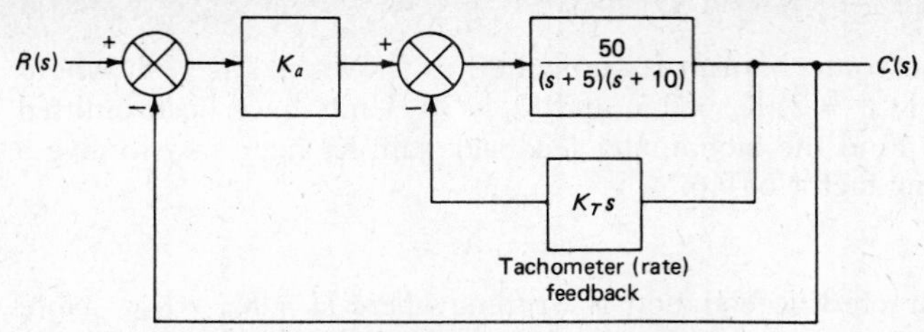

Fig. 7-7 Introduction of rate feedback in system of Fig. 7-6.

$$\text{and} \quad e(\infty) = \frac{r_0}{1 + K_e} = \frac{5}{1 + 20} = \frac{5}{21} \approx 0.24^\circ$$

The error is the same as before, and it can be seen that it does not depend on K_T, the tachometer feedback.

If K_T is set equal to 0.6, the damping factor can be calculated as before.

$$1 + HG(s) = 0$$

$$1 + \frac{50(20)}{s^2 + (50K_T + 15)s + 50} = 0$$

$$s^2 + [50(0.6) + 15]s + 1050 = 0$$

$$s^2 + 45s + 1050 = 0$$

$$\omega_n = \sqrt{1050} \approx 32$$

$$2\zeta\omega_n = 45$$

$$\zeta = \frac{45}{2(32)} = \frac{45}{64} \approx 0.75$$

Table 7-1 summarizes the results. It shows that if gain is increased, the error is reduced; however, the system becomes more unstable. With the introduction of the proper amount of tachometer feedback the error is reduced while maintaining stability.

TABLE 7-1

	No Tachometer Feedback		Tachometer Feedback
	$K_a = 1$	$K_a = 20$	$K_a = 20$
ζ	0.75	0.25	0.75
$e(\infty)$, deg	2.5	0.24	0.24

Example 7-1

A positional servomechanism is configured as shown in Fig. 7-8, where $K_m = 50$, $J = 1$, $F = 2$, $K_p = 0.5$, and $K_a = 30$. Units have been omitted for simplicity. Find the tachometer feedback gain K_T necessary to give a system damping factor of 0.6.

Solution

First, the characteristic equation is written, where $H = K_p + K_T s$. Note also that the input potentiometer K_p does not enter into the calculations since it is not in the loop

$$HG(s) = (K_p + K_T s)\frac{K_a K_m/J}{s(s + F/J)}$$

Substituting the information given, we get

$$HG(s) = (0.5 + K_T s)\frac{1500}{s(s + 2)}$$

$$1 + HG(s) = 1 + \frac{1500(0.5 + K_T s)}{s(s + 2)}$$

Letting $1 + HG(s) = 0$, we see that the characteristic equation is

$$s(s + 2) + 1500(0.5 + K_T s) = 0$$
$$s^2 + 2s + 750 + 1500K_T s = 0$$
$$s^2 + s(2 + 1500K_T) + 750 = 0$$
$$\omega_n = \sqrt{750} = 27.4$$
$$2\zeta\omega_n = 2 + 1500K_T$$

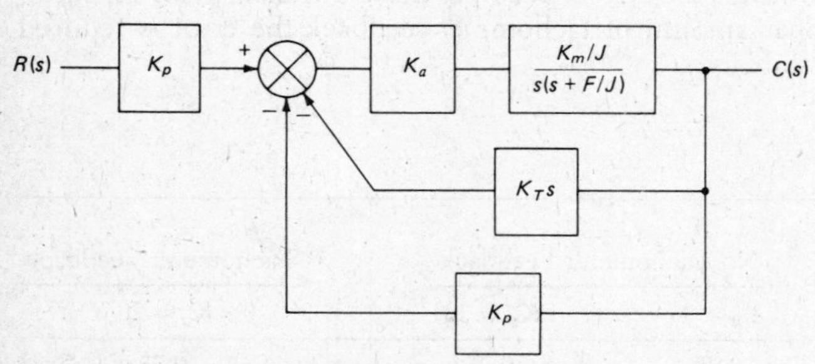

Fig. 7-8 System configuration for Example 7-1.

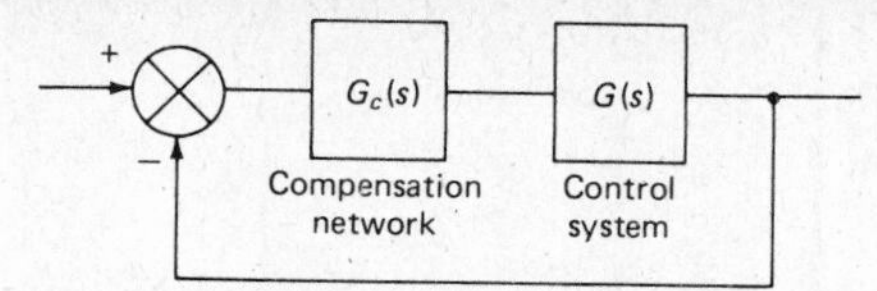

Fig. 7-9 Addition of a compensation network to a control system.

Substituting the desired damping ratio (0.6) into the above equation and solving for K_T, we get

$$2(0.6)(27.4) = 2 + 1500K_T = 32.88$$
$$1500K_T = 30.88 \quad \text{and} \quad K_T = 0.021$$

7-5 COMPENSATION NETWORKS

System performance can be improved by the introduction of compensation networks into the forward loop, as shown in Fig. 7-9. These networks consist of resistors and capacitors in various configurations known as *lag*, *lead*, and *lead-lag networks*.

From Fig. 7-9, the overall open-loop transfer function of the compensated system is

$$G_T(s) = G_c(s)G(s)$$

Using the information from Chap. 6, we can show that (1) the magnitude Bode plot of the compensated system is the sum of the Bode plots for the compensation network and the original control system and (2) that the total phase of the compensated system is the sum of the phase angles of the compensation network and the original control system.

7-6 LAG NETWORKS

Figure 7-10 illustrates a lag network. The transfer function can be shown to be

$$G(s) = \frac{E_o(s)}{E_i(s)} = \frac{\tau s + 1}{\alpha \tau s + 1}$$

where $\alpha = \dfrac{R_1 + R_2}{R_2}$ and $\tau = R_2 C$

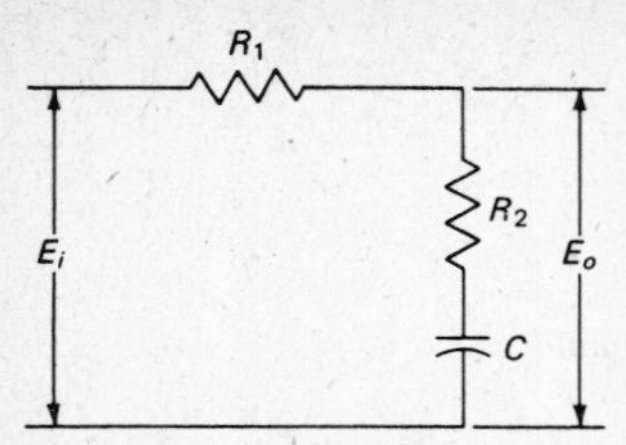

Fig. 7-10 Lag network.

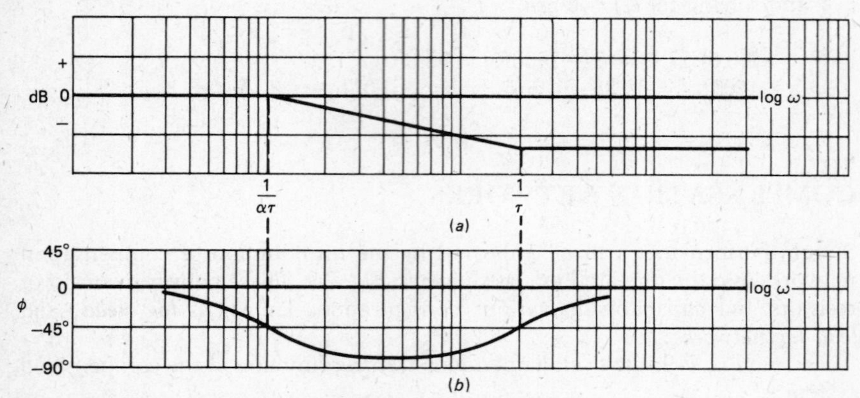

Fig. 7-11 (*a*) Magnitude and (*b*) phase Bode plots for a lag network ($\alpha \geq 10$).

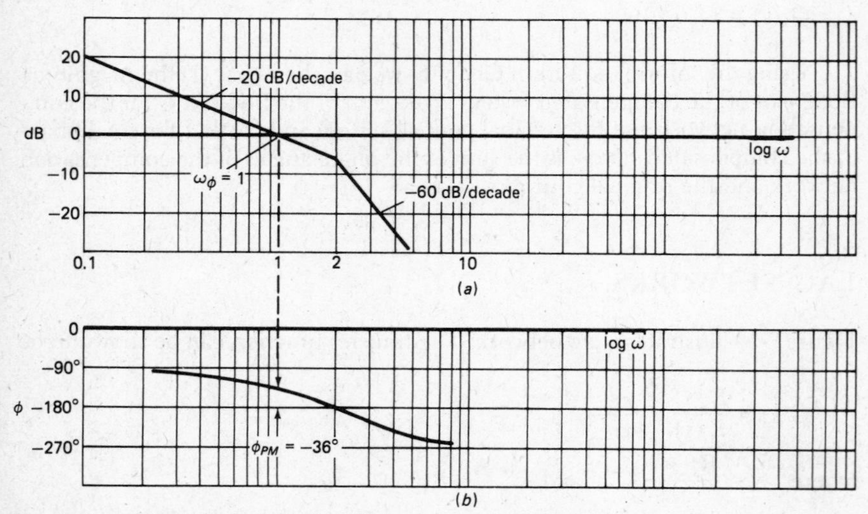

Fig. 7-12 (*a*) Magnitude and (*b*) phase Bode plots for uncompensated system in Example 7-2.

and the value of α is obviously greater than 1. A lag network permits the gain of the original system to be increased without compromising the stability. This reduces the steady-state error. However, since the natural frequency is reduced slightly, the speed of response drops somewhat.

Figure 7-11 shows the magnitude and phase Bode plots for a lag network. Trial and error is generally used to design a compensation network. Various networks are selected until the Bode plot of the compensated system approaches the desired characteristics.

When using a lag network, a typical value of α selected is 10. A value of τ is selected so that the phase margin of the uncompensated system is not appreciably changed. Lag-compensation networks are used when the phase margin is acceptable but the gain is too low to provide small errors. After the lag network is added, the gain can be increased without noticeably changing the stability.

Example 7-2

Consider the transfer function:

$$G(s) = \frac{K}{s(s/2 + 1)^2} \qquad \text{where } K = 1$$

Figure 7-12 shows the Bode plots for this uncompensated system with low gain.

Solution

The phase margin of $-36°$ is measured at $\omega_\phi = 1$ rad/s. To compensate the system with a lag network, select $\alpha = 10$ and $1/\tau \approx 0.1$ rad/s ($1/\tau \approx 0.1\omega_\phi$).

$$\text{Since} \quad G_c = \frac{\tau s + 1}{\alpha \tau s + 1} = \frac{10s + 1}{100s + 1}$$

the new transfer function is

$$G_T(s) = G_c(s)G(s) = \frac{10s + 1}{100s + 1}\,\frac{1}{s(s/2 + 1)^2}$$

The Bode plots for the overall transfer function are shown in Fig. 7-13. Notice that the new phase margin of $-45°$ is close to the old phase margin but the new frequency $\omega_\phi \approx 0.1$ is less than the original phase-margin frequency. While the solid line represents the magnitude Bode plot for $K = 1$, the dotted line shows the effect of increasing K to 10 (20 dB). Even though K has been increased by a factor of 10, the stability has been maintained.

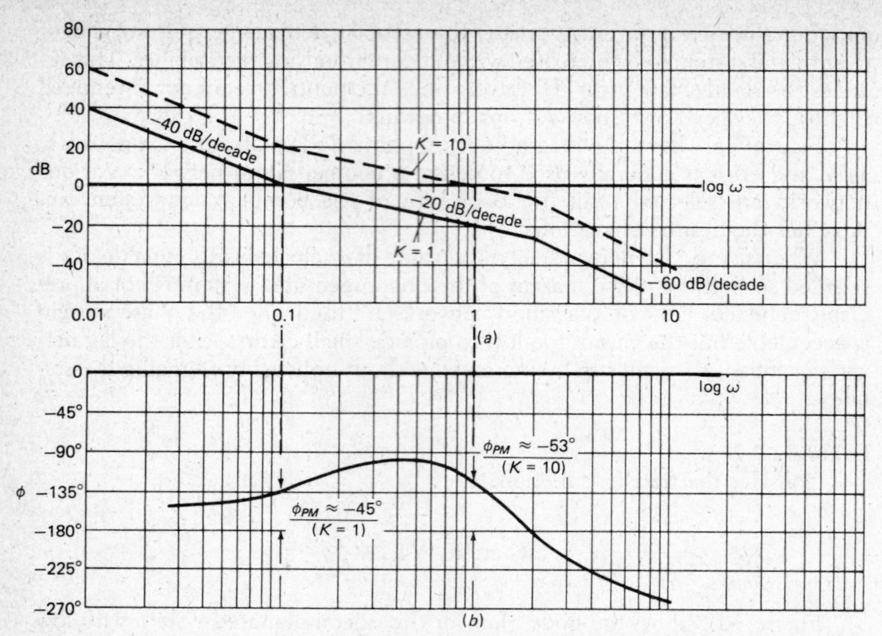

Fig. 7-13 (*a*) Magnitude and (*b*) phase Bode plots for the lag-compensated system in Example 7-2.

7-7 LEAD NETWORKS

The transfer function of the lead network shown in Fig. 7-14 is

$$G_c(s) = \frac{E_o(s)}{E_i(s)} = \alpha \frac{\tau s + 1}{\alpha \tau s + 1}$$

where $\alpha = \dfrac{R_2}{R_1 + R_2} \qquad \tau = R_1 C$

and α is obviously less than 1.

The lead network tends to increase the resonant frequency, resulting in improved response time. Figure 7-15 represents the Bode plot for a lead network. As a general rule, $\alpha = 0.1$ is acceptable. For a stable system, the resonant frequency is increased without appreciably changing the phase margin. The gain can then be increased to reduce the steady-state error, but not as much as in a lag-compensated system. In a system with an unacceptable phase margin, the lead network is used to stabilize the system without causing a large decrease in gain.

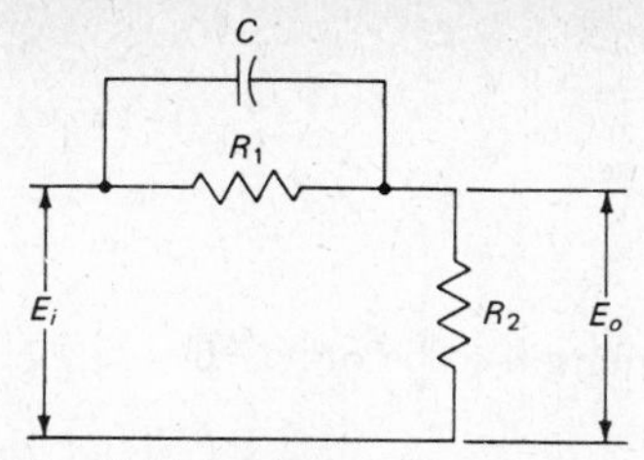

Fig. 7-14 Lead network.

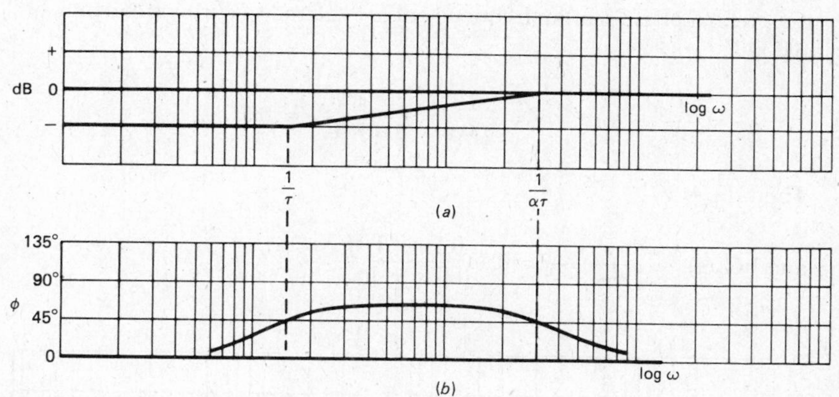

Fig. 7-15 Bode plots for a lead network ($\alpha \leq 0.1$).

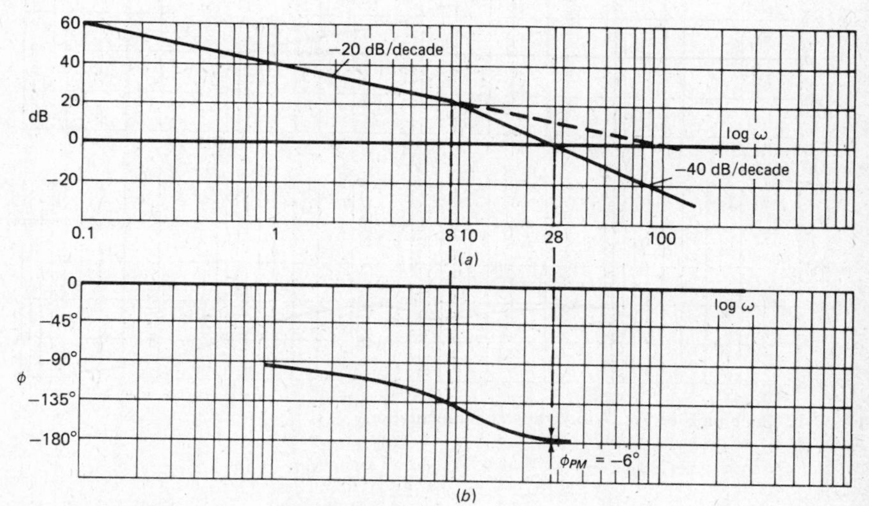

Fig. 7-16 Bode plots for $G(s) = 100/[s(s/8 + 1)]$.

Example 7-3

Consider the transfer function

$$G(s) = \frac{100}{s(s/8 + 1)}$$

The Bode plot in Fig. 7-16 shows the phase margin to be only $-6°$.

Solution

Select a lead network with $\alpha = 0.1$ and an upper break frequency above the original phase-margin frequency of $\omega = 28$ rad/s. With a frequency of 100 rad/s

$$\alpha\tau = \frac{1}{100} \qquad \text{or} \qquad \tau = \frac{1}{100\alpha} = \frac{1}{100 \times 0.1} = 0.1$$

Then $\dfrac{1}{\tau} = 10$

and $G_c(s) = \alpha \dfrac{\tau s + 1}{\alpha\tau s + 1} = 0.1 \dfrac{0.1s + 1}{0.01s + 1}$

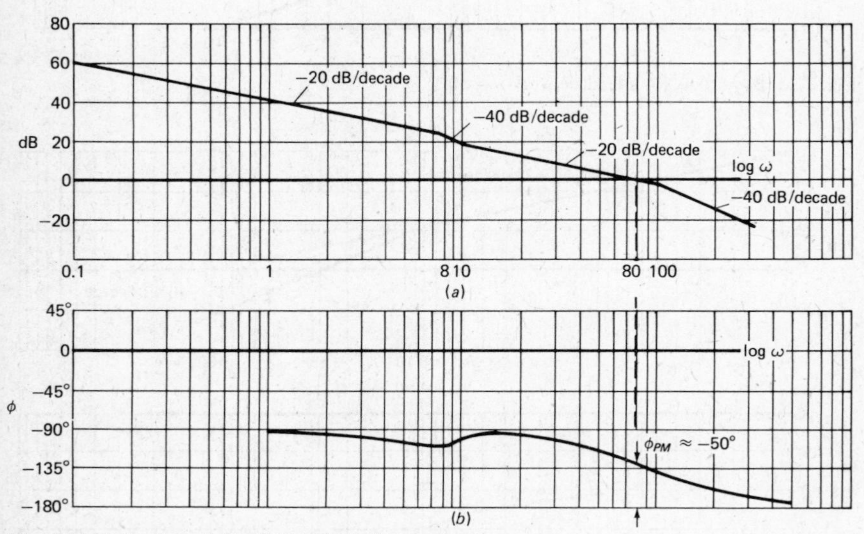

Fig. 7-17 Bode plots of Fig. 7-16 with lead compensation.

Because the term 0.1 reduces the gain of the overall system by a factor of 10, we increase the gain of the original system by 10, or

$$G'(s) = (10)\frac{100}{s(s/8 + 1)}$$

$$\text{Then} \quad G_T(s) = G_c(s)G'(s) = \frac{0.1s + 1}{0.01s + 1}\frac{100}{s(s/8 + 1)}$$

The Bode plot for the compensated system in Fig. 7-17 shows an improvement in phase margin to $-50°$, while the phase-margin frequency has increased from 28 to 80 rad/s.

7-8 LEAD-LAG NETWORKS

The transfer function of the lead-lag network shown in Fig. 7-18 is

$$G_c(s) = \frac{E_o(s)}{E_i(s)} = \frac{(\tau_1 s + 1)(\tau_2 s + 1)}{(\alpha\tau_1 s + 1)(\tau_2 s/\alpha + 1)}$$

where $\tau_1 = R_1C \qquad \tau_2 = R_2C_2$

$$\alpha = \frac{\tau_1 + \tau_2 + \tau_{12} - [(\tau_1 + \tau_2 + \tau_{12})^2 - 4\tau_1\tau_2]^{1/2}}{2\tau_1}$$

and $\tau_{12} = R_1C_2 \qquad \alpha > 1 \qquad \tau_1 > \tau_2$

The lead-lag network can be thought of as a separate lag network and a lead network. Lag compensation permits an increase in gain to reduce steady-state error while a lead compensator permits an increase in response time without

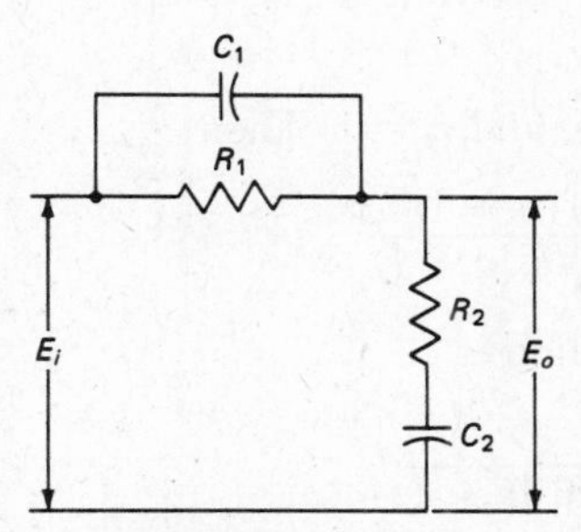

Fig. 7-18 Lead-lag network.

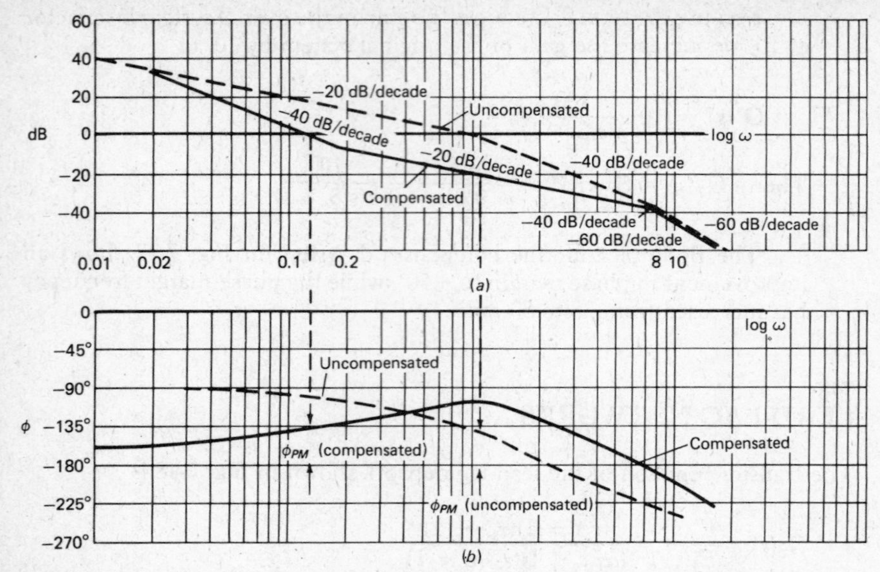

Fig. 7-19 Bode plots for uncompensated (dashed lines) and lead-lag-compensated (solid lines) system in Example 7-4.

loss of stability. Design aids are available to help in the selection of a lead-lag network, but they are beyond the scope of this book. The following example illustrates the use of a lead-lag compensator.

Example 7-4

The uncompensated system transfer function is

$$G(s) = \frac{K}{s(s + 1)(s/8 + 1)} \qquad \text{with} \qquad K = 1$$

Solution

Select a lead-lag network with $\alpha = 10$, $\tau_1 = 5$, and $\tau_2 = 1$. Then

$$G_c(s) = \frac{(\tau_1 s + 1)(\tau_2 s + 1)}{(\alpha\tau_1 s + 1)(\tau_2 s/\alpha + 1)} = \frac{(5s + 1)(s + 1)}{(50s + 1)(0.1s + 1)}$$

The overall transfer function with compensation is

$$G_T(s) = G_c(s)G(s) = \frac{1(5s + 1)}{s(s/8 + 1)(50s + 1)(0.1s + 1)}$$

Figure 7-19 shows the Bode plot for the uncompensated and the compensated system. The phase margin for the uncompensated system is measured at a frequency $\omega = 1$ rad/s. The compensation lowers the frequency and keeps the phase margin approximately constant. If the gain is increased to reduce the error, the phase margin can still be kept within acceptable limits and the speed of response can be increased.

PROBLEMS

7-1. For the system shown in Fig. P7-1, what value of K_T will give a damping factor ζ of 0.7?

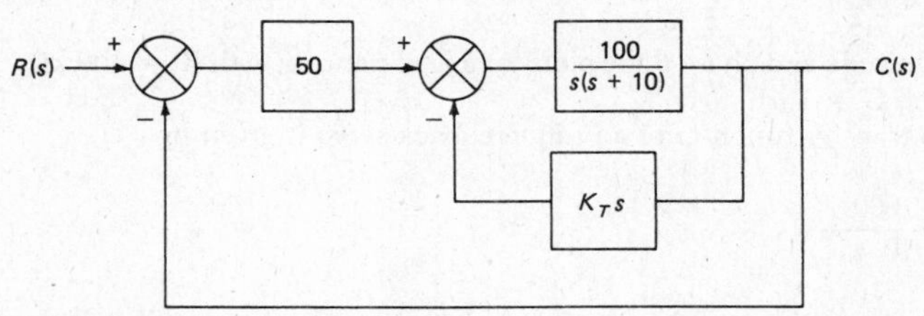

Fig. P7-1

7-2. In the system shown in Fig. P7-2, find K_T for critical damping.

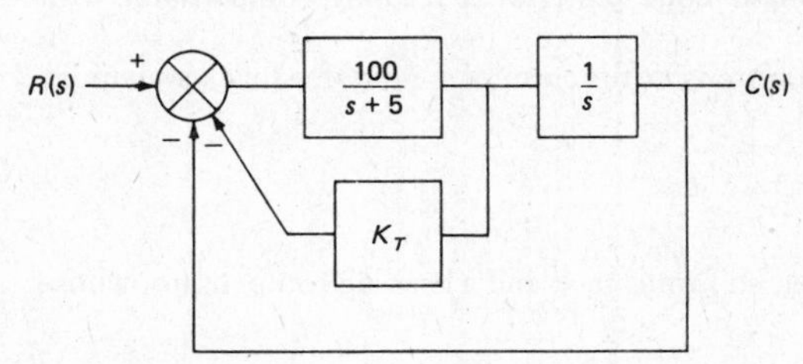

Fig. P7-2

7-3. For the system shown in Fig. P7-3, $r(t) = 10°$ step. Find K_a to obtain a steady-state error of 0.2°. Using the value of K_a just determined, find K_T for critical damping.

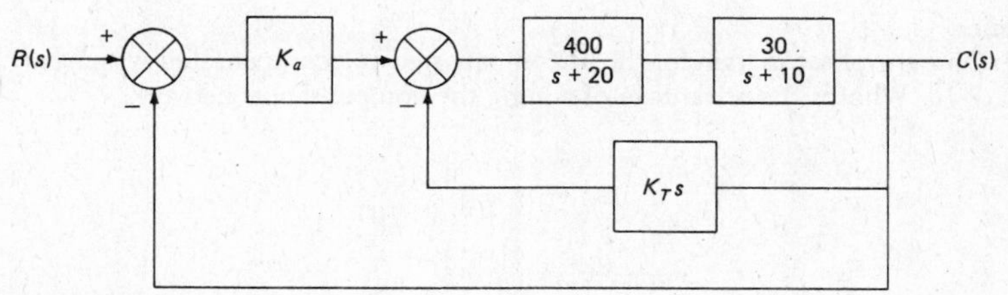

Fig. P7-3

7-4. Sketch the magnitude and phase Bode plots for a lag network with $R_1 = 9\ \text{M}\Omega$, $R_2 = 1\ \text{M}\Omega$, and $C = 20\ \mu\text{F}$.

7-5. The open-loop transfer function of a unity-feedback servosystem is given by

$$G(s) = \frac{6}{s(s/4 + 1)^2}$$

A lag compensator with $\alpha = 10$ and $\tau = 10$ is added to the system. Sketch the magnitude and phase Bode plots for both the uncompensated and compensated systems.

7-6. Determine the gain and phase margins for the uncompensated and compensated systems in Prob. 7-5.

7-7. Sketch the magnitude and phase Bode plots for a lead network with $R_1 = 100\ \text{k}\Omega$, $R_2 = 11\ \text{k}\Omega$, and $C = 1\ \mu\text{F}$.

7-8. The open-loop transfer function of a unity-feedback servo is given by

$$G(s) = \frac{1000}{s(s/10 + 1)}$$

A lead compensator with $\alpha = 0.1$ and $\tau = 0.1$ is added to the system. Sketch the magnitude and phase Bode plots for both the uncompensated and compensated systems.

7-9. Sketch the amplitude and phase Bode plots for a lead-lag compensator with $\alpha = 10$, $\tau_1 = 10$, and $\tau_2 = 5$.

7-10. The uncompensated open-loop transfer function of a unity-feedback system is

$$G(s) = \frac{2}{s(s/2 + 1)(s/10 + 1)}$$

The gain of the system is increased from 2 to 4, and a lead-lag compensator whose transfer function is

$$G_c(s) = \frac{(2s + 1)(s/2 + 1)}{(20s + 1)(s/20 + 1)}$$

is added. Sketch the magnitude and phase Bode plots for both the uncompensated and compensated systems. *Note:* Use the higher gain for the compensated system.

7-11. Determine the phase margin for the uncompensated and compensated systems in Prob. 7-10. What is the advantage of adding the compensation network?

chapter 8

Other Types of Control Systems

A servomechanism was defined as a control system containing electromechanical components. The main element in a servomechanism is an electric motor. Not all control systems use electrical components. Aircraft, for example, use hydraulic systems to control landing gear and control surfaces (ailerons, rudder, wing flaps, etc.).* Hydraulic motors have a greater torque-to-weight ratio than electric motors. Even though a hydraulic pump and reservoir are required, they are shared by many control systems. The overall weight saving is substantial.

8-1 HYDRAULIC CONTROL SYSTEMS

In a hydraulic system the medium for power transmission is a fluid under pressure. The fluid is generally oil since oil also acts as a lubricant. Pneumatic systems are of the same form as hydraulic systems except that the medium is air, which is an advantage when risk of fire or explosion must be minimized. Hydraulic and pneumatic systems are used especially when the fluid under pressure is necessary to perform required functions in a large system such as an aircraft. For equivalent power outputs, these systems require smaller actuators than equivalent electrical systems if the pumps and reservoirs are not included. One example of the mechanical hydraulic system is the power-steering mecha-

* The use of electric motors rather than hydraulic actuators is now being studied for aircraft. Hydraulic lines would be eliminated and replaced with electrical wiring. This type of system is called *fly by wire*.

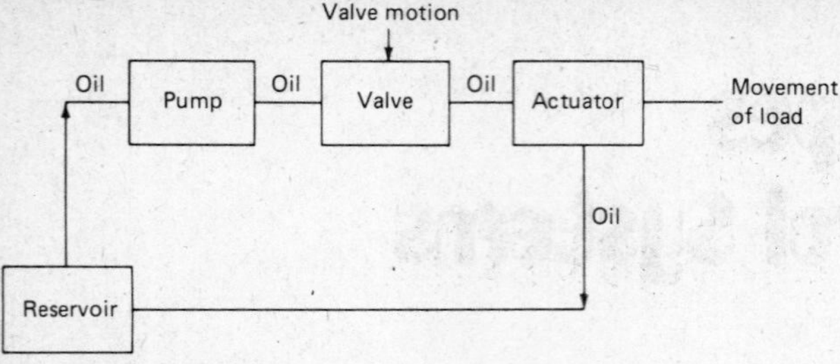

Fig. 8-1 Hydraulic system.

nism in an automobile. The movement of the steering wheel drives a hydraulic valve, which actuates the wheel-turning mechanism through a gearing arrangement.

8-1-1 Basic System Operation

A schematic of a fundamental hydraulic system is shown in Fig. 8-1. A pneumatic system is similar except that no reservoir is required. The valve-actuator arrangement is shown in Fig. 8-2. The valve stem can be positioned manually, electromechanically, or by any other means. For linear operation it must be assumed that valve movement will be limited to very small excursions. In the null

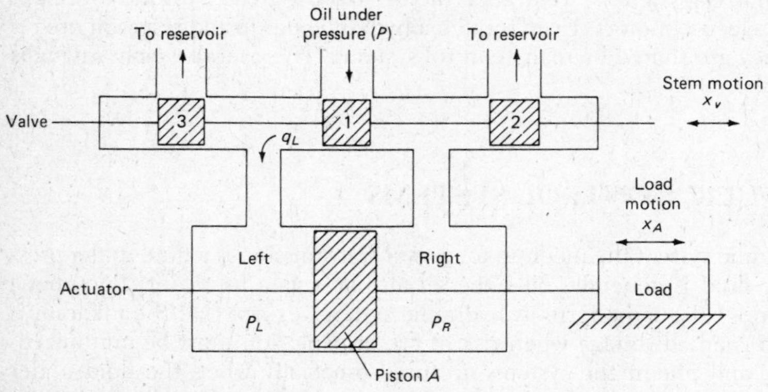

Fig. 8-2 Valve-actuator system:
P = pressure, lb/in²
x = displacement, in
q = flow rate, ft³/min or in³/sec
A = area of piston, in²
M = mass of load slugs
F = force on piston, lb
f = viscous-friction coefficient, lb/(in · s)

position, the input line is blocked so that equal pressure exists on both sides of the actuator piston.

If the valve stem is moved to the right, oil at pressure P enters the actuator cyclinder to the left of the piston. Valve piston 3 still blocks the port to the reservoir, so that the left-hand side of piston A attains a pressure P_L. Valve piston 2 exposes the right-hand side of piston A to the reservoir so that the pressure on the right is rather low (near atmospheric pressure). Assuming incompressibility of the oil and no leakage around the pistons, it follows that the flow rate of oil into the left is proportional to the excursion of the valve and the back pressure to the left of piston A.

8-1-2 Block Diagram for the Valve-Actuator System

Referring to Fig. 8-2, the difference in pressure across the piston for displacement to the right is

$$P_D = P_L - P_R \tag{8-1}$$

The force on the actuating piston becomes

$$F = A(P_L - P_R) = AP_D \tag{8-2}$$

The flow rate into the left side for linear operation is

$$q_L = K_1 x_v - K_2 P_D \tag{8-3}$$

Which merely states that the flow rate increases as the valve stem exposes more of the oil-pressure line to the chamber but decreases as the back pressure increases. The linearized valve characteristics, Eq. (8-3), are plotted in Fig. 8-3.

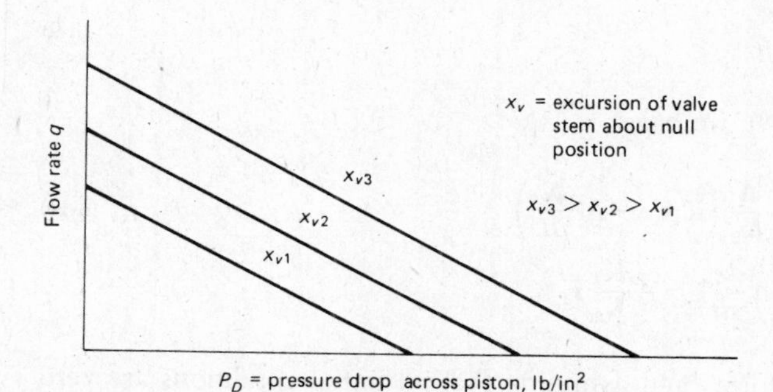

Fig. 8-3 Linearized valve characteristics.

The fluid flowing into the left side must be balanced by the movement of piston A to the right. The rate at which the volume to the left increases is

$$A \frac{dx_A}{dt} \tag{8-4}$$

Then the volume rate increase equals the flow rate of oil, or

$$q_L = A \frac{dx_A}{dt} \tag{8-5}$$

Equating (8-5) and (8-3), we get

$$A \frac{dx_A}{dt} = K_1 x_v - K_2 P_D \tag{8-6}$$

$$\text{or} \quad P_D = \frac{1}{K_2}\left(K_1 x_v - A \frac{dx_A}{dt}\right) \tag{8-7}$$

from Eq. (8-2)

$$F = AP_D$$

Then Eq. (8-7) becomes

$$F = \frac{A}{K_2}\left(K_1 x_v - A \frac{dx_A}{dt}\right) \tag{8-8}$$

But the force developed by the piston is used in overcoming the load

$$F = M \frac{d^2x_A}{dt^2} + f \frac{dx_A}{dt} \tag{8-9}$$

Equating (8-8) and (8-9), we have

$$M \frac{d^2x_A}{dt^2} + f \frac{dx_A}{dt} = \frac{A}{K_2}\left(K_1 x_v - A \frac{dx_A}{dt}\right) \tag{8-10}$$

$$\text{or} \quad M \frac{d^2x_A}{dt^2} + \left(f + \frac{A^2}{K_2}\right) \frac{dx_A}{dt} = A \frac{K_1}{K_2} x_v$$

Taking Laplace transforms and assuming that all initial conditions are zero leads to

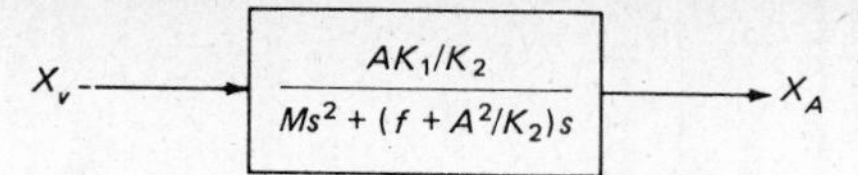

Fig. 8-4 Overall transfer function.

$$\left[Ms^2 + \left(f + \frac{A^2}{K_2}\right)s\right] X_A = A\frac{K_1}{K_2} X_v \qquad (8\text{-}11)$$

The transfer function of interest is

$$\frac{X_A}{X_v} = \frac{AK_1/K_2}{Ms^2 + (f + A^2/K_2)s} \qquad (8\text{-}12)$$

The block diagram of the combined valve-actuator-load system is shown in Fig. 8-4. Of course the entire analysis is the same for the movement to the left, so that Eq. (8-12) is valid independent of direction.

The values of K_1 and K_2 can be determined from characteristics like those in Fig. 8-3, which are either supplied by the manufacturer or determined by test. The area of the piston is specified, and the load value M is a design specification. From Eq. (8-3)

$$q = K_1 x_v - K_2 P_D \qquad (8\text{-}13)$$

and $$K_1 = \left.\frac{\Delta q}{\Delta x_v}\right|_{\Delta P_D = 0} \qquad |K_2| = \left.\frac{\Delta q}{\Delta P_D}\right|_{\Delta x_v = 0} \qquad (8\text{-}14)$$

K_2 is the slope of any characteristic line, and K_1 is the vertical distance between lines (at constant pressure) divided by the change in stem motion.

8-1-3 Block Diagram of the Hydraulic Control System

The schematic of a complete electromechanical hydraulic servo is shown in Fig. 8-5. If the input potentiometer shaft is moved by an amount R, the amplifier is driven by the voltage

$$E = K_p(R - x_A) \qquad (8\text{-}15)$$

The amplifier output V_o excites the solenoid valve winding, which causes the valve stem to move an amount

$$x_v = K_s V_o \qquad (8\text{-}16)$$

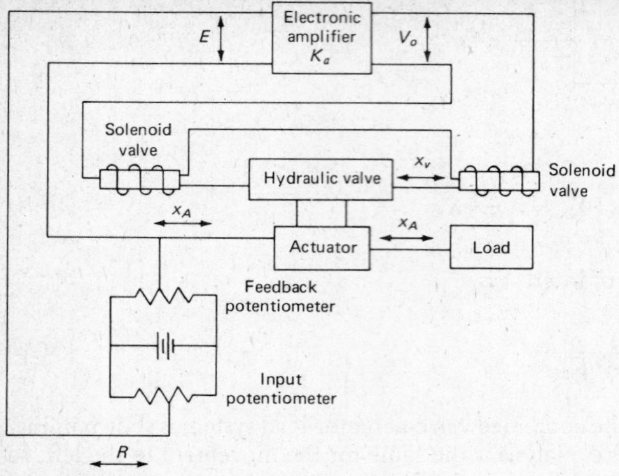

Fig. 8-5 Hydraulic servo.

where K_s is the gain of the solenoid in inches per volt. The movement of the valve causes the load to move by an amount x_A governed by the transfer function [Eq. (8-12)].

The actuator is also coupled to the feedback potentiometer, which moves the same distance x_A. When $x_A = R$, $E = K_p(R - x_A) = 0$. At this point the valve is returned to the null position, and motion ceases. The block diagram is shown in Fig. 8-6, from which the overall transfer function can be obtained. Let

$$\frac{AK_1}{K_2} = K_H \tag{8-17}$$

and $$f + \frac{A^2}{K_2} = f_H \tag{8-18}$$

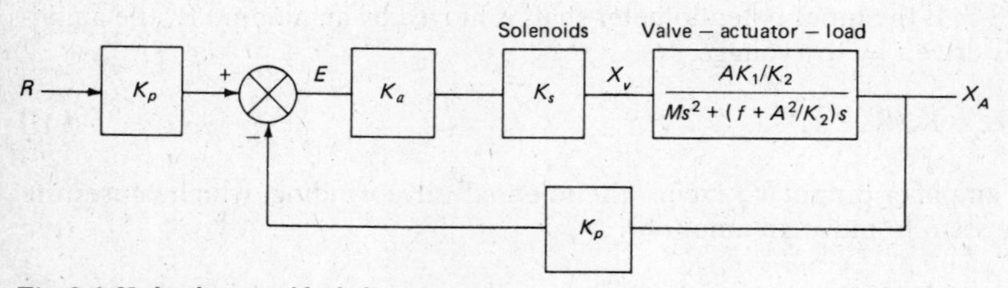

Fig. 8-6 Hydraulic-servo block diagram.

Then $$\frac{X_A}{X_v} = \frac{K_H}{Ms^2 + f_H s} = \frac{K_H/f_H}{s[(M/f_H)s + 1]} \tag{8-19}$$

If we let $M/f_H = \tau_H$, the valve-actuator-load time constant, Eq. (8-19) becomes

$$\frac{X_A}{X_v} = \frac{K_H/f_H}{s(\tau_H s + 1)} \tag{8-20}$$

The overall transfer function is

$$\frac{X_A}{R} = \frac{K_p K_a K_s (K_H/f_H)/[s(\tau_H s + 1)]}{1 + K_p K_a K_s (K_H/f_H)/[s(\tau_H s + 1)]} \tag{8-21}$$

$$\frac{X_A}{R} = \frac{K}{s(\tau_H s + 1) + K} \tag{8-22}$$

where $K = \dfrac{K_p K_a K_s K_H}{f_H}$

The overall system has the normal second-order response and depending on the values of K and τ_H will either be overdamped or underdamped (critical damping is highly unlikely in practice). The analysis of a hydraulic control system is identical to that of a second-order servomechanism. Instead of dealing with dc motor parameters, the equations involve hydraulic-valve, actuator, and solenoid constants.

8-2 HEATING CONTROL SYSTEMS

Home heating control systems are unusual compared with the hydraulic and servo systems discussed so far. Recall that the servo could move the load in either direction. When overshoot occurs, the servo motor reverses and turns the load back in the opposite direction. Most heating systems do not have the ability to cool. If the temperature exceeds the selected value, the heating system merely shuts down and waits for the area to cool. An exception to this is the climate-control system used in some automobiles, which actually cycle between heating and air conditioning.

Figure 8-7 represents a block diagram of a typical home heating system. The desired room temperature is normally set by turning a dial on the thermostat. The thermostat contains a mercury switch, which closes when the room temperature is lower than the setting. The closing of the switch activates the controls necessary to turn the furnace on. The heat generated by the furnace is

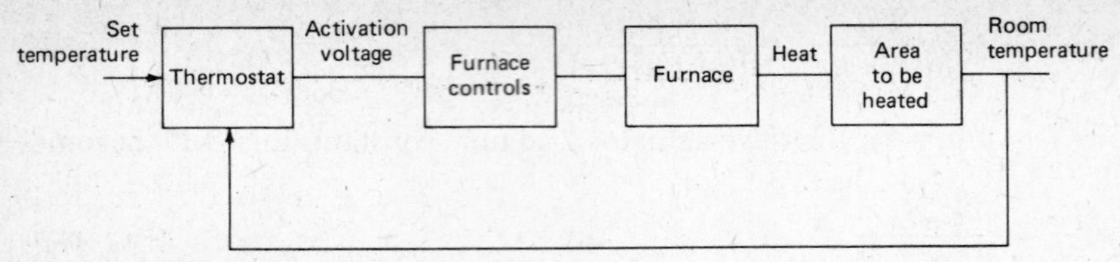

Fig. 8-7 Heating control system.

distributed throughout the area, causing a rise in temperature. When the room temperature reaches or exceeds the setting, the mercury switch opens and halts the heating cycle.

The heating system is not a linear feedback control but an on-off, or bang-bang, system. The amount of heat generated by the furnace is essentially a constant independent of the room temperature. Time is an important factor. If the area is very cold, the furnace just stays on longer. Should the temperature rise above the setting, the furnace merely shuts down until it is called upon again by the thermostat.

This behavior is different from that of the servomechanisms studied thus far, which have been linear systems with the signal varying in proportion to the error, or difference between the set value and the actual value.

8-3 INSTRUMENT SERVOMECHANISMS

Instrument servos are generally used in measurements where continuous recordings of the variable to be observed are required. As an example, if the temperature in a chemical process is to be continuously measured, a temperature transducer can provide an electric signal which is sensed by a recorder. Two basic types of recorders can be used, an open-loop recorder or a closed-loop (servo) recorder. In the open-loop recorder, which is used when the recording pen is to be moved over small distances, the basic movement is a galvanometer

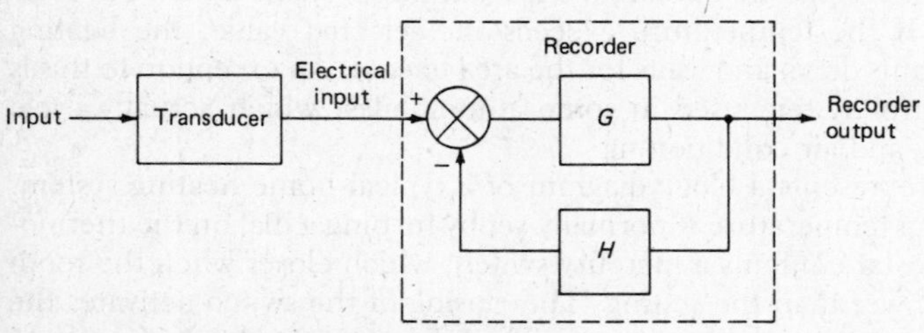

Fig. 8-8 Block diagram for servo recorder.

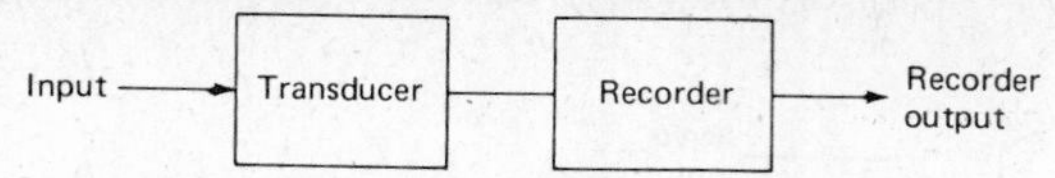

Fig. 8-9 Block diagram for open-loop recorder.

using a stiff restraining spring. In the servo recorder, the basic movement is a motor.

In the open-loop system, the movement of the pen should be strictly proportional to the input, and after initial calibration this may be true temporarily; but the output of these devices is very sensitive to degradation of components and variation in loads. For example, environmental conditions may affect the frictional force between the recording paper and the pen, thereby changing the response of the recorder. Also, since the basic movement for open-loop systems is a galvanometer, which is linear only for small rotation, pen motions are restricted to small excursions.

In a servo recorder, the output (motion of the pen) is compared with the input signal and the difference is amplified and used to drive a motor. Then, when the pen's change in position is proportional to the input, the error signal is reduced to zero. But now any changes in environment cause a change in error signal so that the system can correct itself.

8-3-1 The General-Purpose Recorder

The general-purpose recorder is usually an instrument servo which uses a self-balancing potentiometer to achieve feedback. An example of this is the recorder illustrated in Fig. 8-10, in which a pen is caused to move across recording paper in response to an electric input signal. The input signals must be restricted to low frequency because of the limitations on the mechanical response of the recorder; therefore this type of instrument is sometimes referred to as a dc recorder. The time base for the recorder is provided by a paper-drive motor. The amplifier usually contains voltage- and power-amplification sections. The servomotor driven by the power amplifier is used to move the pen.

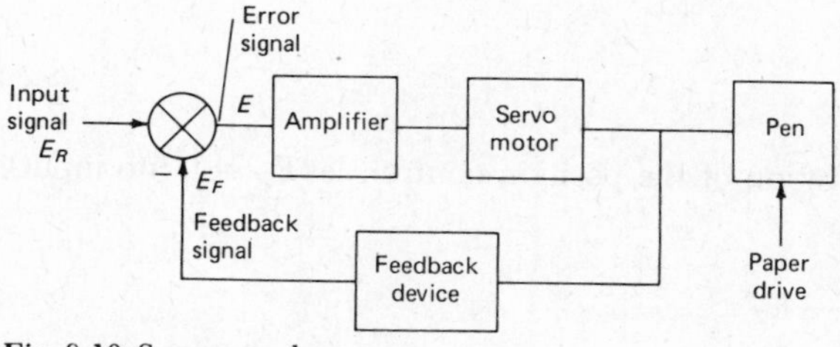

Fig. 8-10 Servo recorder.

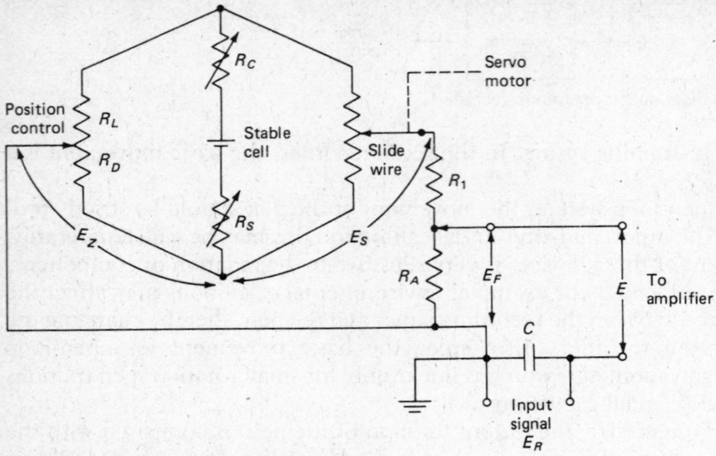

Fig. 8-11 Feedback circuit for servo recorder.

The heart of the instrument is the feedback mechanism, which is basically a Wheatstone bridge. When the pen moves the amount dictated by the input signal, the error signal becomes zero and the pen remains stationary (in the steady state) unless the input changes. That is, the output of the motor, as well as driving the pen, drives a slide-wire resistance. The slide-wire is usually a multiturn potentiometer which is mechanically coupled to the servo motor by the system that moves the pen. The basic feedback circuit is shown in Fig. 8-11. The voltage E_s depends upon the position of the slide-wire and hence represents the mechanism for feedback. The voltage E_Z is a bias which is used to position the pen (zero position). The voltage E_F is

$$E_F = \frac{R_A}{R_1 + R_A}(E_s - E_Z) = K(E_s - E_Z) \tag{8-23}$$

and the error signal is

$$E = E_R - E_F \tag{8-24}$$

To examine the operation of the position control, let $E_R = 0$ (no input); then

$$E = -E_F' = -K(E_s - E_Z) \tag{8-25}$$

Equation (8-25) shows that the servo motor will be driven until

$$E_s = E_Z$$

or the pen will move until the slide-wire voltage equals the bias voltage E_Z.

Now assume that $E_Z = 0$ (under this condition, the mechanical system should position the pen to the zero position on the paper with no input) but that an input E_R is applied. Then

$$E = E_R - KE_s \tag{8-26}$$

or the motor will drive the pen until the slide-wire output equals the input E_R. In other words, the pen will move by an amount proportional to E_R. The factor K, due to the resistors R_A and R_1, governs the sensitivity of the recorder. For example, if the maximum slide-wire voltage were 1 V and $K = 0.1$, the maximum input voltage (to cause full deflection of the pen) would be limited to 100 mV. This maximum input is also called the *span*. By providing taps on R_A, the sensitivity can be varied in fixed steps. The rheostat R_s, since it affects the maximum slide-wire voltage, can be used as a vernier. The rheostat R_c is of low value and is used for calibration.

The capacitor C is used to filter the input signal so that fast changes are not applied and damping is improved. Actually, the servo system is nonlinear, since for anything other than very small inputs, the amplifiers are driven into saturation. In saturation, the motor turns at constant speed until the error signal (input minus feedback) is small enough for linear operation. The nonlinear operation, in conjunction with the filter capacitor, should provide fast response with very little oscillation if any. This is not possible in a linear system, where fast response requires low damping and results in overshoot and considerable oscillation.

8-4 OPEN-LOOP SYSTEM

In an open-loop system, the output signal is not automatically measured and fed back for comparison with the input signal. Open-loop systems depend on prearranged settings to accomplish a desired result. In other words, the system is calibrated to yield desired results. For example, suppose a motor is to be operated at various speeds; Fig. 8-12 illustrates the calibration procedure.

The potentiometer is turned until the tachometer reads the desired speed. A marking is then made on the potentiometer dial corresponding to this speed. The process is continued until all desired values of speed are marked on the

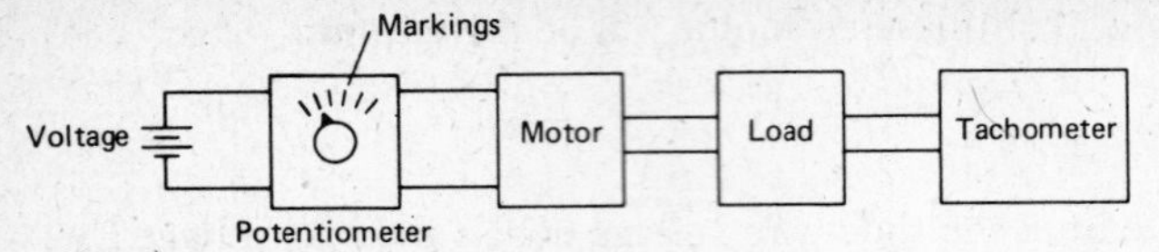

Fig. 8-12 Calibrating a motor for speed.

dial. As long as the load on the motor is not changed, the motor can be made to turn at the required speed by turning the potentiometer to the desired value. If the load changes or the condition of the motor bearings degrades, the speed will not correspond to the calibrated markings. Speed can be kept constant if an operator constantly watches the tachometer and adjusts the potentiometer manually. In a feedback system, automatic correction of voltage takes place as speed tries to change.

8-4-1 Numerical Control Systems

Numerically controlled machines are generally open-loop systems which employ stepper motors (see Chap. 9) to sequence a machine tool through various operations. Operations include moving a part (vertically, horizontally, and laterally) and controlling speed, drilling, punching, grinding, and milling.

The operations to be performed must be defined mathematically so that they can be symbolically programmed onto paper tape, magnetic tape, or

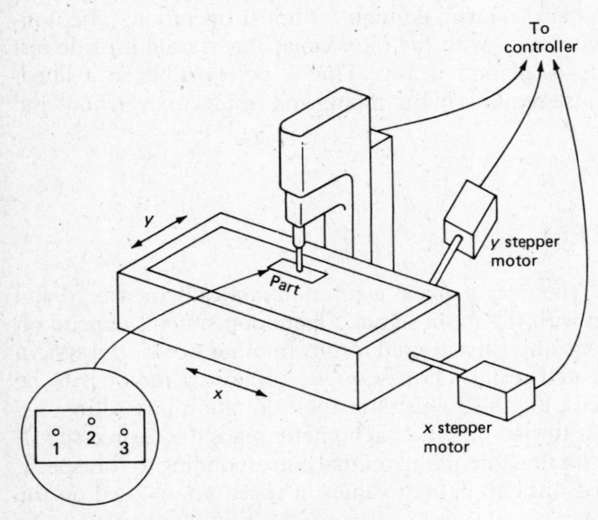

Fig. 8-13 Numerically controlled machine.

punched cards. The program is then processed by a computer to produce a numerically coded tape. This tape is then fed into a machine control unit which produces the signals controlling and operating the machine tools.

A simple numerically controlled machine can be a drill press with a movable table to position the part in the x and y position, as shown in Fig. 8-13. The object is to drill three holes in the part at specific locations as shown in the inset. The machine is programmed so that the proper logic signals are fed to the x and y stepper motors to position the table to locate the spot for hole 1 under the drill bit. The program also causes the drill to turn on and engage the part for a certain amount of time. The drill then reverses and the table is sequenced to line up for hole 2. The process is repeated for holes 2 and 3.

PROBLEMS

8-1. What is the advantage of a hydraulic motor over an electric motor?
8-2. When should a hydraulic system be selected over an electric system?
8-3. Explain the basic operation of the valve-actuator system in a hydraulic servo.
8-4. The physical constants of a hydraulic servo are

$$M = 50 \text{ slugs} \quad A = 0.02 \text{ in}^2 \quad K_a = 100 \quad K_p = 1 \text{ V/rad}$$
$$K_1 = 50 \quad K_2 = 0.001 \quad K_s = 0.001 \quad F = 4.6$$

(a) Prepare a block diagram for the system using numerical values for all transfer functions.
(b) Determine the overall transfer function X_A/R.

8-5. The characteristics of a hydraulic valve are shown in Fig. P8-5. Find the constants K_1 and K_2 (give units).

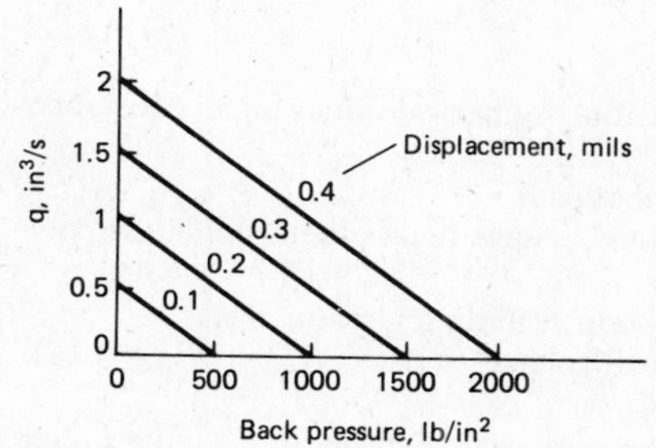

Fig. P8-5

8-6. A hydraulic valve with $K_1 = 10^2$ and $K_2 = 10^{-1}$ drives an actuator whose piston area is 1 in². The load mass is 7.5 slug, and the friction coefficient is 5 lb/(in · s).

(a) Find the transfer function $X_A/X_{v_{in}}$.
(b) What is the time constant of this system?
(c) If x_v is a step, as shown in Fig. P8-6, solve for X_A as a function of s.

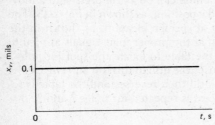

Fig. P8-6

8-7. The system in Prob. 8-6 is used in the servo shown in Fig. 8-5. The input and feedback potentiometers both have a stroke of 2 in and are excited by 5 V. The amplifier gain is 30, and the response of the solenoid is shown in Fig. P8-7.

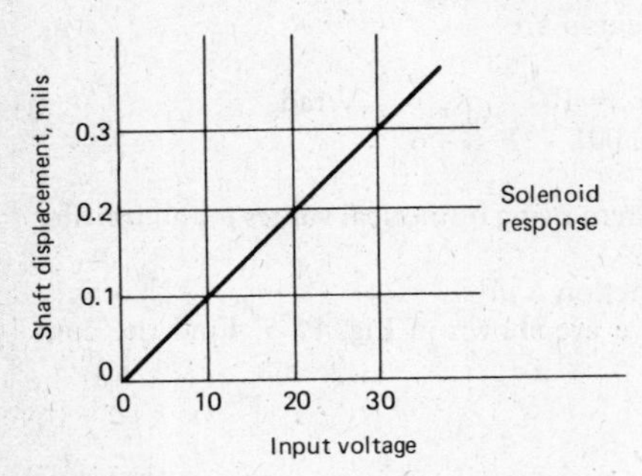

Fig. P8-7

(a) Sketch the closed-loop diagram indicating numerical values for all constants.
(b) Write the overall transfer function.
(c) Is the system underdamped or overdamped?

8-8. What characteristics of most home heating systems makes them nonlinear control systems?

8-9. What is the main disadvantage of open-loop recording instruments?

8-10. In Fig. 8-11, R_1 = 90 kΩ, R_A = 10 kΩ, cell voltage = 1.1 V, R_s = 0, R_c = 10 Ω, battery current = 10 mA.
(a) If R_D = 0, what value of E_R will cause full-scale deflection of the pen, i.e., what is the span?
(b) What happens if E_R is negative with R_D = 0?
(c) If R_L = 0.8 kΩ and R_D = 0.2 kΩ, what is the span of the recorder?

8-11. Repeat Prob. 8-10 if R_A = 20 kΩ and R_1 = 80 kΩ.

chapter 9

Digital Servos and Components

In recent years there has been an increase in the use of the digital computer as a means of controlling servomechanisms. The advent of the economical and simple-to-use microprocessor and the tremendous demand for computer peripherals has also brought digital controls in contact with the servomechanism. As a result, the digital servo has become just as important as the linear servomechanisms discussed in the previous chapters, if not more important.

9-1 DC STEPPER MOTOR

Perhaps the single most important electromechanical component used in digital servos is the *stepper motor*. It is a device whose input is an electrical pulse and whose output is the rotation of its shaft. For each pulse the motor receives, its shaft rotates through a fixed precise angle. Depending on the pattern of pulses applied, the stepper motor can be used to control the position and/or velocity of a load accurately. The significance of this type of control is as follows:

1. The error is the single step error (usually less than 5 percent of one step) and therefore is noncumulative.
2. The need for feedback to determine the shaft position is eliminated.
3. The need for converting from digital to analog signals when interfaced with a digital computer is eliminated.

The motor is controlled by drive circuitry, which dictates the motor's response, and a dc power supply.

9-1-1 Definitions

Step Angle (SA) This is the specific angular rotation in degrees the shaft moves each time its winding polarity is changed. It is accomplished by a single input pulse. It is expressed in deg/step, or simply deg.

Steps per Revolution (SPR) This represents the total number of steps required by a motor's shaft for it to rotate a full 360°

$$\text{SPR} = \frac{360°}{\text{SA}} \tag{9-1}$$

Steps per Second (SPS) This is the number of angular steps the motor goes through in 1 s. SPS is comparable to speed in revolutions per minute of the ac or dc motor described in Chap. 2

$$\text{SPS} = \frac{(\text{r/min})(\text{SPR})}{60} \tag{9-2}$$

or $$\omega = 2\pi \frac{\text{SPS}}{\text{SPR}} \text{ rad/s} \tag{9-3}$$

Step Accuracy This is the position-accuracy tolerance, generally expressed as a percentage of a single step angle.

Holding Torque With the motor shaft at zero speed (standstill) and rated power applied, holding torque is the minimum amount of external torque applied to the shaft which will break it away from its holding position.

Residual Torque This is the torque present at standstill with power off. It is present only in permanent-magnet rotor-type motors.

Step Response This is the time for a motor to move through a single step. It is a function of the motor's torque-to-inertia ratio and the characteristics of the drive circuitry.

Torque-to-Inertia Ratio (TIR) This is figure of merit for a stepper motor. The higher the TIR the better the step response will be

$$\text{TIR} = \frac{\text{holding torque (oz} \cdot \text{in)}}{\text{rotor inertia (oz} \cdot \text{in} \cdot \text{s}^2)} \tag{9-4}$$

Drive Circuitry This general term describes the electronics that control the motor, consisting usually of a power supply, sequencing logic, and power switching components.

9-1-2 Typical Construction and Operation

The stator of the motor has several poles (frequently eight), whose polarities are changed by electronic switches. The net result of the switching is to rotate the average north and south poles of the stator. The rotor's north pole will line up with the stator's south pole. The rotor's magnetism can be generated by external excitation or a permanent magnet. The latter will be treated here. As the average stator field rotates through steps, the rotor will follow it in a similar stepwise fashion. In order to obtain better resolution, small teeth are machined on the rotor and stator. These teeth do not come in contact with each other but serve as low-reluctance paths (Fig. 9-1).

If a stepper motor has four stator windings (eight poles), it might be wired as shown in Fig. 9-2. This wiring technique is known as the *four-step switching sequence*. Note that there are four possible switch positions, hence the configuration name. Windings A and A′ are not energized at the same time since they create fields that are equal and opposite.

As the input pulses are applied, the switches go through their four-step sequence. The sequence repeats over and over again, as shown in Fig. 9-2. Each time four pulses are applied and the switches go through their sequence, the rotor rotates one full tooth pitch. That is, the rotor turns so that each tooth now lines up with the next succeeding stator tooth. If there are 100 teeth on the

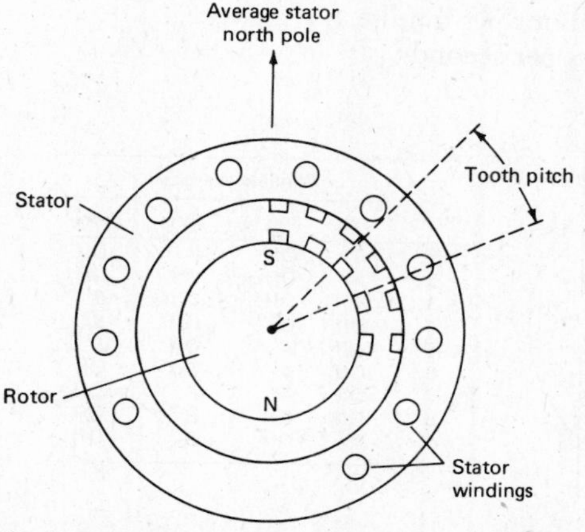

Fig. 9-1 Typical permanent-magnet stepper-motor construction.

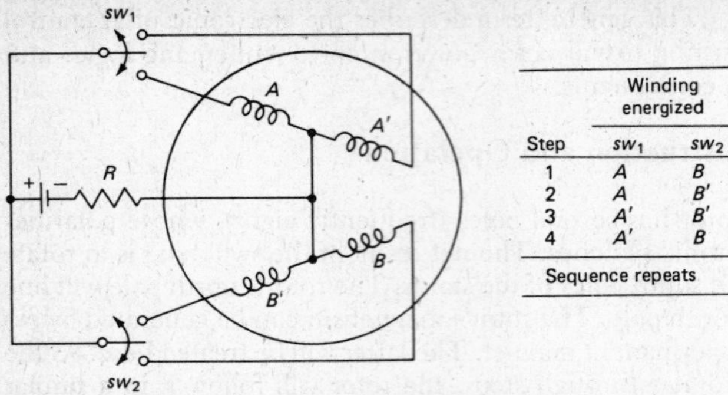

Step	Winding energized sw_1	Winding energized sw_2
1	A	B
2	A	B'
3	A'	B'
4	A'	B
Sequence repeats		

Fig. 9-2 Eight-pole stator wired for four-step sequence.

rotor, it would take 4 steps for each tooth, or 400 steps, for the rotor to make a full revolution. This can be described by

$$\text{SPR} = (\text{SS})(N_r) \tag{9-5}$$

where N_r is the number of the rotor teeth and SS represents the step sequence.

Example 9-1

A stepper motor wired for a four-step sequence has 180 rotor teeth. It is fed a pulse train with a 500-Hz frequency. Find:

(a) The steps per revolution
(b) The step angle
(c) The rotor speed in revolutions per minute
(d) The rotor speed in radians per second

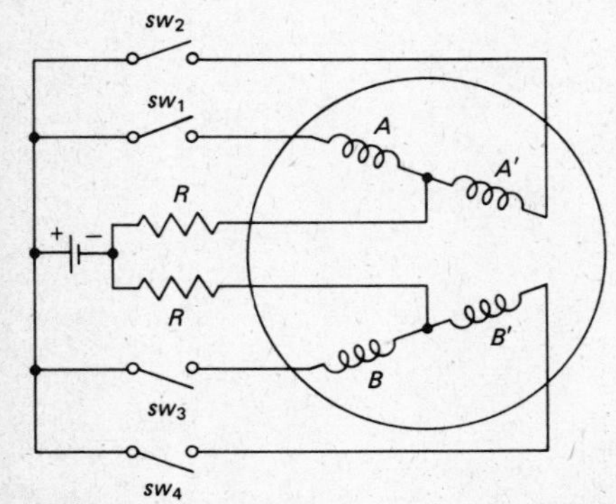

Step	Winding energized sw_1	sw_2	sw_3	sw_4
1	A	Off	B	Off
2	A	Off	Off	Off
3	A	Off	Off	B'
4	Off	Off	Off	B'
5	Off	A'	Off	B'
6	Off	A'	Off	Off
7	Off	A'	B	Off
8	Off	Off	B	Off
1	A	Off	B	Off

Fig. 9-3 Eight-pole stator wired for eight-step sequence.

Solution

(a) Using Eq. (9-5), we get

$$\text{SPR} = (\text{SS})(N_r) = 4(180) = 720 \text{ steps/r}$$

(b) Using Eq. (9-1), we see that the single-step angle is given by

$$\text{SA} = \frac{360°}{\text{SPR}} = \frac{360°}{720} = 0.5°/\text{step}$$

(c) From Eq. (9-2)

$$\text{r/min} = \frac{60(\text{SPS})}{\text{SPR}} = \frac{60(500)}{720} = 41.67 \text{ r/min}$$

(d) Applying Eq. (9-3), we get

$$\omega = 2\pi \frac{500}{720} = 4.36 \text{ rad/s}$$

An alternate way of wiring an eight-pole stator is shown in Fig. 9-3. Here four switches are used instead of two, as in Fig. 9-2. This permits both windings A and A′ (or B and B′) to be deenergized at the same time.

A simpler design is the three-lead motor shown in Fig. 9-4. Although there are only two windings, the switching is more complicated and the power must be supplied from a dual voltage supply. In addition, the four-winding motor develops more torque than a comparable two-winding motor at high speeds.

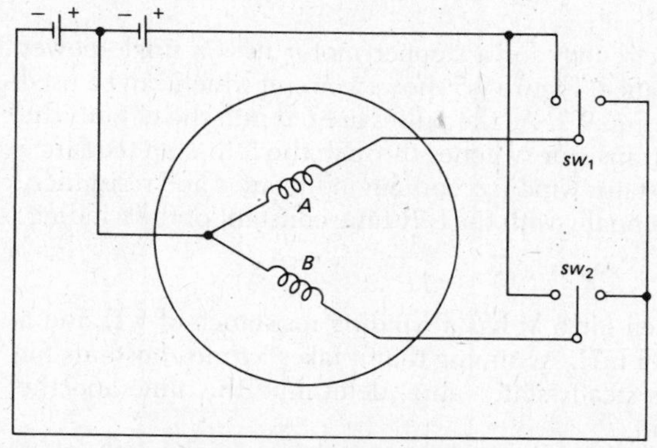

Fig. 9-4 Four-pole motor with four-step sequence.

Example 9-2

The motor shown in Fig. 9-3 has 90 rotor teeth. It is desired to have the motor rotate 100° in 0.25 s. How many pulses must be applied to the motor and at what frequency? Assume zero acceleration time.

Solution

The motor in Fig. 9-3 has an eight-step sequence; therefore, using Eq. (9-5) and (9-1), we have

$$\text{SPR} = 8(90) = 720 \text{ steps/r}$$

$$\text{SA} = \frac{360^\circ}{\text{SPR}} = \frac{360}{720} = 0.5^\circ/\text{step}$$

Thus to rotate 100°, the motor must receive

$$\frac{100^\circ}{0.5^\circ/\text{step}} = 200 \text{ steps or } 200 \text{ pulses}$$

$$\frac{200 \text{ pulses}}{0.25 \text{ s}} = 800 \text{ pulses/s}$$

The answer is that 200 pulses at a rate of 800 Hz are needed.

9-1-3 Drive Circuitry

This part of a stepper-motor control system is just as important as the motor itself. It is the drive circuitry which causes current to flow in the stator windings. Motor torque, and hence acceleration, is proportional to current. The motor's step response therefore depends on the speed at which current in its stator windings rises and falls.

A simple type of drive circuitry for a stepper motor uses a single power supply and no external resistance. Figure 9-5 shows a circuit which can be used to drive the motor shown in Fig. 9-2. As the pulses are fed into the circuit, the gating and flip-flops put the transistor switches through the four-step sequence shown in Fig. 9-2. Since the stator windings contain inductance and resistance, the current I_s will rise exponentially with the L/R time constant of the winding.

Example 9-3

A stepper motor designed for 6 V has a winding resistance of 5 Ω and a winding inductance of 15 mH. Assuming that it takes 5 time constants for the current to reach its steady-state value, determine this time and the value of the final current.

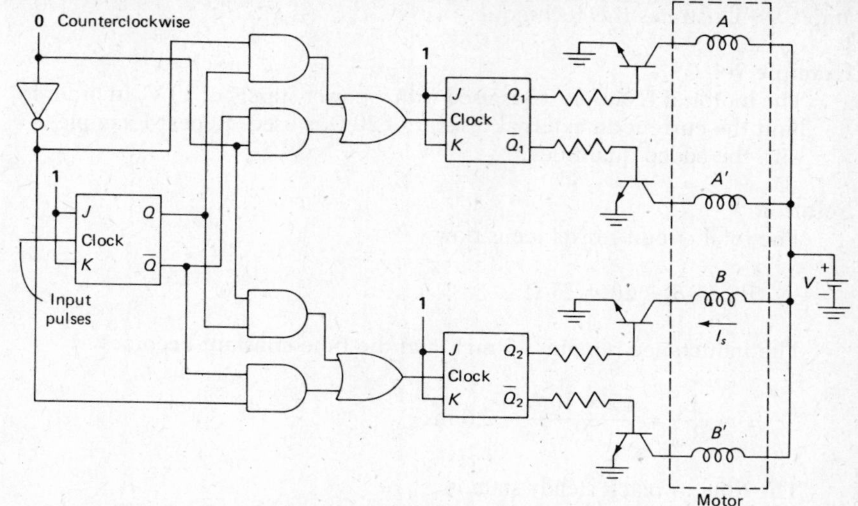

Fig. 9-5 Simple stepper-motor drive circuit. Boldface **1** and **0** represent the binary digits.

Solution

The time constant L/R is

$$\tau = \frac{L}{R} = \frac{15 \times 10^{-3}}{5} = 3 \text{ ms}$$

The time to reach steady state is

$$t = 5\tau = 5(3 \text{ ms}) = 15 \text{ ms}$$

The final value of current will be

$$I = \frac{V}{R} = \frac{6}{5} = 1.2 \text{ A}$$

Example 9-3 indicates that it takes 15 ms for the motor current (or torque) to reach its full value. This may or may not be fast enough in a given application. One way to speed the response time up is to overexcite the motor; i.e., a higher voltage is used for the supply. This, however, will give rise to a larger current, which would exceed the rating of the motor. In order to avoid this problem, some means of current limiting must be employed. The simplest way to limit the current is to place an external resistor in series with the supply. Example 9-4 illustrates this technique.

Example 9-4

The motor of Example 9-3 is used with a power supply of 30 V. In order to limit the current an external resistor of 20 Ω is used. Repeat Example 9-3 with the added quantities.

Solution

The total circuit resistance is now

$$R_{tot} = 5 + 20 = 25\ \Omega$$

The inductance remains 15 mH, and the time constant becomes

$$\tau = \frac{L}{R_{tot}} = \frac{15 \times 10^{-3}}{25} = 0.6 \text{ ms}$$

The time to reach steady state is

$$t = 5\tau = 5(0.6 \text{ ms}) = 3 \text{ ms}$$

and the final value of current is

$$I = \frac{V}{R_{tot}} = \frac{30}{25} = 1.2 \text{ A} \qquad \text{as before}$$

It should be noted that although the objective (a shorter response time) was accomplished quite simply, there is a major disadvantage to series resistance limiting. In Example 9-4 the external resistance is 4 times the motor resistance. This in effect means that 80 percent of the power supplied is dissipated externally from the motor. This in turn causes a less efficient system.

Another method of limiting the current is the chopper technique. Here, a high voltage is again used to overexcite the motor, but the voltage is switched off and on repeatedly so that the current does not exceed a specified limit. The switching gives rise to an average current in the motor winding and continues until the winding is deenergized. The advantage here is the higher efficiency obtained; however, the drive circuitry is much more complex.

Another method is the dual-voltage technique. As the name implies, two voltages are used. A high voltage is used initially to overexcite the motor. At the instant the current reaches a specified limit, the high voltage is switched out and a lower voltage is switched in. The lower voltage value is such as to maintain the current at its value when the switching occurred. Here again, even though increased efficiency is obtained, the drive circuitry is complex. In addition, this technique requires two power supplies.

9-1-4 Motor Selection

There are many criteria used in the selection of a stepper motor. To make the optimum selection would require an extensive discussion of the mechanical, load, and electronic drive requirements as well as the economics involved. A brief discussion will be given here of the torque requirements of the motor to be chosen.

If a motor is required to come up to a given speed (SPS) in a given time t, the angular acceleration can be obtained from

$$\alpha = \frac{\text{SPS}}{t}\frac{\text{SA}}{57.3} \quad \text{rad/s}^2 \qquad (9\text{-}6)$$

where SA divided by 57.3 represents the step angle in radians.

Example 9-5

A stepper motor is used to drive a constant load torque of 50 oz · in. The step angle is 1.8°, and the load inertia is 2.0 oz · in · s². If the load is to be accelerated to 100 steps per second in 0.5 s, find the minimum rated torque needed, neglecting the motor's rotor inertia.

Solution

From Eq. (9-6) the acceleration is

$$\alpha = \frac{100}{0.5}\frac{1.8}{57.3} = 6.28 \text{ rad/s}^2$$

The torque required to accelerate the load can be obtained using $t_{ac} = J\alpha$ [Eq. (2-14)]:

$$t_{ac} = 2(6.28) = 12.56 \text{ oz} \cdot \text{in}$$

The total torque required is the sum of the accelerating torque and the load torque. The motor selected must produce a torque of 62.56 oz · in (50 + 12.56) at a speed of 100 steps per second. This can be found by consulting various torque-speed curves. The rotor inertia would have to be negligible compared with the load inertia of 2 oz · in · s².

In some applications where no acceleration time is permitted at high stepping rates (greater than 50 steps per second) the angular acceleration can be approximated by

$$\alpha = \frac{(SPS)^2}{2}\frac{SA}{57.3} \quad \text{rad/s}^2 \tag{9-7}$$

Example 9-6

A motor using a step angle of 1.8° must drive the load of Example 9-5 at 100 steps per second. No acceleration time is allowed. Neglecting the motor's inertia, determine the torque rating required.

Solution

Since no acceleration time is allowed, Eq. (9-7) is used to determine the acceleration

$$\alpha = \frac{(100)^2}{2}\frac{1.8}{57.3} = 157.07 \text{ rad/s}^2$$

Again using Eq. (2-14), we find that the torque needed to accelerate the total inertia is

$$t_{ac} = 2(157.07) = 314.14 \text{ oz} \cdot \text{in}$$

Thus, the total torque required at 100 steps per second is

$$314.14 + 50 = 364.14 \text{ oz} \cdot \text{in}$$

It should be noted that Example 9-5 is a typical calculation that would be used in a velocity control system while Example 9-6 is a calculation that would be used in a positional servomechanism.

9-2 SHAFT ENCODER

This electromechanical device senses rotation mechanically and converts it into an electrical signal. There are two basic types of shaft encoders, incremental and absolute.

9-2-1 Incremental Shaft Encoder

The incremental shaft encoder is one whose output is a pulse for each incremental change in its shaft position. This can be accomplished in several ways.

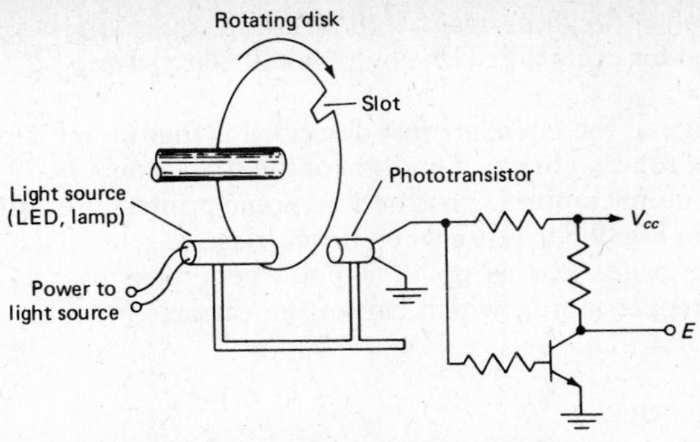

Fig. 9-6 Simple optical incremental shaft encoder.

Two common ways to build an incremental encoder are optically or with mechanical contacts. Figures 9-6 and 9-7 show both techniques for an encoder which senses a full 360° rotation of its shaft. In Fig. 9-6 since the disk prevents light from reaching the phototransistor, E is held low. However, each time the slot passes the light source, the phototransistor will fire for as long as the light is on it. During the time when the phototransistor is on, E will be high. A pulse is thus generated each time the slot passes the light source. Obviously, if the disk has four equally spaced slots, a pulse will be generated for every 90° rotation of the shaft. In general, for n equally spaced slots, a pulse is generated for every $(360/n)°$ rotation of the shaft

$$\text{Pulse angle} = \frac{360}{n} \tag{9-8}$$

In Fig. 9-7, as the disk rotates, the transistor base is grounded each time the tab makes contact with the brush. This causes E to go high. A pulse is gen-

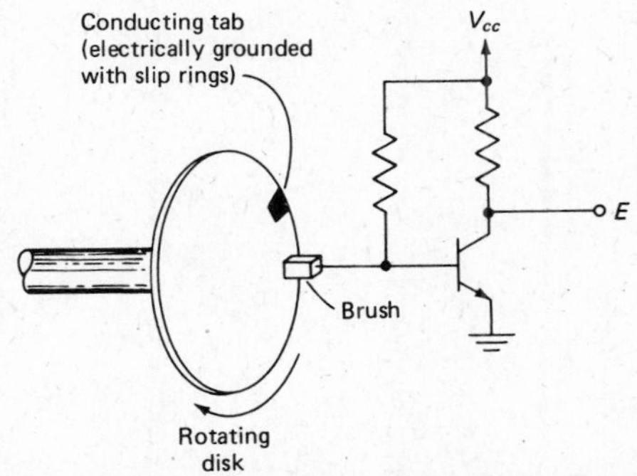

Fig. 9-7 Simple mechanical incremental shaft encoder.

erated for each 360° revolution of the disk. Again, if there are 18 equally spaced tabs, a pulse will be generated for every 20°. This could be calculated using Eq. (9-8).

The obvious disadvantage of the encoders just described is that no information about the direction of rotation or exact position of the shaft is obtained. This can be corrected with a more complex construction. Some printers use an encoder like the one shown in Fig. 9-6 to sense when the carriage or print head has reached the end of the page. At this point, a pulse generated by the encoder is used to drive a stepper motor, which causes the carriage (or print head) to return.

9-2-2 Absolute Shaft Encoder

An absolute encoder has the advantage over an incremental encoder of enabling the user to determine the actual position of the shaft at all times. It too consists of a disk, but in this case, the disk is divided into several concentric circles each of which represents 1 bit of digital information. The outermost circle represents the least significant bit (LSB) while the innermost circle represents the most significant bit (MSB). Each circle is divided into portions of conducting and nonconducting surfaces. For each bit there is a brush which makes contact with the circle, thus determining whether the bit is high (**1**) or low (**0**). (The two binary digits are set in boldface for clarity.) A simple 3-bit binary shaft encoder is shown in Fig. 9-8. The dark portions of each circle can be connected through slip rings to a 5-V supply, and the light portions can be grounded (5 V would thus represent a **1** and 0 V would represent a **0**).

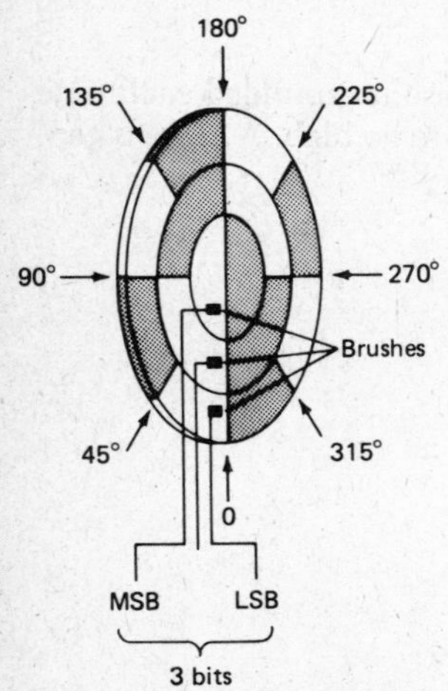

Fig. 9-8 Three-bit binary shaft encoder.

Notice that as the disk turns counterclockwise the bits will count from **000** to **111** in binary. The fact that there are 3 bits means that there are eight (2^3) possible states. Dividing 360° by 8 means that each binary count from **000** to **111** represents an angular rotation of 45°. The resolution of the 3-bit encoder is therefore 45°. If the encoder has an output of **011**, the shaft rotation will be an angle of 135 to 180°. To improve the resolution, more bits are required.

Example 9-7

An 8-bit binary shaft encoder is rotated 70°. Determine the resolution of the encoder and the encoder output for the given rotation.

Solution

If n is the number of bits, the resolution is given by

$$\text{Resolution} = \frac{360}{2^n} \tag{9-9}$$

$$\text{Then} \quad \text{Resolution} = \frac{360}{2^8} = \frac{360}{256} = 1.41^\circ$$

The output will therefore increase by a count of 1 for each 1.41° of rotation. For 70° rotation this represents

$$\frac{70}{1.41} = 49.65$$

or 49 steps; this converts into **00110001** in binary, which will be the encoder output.

One of the disadvantages of a binary encoder arises from the fact that between any two steps several bits may change state. If the brushes are slightly misaligned, a tremendous error can result. As an illustration consider the encoder of Fig. 9-8. If the shaft were displaced 181° from its zero position, the output would read **100**. However, if the MSB brush were slightly misaligned, the output might read 000, giving rise to an error corresponding to four encoder steps. This misalignment problem is somewhat remedied by the use of the Gray code on the disk rather than the binary code. Although misalignment errors still may occur, they will not be as severe as with the binary encoder. The error is minimized because between any two succeeding Gray states only 1 bit changes. Table 9-1 compares the binary and Gray codes.

A diagram of a 3-bit Gray shaft encoder is shown in Fig. 9-9. Although the binary-encoder output can be fed directly into a computer, the Gray code can easily be converted into binary with a minimum of hardware.

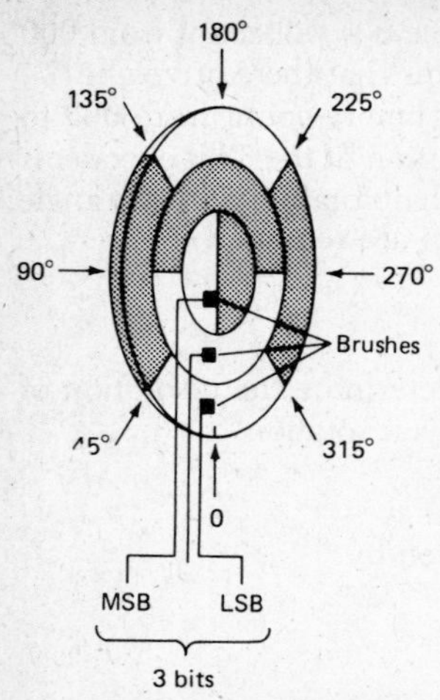

Fig. 9-9 Three-bit Gray shaft encoder.

TABLE 9-1
Comparison of Binary and Gray Codes

Decimal	Binary	Gray	Decimal	Binary	Gray
0	0000	0000	6	0110	0101
1	0001	0001	7	0111	0100
2	0010	0011	8	1000	1100
3	0011	0010	9	1001	1101
4	0100	0110	10	1010	1111
5	0101	0111			

If the misalignment problem considered for the binary encoder is examined here, the error will be found to be much smaller. When the shaft is turned 181°, the Gray output will be **110.** However, due to the misalignment considered before, the output would be **010,** which corresponds to an error of one encoder step. For the misalignment considered, the error for the binary encoder was 4 times as great.

9-3 SOLENOID

The solenoid is an electromechanical component found in many digital servomechanisms. It is a device which converts electrical energy into mechanical motion. An applied voltage causes a current to flow in a coil of wire mounted in a

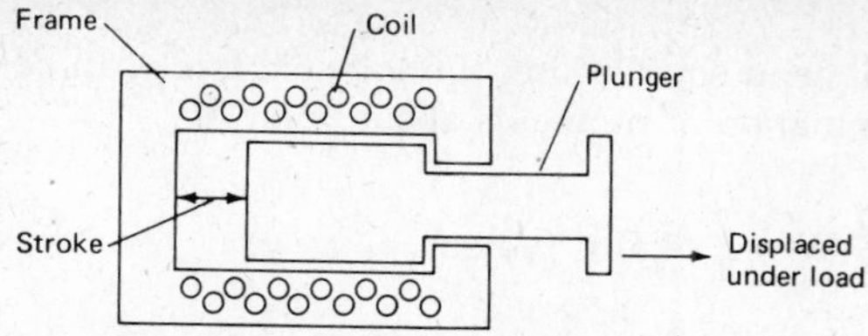

Fig. 9-10 Simplified diagram of a linear solenoid.

metallic frame. The current creates a magnetic field which tends to pull an armature (metal plunger) into the coil until it seats itself firmly against the frame.

Figure 9-10 is a simplified picture of a solenoid. It shows the plunger under load displaced to the right. When the coil is energized, the magnetic force developed pulls the plunger (and the load since it is attached) to the left until it hits the frame. The distance the plunger moves is called the *stroke* of the solenoid. The force exerted by the plunger is specified for a given solenoid at a given voltage.

In ac solenoids a high current is drawn when the coil is energized. This current decreases as the plunger closes and settles at a low holding current when the plunger is seated. If the plunger is overloaded or obstructed, it will not close all the way. This means that a large current will continue to flow in the coil. Eventually, the coil insulation will fail and the coil will short out. Since the current drawn at the start of an operating cycle is high, a solenoid designed for intermittent duty may overheat and short out when subjected to continuous duty.

In dc solenoids, the current remains constant throughout the stroke, and therefore the problems discussed above do not exist.

Example 9-8

A solenoid is used to lift the head of a printer. The head assembly weighs 4.5 oz and is to be raised 0.5 in. A simplified diagram indicating the linkage is shown in Fig. 9-11. Determine the force and stroke specifications of the solenoid.

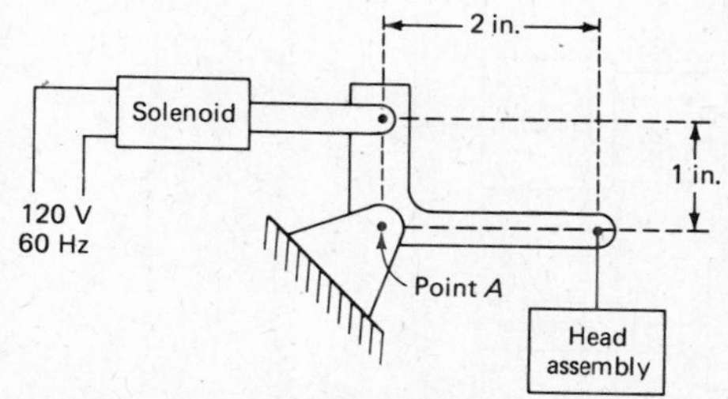

Fig. 9-11 Simplified print-head lift mechanism.

Solution

Since the head assembly will be in equilibrium, the solenoid force can be calculated by taking the summation of moments about point A:

$$1F = 2W \qquad F = \frac{2}{1}W = \frac{2}{1}(4.5) = 9.0 \text{ oz}$$

In a similar fashion, the stroke can be approximated by

$$\text{Stroke} \times 2 = \text{lift distance} \times 1$$
$$\text{Stroke} = 0.5(\tfrac{1}{2}) = 0.25 \text{ in}$$

A solenoid with a stroke of 0.25 in should be selected, but the solenoid should also provide a minimum force (when the plunger is seated) approximately 20 percent greater than the 9 oz calculated. This will ensure that the plunger seats itself under conditions of reduced line voltage and increased opposition due to a buildup of friction.

9-4 SYSTEM CONFIGURATIONS

There are two basic types of digital servomechanisms. The first involves the use of digital signals to control an analog system (Fig. 9-12). The second involves the use of digital signals to control an incremental system (Fig. 9-13).

The following definitions apply to components in Figs. 9-12 and 9-13 which have not yet been discussed.

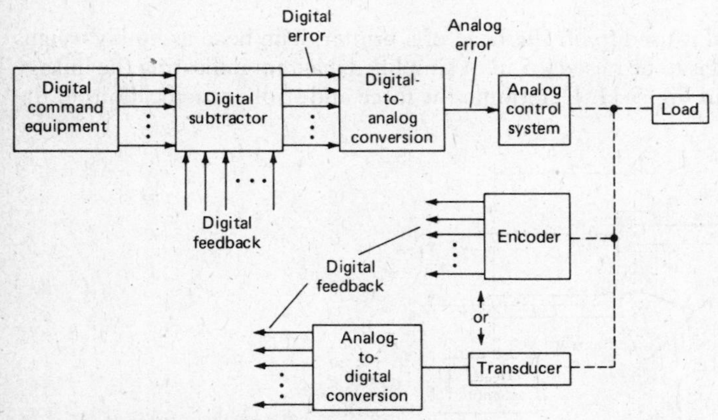

Fig. 9-12 Digitally controlled analog system.

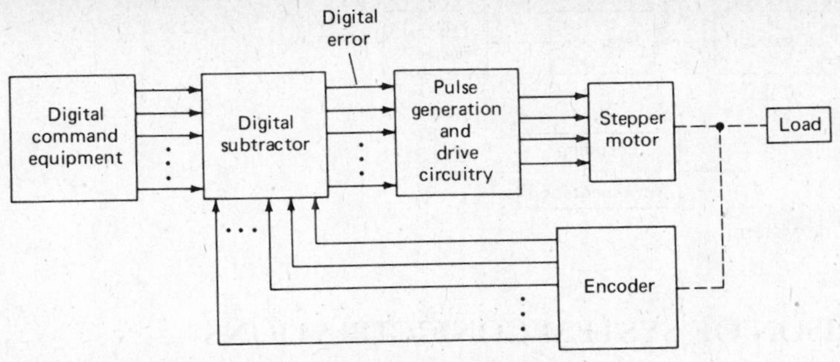

Fig. 9-13 Digitally controlled incremental (digital) system.

Digital Command Equipment This encompasses several peripheral devices which supply input information (usually in binary form) for a control system. These devices include magnetic tape or disk, floppy disk, paper tape, or card reader. A computer itself may even be used to supply the input to a system.

Digital Subtractor This consists of logic circuitry that takes the difference between the digital input and the digital feedback signals.

D/A and A/D Converters These represent components which convert digital signals into an equivalent analog signal and vice versa. Generally, they are needed when a digital computer is interfaced with an analog system. Because of their importance, they are given a more thorough treatment in Chap. 10.

It should be noted that if the encoders used in Fig. 9-12 or 9-13 produce an output in a code other than binary, e.g., Gray code, an additional component is needed to convert from that code into binary.

When a digital computer is used in a control system, it can be used either inside (on-line system) or outside (off-line system) the loop. An *on-line system* is one in which the computer continuously monitors and corrects the system automatically by containing the proper interface equipment. In an *off-line system* an operator receives information from the computer, as well as feeding information into the computer, and manually takes corrective steps. General block diagrams for both systems are shown in Figs. 9-14 and 9-15.

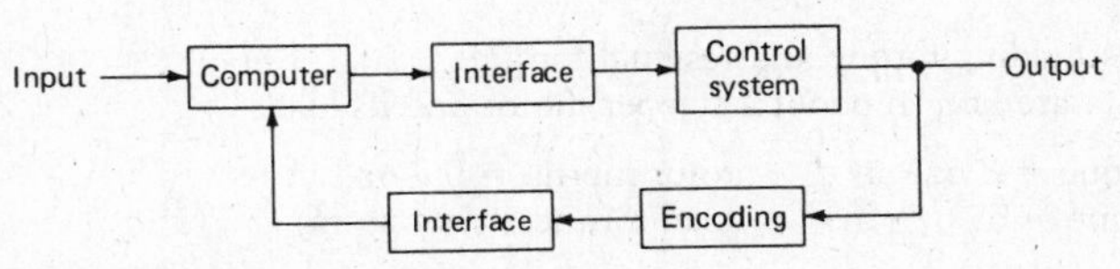

Fig. 9-14 On-line system.

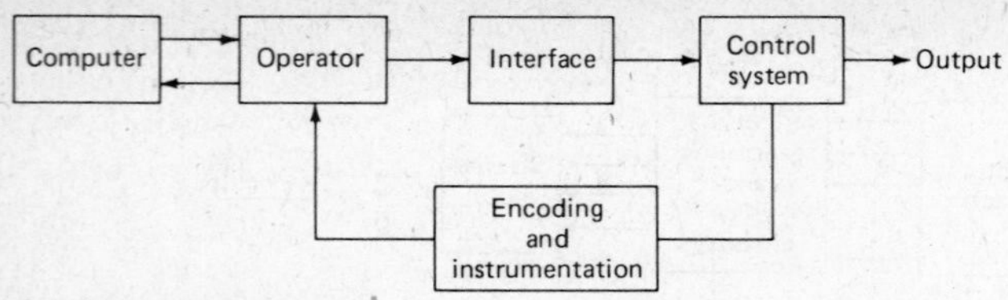

Fig. 9-15 Off-line system.

9-5 COMPARISON OF SYSTEM CONFIGURATIONS

Stepper motors always exhibit a damped oscillation (*ringing*) when commanded to a specific position or to perform continuous stepping. This ringing can be reduced to a great extent. One technique is to use a mechanical damper mounted on the shaft either internally or externally. Another technique utilizes an internal silicone viscous damping fluid with which a response close to critically damped can be obtained. In both cases the step response is increased.

Analog servos, on the other hand, can be electronically damped to eliminate overshoot and ringing completely. Furthermore, with the use of other compensation techniques (Chap. 7) an increase in step response can be virtually eliminated.

In addition, stepper motors are limited to approximately 2000 steps per second because of the limited power slew rate into the motor's field windings (inductive load). On the other hand, analog servos are limited in speed by the control-signal slew rate, which is much greater. They can obtain speeds on the order of 50,000 steps per second.

The main advantage of the stepper motor is the *cost* and *simplicity* of the system. If the limited speed, slight overshoot, and ringing are acceptable, an incremental system would be the choice over an analog system.

PROBLEMS

9-1. Give a general description of a stepper motor. Include in your discussion input, output, error, interfacing advantages, and its use as an open-loop controller.

9-2. A stepper motor makes 200 steps per revolution. It is running at 500 steps per second. Find (**a**) step angle, (**b**) speed in revolutions per minute, and (**c**) speed in radians per second.

9-3. Explain the terms holding torque and residual torque.

9-4. The following data are known about a stepper motor and its load:

Residual torque = 8 oz · in　　　rotor inertia = 0.5 oz · in · s^2
Holding torque = 65 oz · in　　　load torque = 50 oz · in

What is its torque-to-inertia ratio?

9-5. The motor of Prob. 9-2 has 25 rotor teeth. What switching sequence is it wired for?

9-6. A stepper motor wired for a four-step sequence has 50 rotor teeth. It is fed a 2-kHz pulse train. Find:
(a) Steps per revolution
(b) Step angle
(c) Speed in revolutions per minute
(d) Speed in radians per second

9-7. The motor of Prob. 9-6 is used in an open-loop position servomechanism. Describe the input necessary to make it turn 90° in 100 ms.

9-8. A stepper motor with a step angle of 0.9° is fed a 4-kHz pulse train for 50 ms. What angle will it turn through?

9-9. A stepper motor using a 12-V supply draws 2 A. If it takes 25 ms to reach 2 A after the voltage is switched on, find the motor's resistance and inductance.

9-10. The motor of Prob. 9-9 is used with a 50-V supply. Find:
(a) The external resistance necessary to limit the current to 2 A
(b) The time it will now take to reach 2 A

9-11. A stepper motor is to be connected directly to a load of 10 in · lb. The application requires a load resolution (step angle) of 0.9°. The load must be driven at 500 steps per second and must attain this speed in 250 ms. The load inertia is 1.6 in · lb · s^2. Assuming that the motor's rotor inertia is negligible, define the torque-speed requirement.

9-12. Repeat Prob. 9-11 if the motor is coupled to the load with gears. The gear ratio is equal to $\frac{1}{2}$.

9-13. A motor with a 2° step angle is to drive a 20-oz · in load through 360° in 1 s. No acceleration time is allowed. The motor is coupled directly to the load. The motor's rotor inertia is negligible. The load inertia is 1.4 oz · in · s^2. What is the motor's torque-speed requirement?

9-14. Repeat Prob. 9-13 if the motor is connected to the load with gears. The gear ratio is 2.

9-15. Explain the difference between incremental and absolute shaft encoders.

9-16. A 6-bit binary shaft encoder is turned to its 90° counterclockwise position. What should its output be? What is the resolution of the encoder?

9-17. A 4-bit Gray shaft encoder has the number 0101 at its output. Through what angle has it turned with respect to its zero? What is the encoder's resolution?

9-18. Describe the effects of an overloaded or obstructed solenoid if it is (a) ac type and (b) dc type?

9-19. What is the difference between on-line and off-line computer control systems?

chapter 10

Data Acquisition and Interface Components

Because of its speed and accuracy, the digital computer is a powerful tool in the area of control systems. It can be used to receive data and interpret it (off-line) or it can be used within a system (on-line) to receive data and automatically supply control signals. Most digital computers can understand and communicate in the digital language known as the binary code. In order to interface digital computers with the analog world, special components are necessary which can convert information from analog to digital form (A/D converter) and also from digital to analog form (D/A converter).

10-1 DIGITAL-TO-ANALOG CONVERTER

A D/A converter (also called a DAC) is a device which produces at its output terminal an analog voltage or current proportional to the digital signal present at its input. The input is supplied in parallel* form from a digital computer or any circuit producing an appropriate digital code. The digital code must have the same number of bits as the converter. Popular converters have from 4 to 18 bits.

* Parallel-data transfer is accomplished by using one line for every bit. The entire digital word is transmitted with one clock pulse, but eight lines are needed for an 8-bit word. Serial-data transfer uses only one line, but eight clock pulses are needed to transmit an 8-bit word, i.e., 1 bit at a time.

10-1-1 Definitions

Full-Scale Voltage (FSV) This represents the maximum possible output voltage for a D/A converter. It is the output that corresponds to the maximum input regardless of the input code used.

Resolution The resolution of a D/A converter is the smallest change it can produce in its output voltage. Ultimately this depends on the number of bits used for the input. For this reason a D/A converter with an 8-bit binary input is said to have a resolution of 8 bits. Resolution is also expressed in the following ways for an n-bit converter:

1. As a voltage, resolution is simply the voltage output corresponding to the least significant bit (LSB).

$$\text{Resolution} = \frac{E_R}{2^n} \quad \text{for } R\text{-}2R \text{ ladder*} \tag{10-1}$$

$$\text{Resolution} = E_R \frac{R_f}{R} \quad \text{for weighted resistor*} \tag{10-2}$$

where E_R is the analog reference voltage. It should be noted that in Eq. (10-1) and (10-2) the resolution in volts is the voltage corresponding to the LSB of the converter. For binary coded decimal† this is the LSB of the least significant digit (LSD).

2. As a decimal (1 part in 2^n)

$$\text{Resolution} = \frac{1}{2^n} \quad \text{for pure binary} \tag{10-3}$$

$$\text{or} \quad \text{Resolution} = \frac{1}{10^d} \quad \text{for BCD} \tag{10-4}$$

where d is the number of decimal digits.

3. As a percentage

$$\text{Resolution, \%} = \frac{1}{2^n}(100) \quad \text{for pure binary} \tag{10-5}$$

$$\text{or} \quad \text{Resolution, \%} = \frac{1}{10^d}(100) \quad \text{for BCD} \tag{10-6}$$

* Ladder and weighted-resistor networks are discussed in Sec. 10-1-2.
† Binary coded decimal (BCD) is a binary representation of a decimal number (see Fig. 10-6).

4. In parts per million (ppm)

$$\text{Resolution, ppm} = \frac{10^6}{2^n} \quad \text{for pure binary} \tag{10-7}$$

$$\text{Resolution, ppm} = \frac{10^6}{10^d} \quad \text{for BCD} \tag{10-8}$$

Monotonicity This property means that the output never decreases for an increase in the input; i.e., the slope of the input-output plot does not change polarity.

Linearity This specifies the maximum deviation from a best straight line drawn through the input-output plot. Standard linearity is $\pm\frac{1}{2}$LSB. Standard linearity guarantees that a converter is monotonic. It should be noted, however, that a monotonic converter does not have to meet standard linearity. Differential linearity is the maximum deviation between any two steps. For a linearity of $\pm\frac{1}{2}$LSB, the differential linearity would be one LSB.

Scale Factor This is the slope of a best straight line drawn through the input-output plot. The ideal scale factor (**ISF**) is equivalent to the resolution expressed in volts per step.

Accuracy This specifies the error in output voltage with respect to the theoretical output expressed as a percentage of FSV.

Settling Time Even though the analog output increases or decreases in steps, each step can have an overshoot and ensuing damped oscillation. The settling time is the time between application of the input and the output settling to within a specified band (generally its linearity, that is, $\pm\frac{1}{2}$LSB) of the final value.

Glitches These are high-frequency spikes (sometimes very large) which occur in the output for some changes in input. The turn-on and turn-off times of transistors (hence each bit) are different. When the binary input is 0011 and increasing, the next input will be 0100. If it is assumed that the turn-on time is greater than the turn-off time, the input will assume a temporary erroneous state of 0000. The converter output will jump toward 0 V until the input becomes 0100, at which point the output will rise to its proper value.

VDAC An abbreviation sometimes used for a D/A converter whose output is a voltage.

IDAC An abbreviation sometimes used for a D/A converter whose output is a current.

Multiplying D/A Converter (MDAC) If the reference voltage used for a D/A converter can be varied during its operation, the analog output is directly proportional to the product of the reference voltage and the digital input. Such a converter is called *a multiplying D/A converter* (MDAC).

MIDAC This refers to a MDAC whose output is a current.

Two-Quadrant Multiplication This refers to the fact that the output polarity in a MDAC can change. It can be done in two ways: In bipolar digital the digital input word controls output polarity (in this case offset binary is used), and in bipolar analog the analog reference input controls output polarity.

Four-Quadrant Multiplication This is a combination of the two forms of two-quadrant multiplication. In other words, output polarity is controlled by either the analog reference or by the offset binary digital input word.

10-1-2 Types of D/A Converters

Binary *R*-2*R* Ladder DAC. Figure 10-1 shows a 4-bit binary *R*-2*R* ladder converter. It only uses resistors of magnitude *R* and 2*R*. If better resolution is required, the number of bits must be increased. A grounded switch represents a 0 input, while a switch connected to the reference represents a 1 input. The MSB is the one closest to the OP-AMP and the LSB is the farthest. If the bits are numbered starting with the MSB, the MSB will be numbered 1 while the

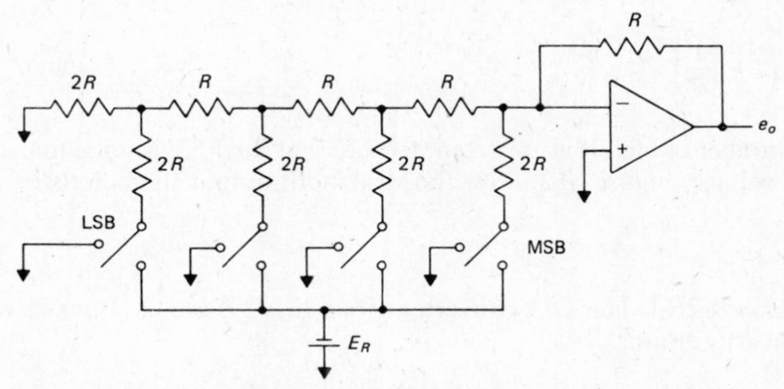

Fig. 10-1 Four-bit binary *R*-2*R* ladder D/A converter.

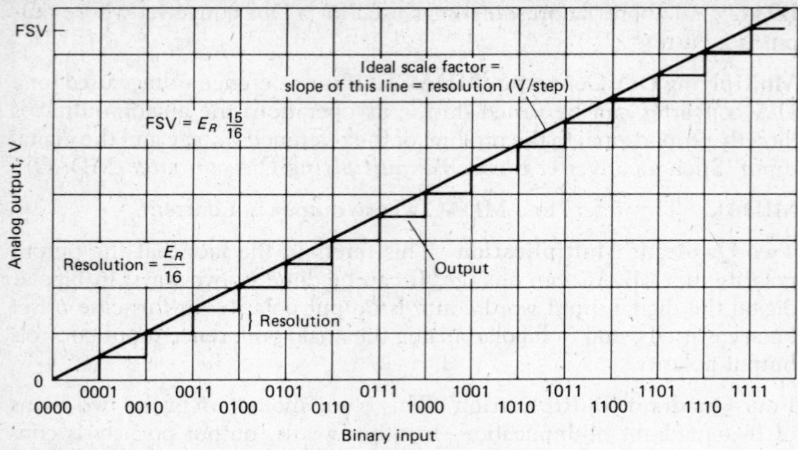

Fig. 10-2 Ideal input-output plot of a 4-bit D/A converter.

LSB will be numbered n for an n-bit converter. If all bits are 0, the output will be zero. For any bit, the output is given by

$$e_o = \frac{E_R}{2^n} \tag{10-9}$$

where n represents the bit number.

The output for more than one bit is the sum of the individual bit outputs. FSV, which corresponds to the condition of all inputs 1, is given by

$$\text{FSV} = \frac{E_R}{2^1} + \frac{E_R}{2^2} + \cdots + \frac{E_R}{2^n}$$

$$\text{or} \quad \text{FSV} = E_R\left(1 - \frac{1}{2^n}\right) \tag{10-10}$$

where n is the number of bits. It is important to note that the FSV is not equal to the reference voltage. Figure 10-2 shows the ideal input-output characteristic for a 4-bit DAC.

Example 10-1

An 8-bit binary R-$2R$ ladder D/A converter uses a 12-V reference. It meets standard linearity. Find:

(a) The FSV
(b) The resolution (in all forms)

(c) The ideal scale factor
(d) The output for the input 00110001
(e) The maximum deviation in volts that can be expected from the output in part (*d*)

Solution

(a) From Eq. (10-10)

$$FSV = 12\left(1 - \frac{1}{2^8}\right) = 11.95 \text{ V}$$

(b) From Eqs. (10-1), (10-3), (10-5), and (10-7), respectively,

$$\text{Resolution} = \frac{12}{2^8} = 0.047 \text{ V}$$

$$\text{or} \quad \text{Resolution} = 1 \text{ part in } 2^8 = \frac{1}{2^8} = \frac{1}{256} = 0.0039$$

$$\text{or} \quad \text{Resolution} = \frac{1}{2^8}(100) = 0.39\%$$

$$\text{or} \quad \text{Resolution} = \frac{10^6}{2^8} = 3906 \text{ ppm}$$

(c) Ideal scale factor = resolution (V/step)

$$ISF = 0.047 \text{ V/step}$$

(d) Using Eq. (10-9) for bit numbers 3, 4, and 8, we get

$$e_o = \frac{12}{2^3} + \frac{12}{2^4} + \frac{12}{2^8} = 1.5 + 0.75 + 0.047 = 2.297 \text{ V}$$

(e) The maximum deviation is specified by the linearity, which in this case is $\pm\frac{1}{2}$LSB. The voltage of the LSB also corresponds to the resolution in volts. Therefore, the maximum deviation is

$$\pm\tfrac{1}{2}(0.047) = \pm 0.0235 \text{ V}$$

and e_o can be 2.297 ± 0.0235 V.

Weighted-Current-Source D/A Converter. Figure 10-3 is a simplified diagram of this type of converter. In this case the output is a current (I_o).

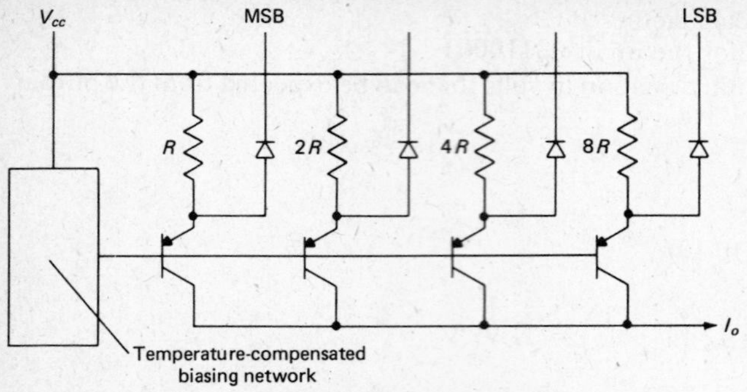

Fig. 10-3 Four-bit weighted-current-source D/A converter.

Current-output D/A converters are faster than voltage-output types since the output is not limited by the slew rate* of an OP-AMP. It should be noted that if a voltage is desired in Fig. 10-3, the output can be fed into an OP-AMP or tied through a resistor to ground. The former, as mentioned before, will slow down the response while the latter may introduce loading problems.

Weighted-Resistor DAC. Figure 10-4 shows this type of converter. Note that in this configuration the bits are weighted, by each having a different input resistance to the amplifier. The LSB has an input resistor equal to R. Each succeeding bit has an input resistor equal to one-half the previous bit. The output voltage, then, for each succeeding bit, will be twice the voltage for the preceding bit (see Sec. 2-15).

The output voltage for several bits is the sum of the outputs corresponding to each bit. The FSV is the output when all bits are on. It is given by

$$\text{FSV} = E_R \frac{R_f}{R}(2^n - 1) \tag{10-11}$$

Example 10-2

A 4-bit weighted-resistor DAC is shown in Fig. 10-5. It has standard linearity. Find:

(a) The FSV
(b) The resolution (in all forms)
(c) The ISF
(d) The input when the output is 4.4 V

* Slew rate is the maximum rate at which the output of an OP-AMP can change. It is usually specified in volts per microsecond.

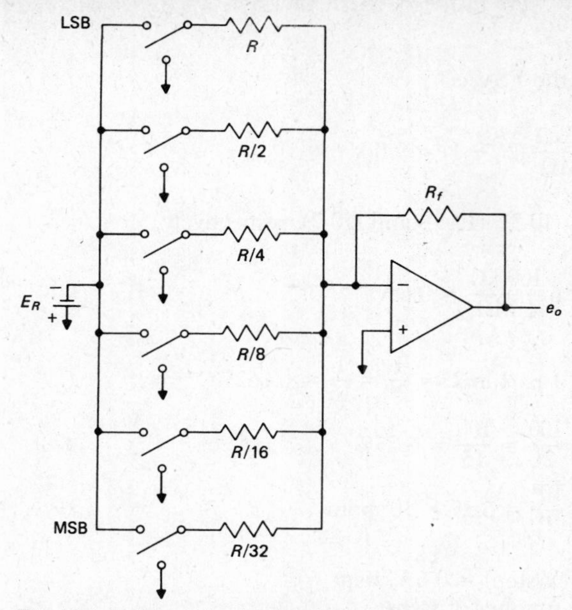

Fig. 10-4 Six-bit weighted-resistor D/A converter.

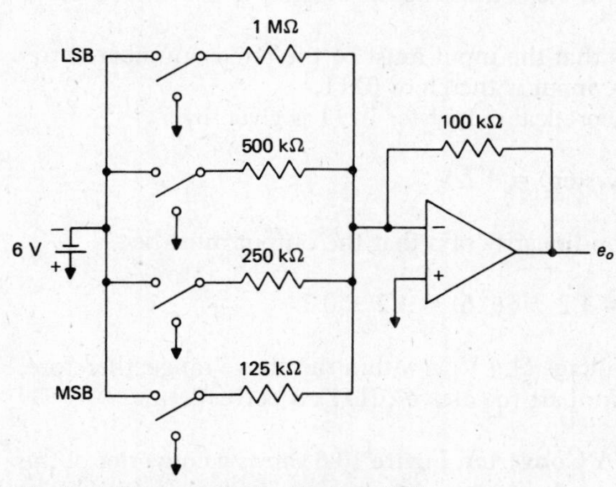

Fig. 10-5 D/A converter for Example 10-2.

Solution

(a) From Eq. (10-11) the FSV is

$$\text{FSV} = 6\frac{100\ \text{k}\Omega}{1\ \text{M}\Omega}(2^4 - 1) = 9.0\ \text{V}$$

(b) From Eqs. (10-2), (10-3), (10-5), and (10-7), respectively,

$$\text{Resolution} = 6\frac{100\ \text{k}\Omega}{1\ \text{M}\Omega} = 0.6\ \text{V}$$

$$\text{or}\quad \text{Resolution} = 1 \text{ part in } 2^4 = \frac{1}{2^4} = \frac{1}{16} = 0.0625$$

$$\text{or}\quad \text{Resolution} = \frac{100}{2^4} = \frac{100}{16} = 6.25\%$$

$$\text{or}\quad \text{Resolution} = \frac{10^6}{2^4} = 6.25 \times 10^4\ \text{ppm}$$

(c) ISF = resolution (V/step) = 0.6 V/step

(d) Since e_o = ISF × number of steps

$$\text{Number of steps} = \frac{e_o}{\text{ISF}} = \frac{4.4}{0.6} = 7.333$$

Since the number of steps must be an integer, it is rounded off to seven steps.

Seven steps means that the input must be the binary number corresponding to 7. The input is therefore **0111**.

As a check, the theoretical output for **0111** is given by

$$e_o = (7\ \text{steps})(0.6\ \text{V/step}) = 4.2\ \text{V}$$

The specification on linearity says that the output must be

$$e_o = 4.2 \pm \tfrac{1}{2}\text{LSB} = 4.2 \pm \tfrac{1}{2}(0.6) = 4.2 \pm 0.3$$

The given output voltage (4.4 V) is within the above range; therefore, the answer found for part (*d*) above (**0111**) is correct.

***R*-2*R* Ladder BCD D/A Converter.** Figure 10-6 shows a converter of this type. In the decimal number system the largest value for any digit is 9. The

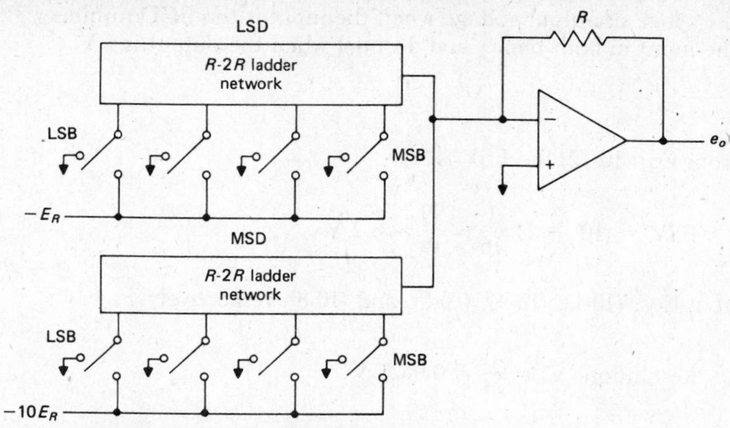

Fig. 10-6 Two-digit R-2R ladder BCD D/A converter.

binary number **1001** is equal to 9, therefore 4 bits are needed to represent each decimal digit (3 bits can only go as high as **111**, which is the decimal number 7). The converter in Fig. 10-6 has two 4-bit R-2R ladders. Its input is a two-digit BCD number. The largest decimal input is therefore 99.

If the largest number is 99, there are 100 possible states (0 to 99) the input can assume. If the input was pure binary, the total number of states would be $2^8 = 256$. It is for this reason that better resolution is obtained for a given number of bits using pure binary as opposed to BCD [see Eqs. (10-3) and (10-4)].

Referring to Fig. 10-6, we can see that the digits are weighted by increasing the analog reference by a factor of 10. If two identical binary numbers are fed into each of the ladder networks, the output will be the sum of two voltages. The one due to the MSD will be 10 times greater than that due to the LSD.

The FSV for an R-2R ladder BCD D/A converter is given by

$$\text{FSV} = (10^d - 1)\frac{E_R}{16} \tag{10-12}$$

where d is the number of decimal digits.

Example 10-3

A two-digit R-2R ladder BCD D/A converter meets standard linearity. Its analog reference voltage E_R is 1 V. Find:

(a) The FSV
(b) The resolution (in all forms)

(c) The range of output voltage when the input is the BCD number 73
(d) The input in both binary and decimal when the output is 3 V

Solution

(a) From Eq. (10-12) the FSV is

$$\text{FSV} = (10^2 - 1)\frac{1}{16} = \frac{99}{16} \approx 6.2 \text{ V}$$

(b) From Eqs. (10-1), (10-4), (10-6), and (10-8), respectively,

$$\text{Resolution (V)} = \frac{1}{2^4} = 0.0625 \text{ V}$$

$$\text{or} \quad \text{Resolution} = \frac{1}{10^2} = 0.01$$

$$\text{or} \quad \text{Resolution} = \frac{100}{10^2} = 1\%$$

$$\text{or} \quad \text{Resolution} = \frac{10^6}{10^2} = 10{,}000 \text{ ppm}$$

(c) The output is the input multiplied by the scale factor $\pm\frac{1}{2}$LSB

$$e_o = 73\ (0.0625 \text{ V/step}) \pm \tfrac{1}{2}(0.0625) = 4.56 \pm 0.0313 \text{ V}$$

(d) The input is the output divided by the scale factor

$$\text{Input} = \frac{3 \text{ V}}{0.0625 \text{ V/step}} = 48 \text{ steps}$$

The input is the decimal number 48. The binary input is the equivalent of 4 for the MSD and 8 for the LSD

$$\underbrace{0\ 1\ 0\ 0}_{\text{MSD}} \quad \underbrace{1\ 0\ 0\ 0}_{\text{LSD}}$$

Weighted-Resistor BCD D/A Converter. The weighted-resistor BCD converter is similar in construction to the weighted-resistor converter for pure binary (see Fig. 10-4). It has a basic difference due to its coded input. The converter must have 4 bits for every decimal digit it converts.

In Fig. 10-7 the bank of resistors connected to E_R represents the LSD of the two-digit decimal number (the ones digit). The bank of resistors connected to $10E_R$ represents the MSD of the two-digit decimal number (the tens digit). By connecting each succeeding digit to a reference 10 times greater than the previous one, each digit will produce an output 10 times greater than the previous one. This is the identical property of the decimal number system [30 = 10(3)].

For the converter in Fig. 10-7 the FSV is given by

$$\text{FSV} = (10^d - 1)E_R \frac{R_f}{R} \tag{10-13}$$

where d is the number of decimal digits.

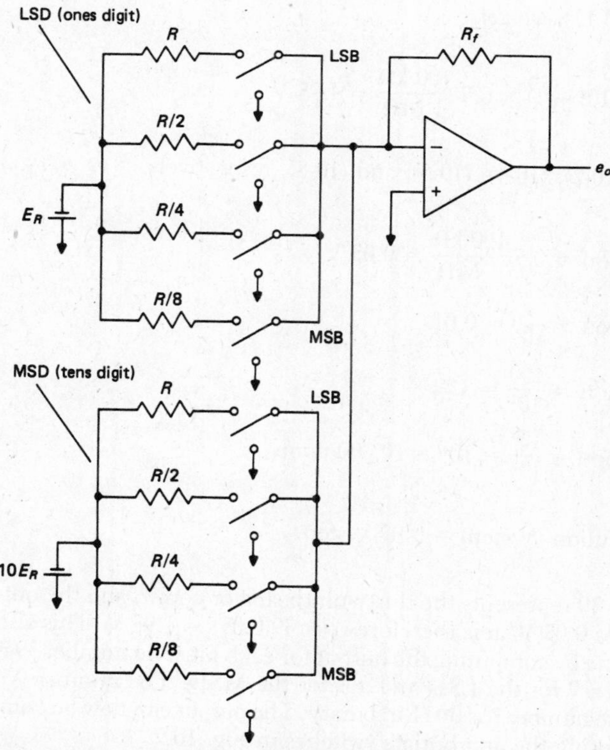

Fig. 10-7 Two-digit weighted-resistor BCD D/A converter.

Example 10-4

A two-digit weighted-resistor BCD D/A converter meets standard linearity. It is configured as in Fig. 10-7 with $E_R = 0.5$ V, $R = 1$ MΩ, and $R_f = 100$ kΩ. Find:

(a) The FSV
(b) The resolution (in all forms)
(c) The ISF
(d) The output for the decimal input 39
(e) The maximum deviation in volts that can be expected from the output in part (*d*).

Solution

(a) Using Eq. (10-13), we get

$$\text{FSV} = (10^2 - 1) \times 0.5 \frac{100 \text{ k}\Omega}{1 \text{ M}\Omega} = 4.95 \text{ V}$$

(b) From Eqs. (10-2), (10-4), (10-6), and (10-8), respectively,

$$\text{Resolution} = 0.5 \frac{100 \text{ k}\Omega}{1 \text{ M}\Omega} = 0.05 \text{ V}$$

$$\text{or} \quad \text{Resolution} = \frac{1}{10^2} = 0.01$$

$$\text{or} \quad \text{Resolution} = \frac{100}{10^2} = 1\%$$

$$\text{or} \quad \text{Resolution} = \frac{10^6}{10^2} = 10^4 = 10{,}000 \text{ ppm}$$

(c) ISF = resolution (V/step) = 0.05 V/step

(d) the number 39 represents the thirty-ninth step or count, and the output increases 0.05 V/step; therefore $e_o = 39(0.05) = 1.95$ V. This also could be done by computing the output for each bit. The number 39 is made up of a 9 for the LSD and a 3 for the MSD. The number 9 is 1001 and the number 3 is 0011 in binary. The output can now be computed by closing the appropriate switches in Fig. 10-7:

$$e_o = 10E_R \overbrace{\left(\frac{R_f}{R} + \frac{R_f}{R/2}\right)}^{3} + E_R \overbrace{\left(\frac{R_f}{R} + \frac{R_f}{R/8}\right)}^{9}$$
$$= 5\left(\frac{100\ \text{k}\Omega}{1\ \text{M}\Omega} + \frac{100\ \text{k}\Omega}{0.5\ \text{M}\Omega}\right) + 0.5\left(\frac{100\ \text{k}\Omega}{1\ \text{M}\Omega} + \frac{100\ \text{k}\Omega}{0.125\ \text{M}\Omega}\right)$$
$$= 5(0.1 + 0.2) + 0.5(0.1 + 0.8)$$
$$= 5(0.3) + 0.5(0.9) = 1.5 + 0.45 = 1.95\ \text{V}$$

(e) the maximum deviation in volts is given by the linearity in volts

$$\text{LSB of LSD} = E_R \frac{100\ \text{k}\Omega}{1\ \text{M}\Omega}$$
$$\text{LSB} = 0.5(0.1) = 0.05$$

Note that this is the same as the resolution in volts:

$$\pm\tfrac{1}{2}\text{LSB} = \pm 0.025\ \text{V}$$

Thus $e_o = 1.95 \pm 0.025$ V

The weighted-resistor converter shown in Fig. 10-7 can be configured so that only one analog reference is needed. This is accomplished by making each input resistor for the tens digit one-tenth the resistance of the corresponding resistor for the ones digit. In this manner the output for a given bit in the tens digit will be 10 times the output for the corresponding bit in the ones digit. The circuit is shown in Fig. 10-8. The disadvantage of this converter is the large number of different resistors required.

Companding D/A Converter. The companding converter uses a special digital code that enables it to convert signals having a large dynamic range. The difference between this converter and those already discussed is that it requires only 8 bits to achieve the range of 13 bits in a standard D/A converter while maintaining a small resolution. The resolution in this case is a function of the input rather than the reference voltage. This is of importance in communications and control systems which are interfaced with 8-bit microprocessors. The input-output relationship is logarithmic instead of linear. It is a piecewise linear approximation of Eq. (10-14), which is a statement of the Bell μ-255 logarithmic law

$$e_o = 0.18 \ln (1 + 255 e_{\text{norm}}) \tag{10-14}$$

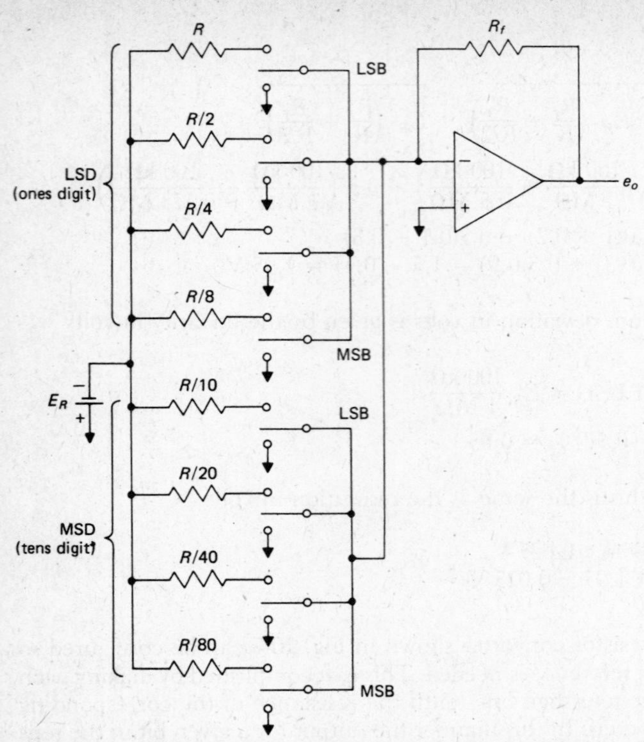

Fig. 10-8 Two-digit weighted-resistor BCD D/A converter with one analog reference.

where e_o is the analog output voltage and e_{norm} is the normalized input-signal level, i.e., the input divided by the full-scale value of the input

$$-1 < e_{\text{norm}} < 1$$

The 8 bits are used as follows:

1. One bit is used for the sign.
2. Three bits are used to denote eight (000 to 111) different chords.
3. Four bits are used to denote 16 (0000 to 1111) different steps on each chord.

In the first chord (000) each of the 16 steps has a specific value, say 1 mV. In the second chord (001) each of the 16 steps now has the value of 2 mV; in the third chord (010) each step is 4 mV; and so forth. The eighth chord (111) has a step size of 128 mV. The input-output plot is shown in Fig. 10-9.

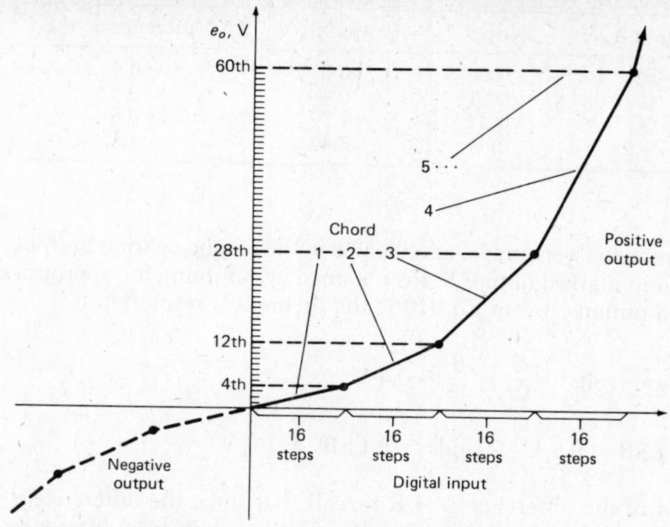

Fig. 10-9 Input-output plot for a companding D/A converter.

The D/A converters discussed here are but a few of the many types available. They are a representative sample, however, and illustrate the calculations used in D/A conversion.

10-1-3 Testing D/A Converters

The two parameters which are most important to check on a D/A converter are the settling time and its accuracy.

The most complete test for accuracy would be to test each of the output levels and determine the scale factor from the best straight line drawn through them. This becomes very tedious, however, as the number of bits increases. For an 8-bit converter there are $256(2^n)$ levels to be checked. A more practical test is one which simply checks the linearity and monotonicity of the converter. This can be accomplished by making n tests for an n-bit converter. It involves comparing the output for a given bit with the sum of the outputs of all the less significant bits. As long as the difference in this comparison is less than two LSBs the converter meets standard linearity and hence monotonicity. As long as the difference is always positive, the converter is monotonic but not necessarily linear. The latter test supersedes the former test.

Example 10-5

A 4-bit D/A converter with $E_R = 8$ V uses an R-$2R$ ladder network. It is tested for accuracy and the results are:

Input	Output A, V	Input	Output B, V	Difference A − B
0001	0.4	0000	0.0	0.4
0010	0.9	0001	0.4	0.5
0100	2.1	0011	1.3	0.8
1000	4.5	0111	3.4	1.1

Solution

Four measurements were made, as indicated in the second column (output A). The column marked output B was obtained by summing the appropriate measured outputs. From Eq. (10-1) the theoretical resolution is

$$\text{Resolution (volts)} = \frac{8}{2^4} = \frac{8}{16} = 0.5 \text{ V}$$

Therefore LSB = 0.5 V and 2 LSBs = 1.0 V

Examination of the difference A − B reveals that since the difference is always positive, the converter is monotonic and that since a difference exists (1.1 V) which is greater than two LSBs (1.0 V), the converter does not meet standard linearity.

Testing the settling time is not quite as simple. For a large number of bits this requires measuring a very small change in output (millivolts) when the output itself can be as much as 10 V. The time for this change is on the order of nanoseconds for fast DACs. Insofar as the test can be performed, it is not simple enough to be briefly covered, but it should be noted that worst-case tests are usually made. These tests include transitions from 000 · · · to 111 · · · and 0111 · · · to 1000 · · · . The latter consists of worst-case glitch which may lengthen the settling time.

As a final note, it should be mentioned that there are elaborate test systems available for D/A converters. Depending on the application of the converter and its environment, the cost of such a test system may or may not be justified.

10-2 ANALOG-TO-DIGITAL CONVERTER

An A/D converter is a device that produces at its output a digital signal, usually in binary, proportional to the analog voltage present at its input. A/D converters are much more complicated in design and operation than D/A converters. The output of a D/A converter follows changes in its input almost instantaneously. The A/D converter, on the other hand, takes a finite amount of time to perform a conversion. During this time the input to the converter must be held constant for an accurate conversion to take place. Many A/D converters use a D/A converter within them to perform the conversion. Because they are more complex, A/D converters cost more than comparable D/A converters.

10-2-1 Definitions

Quantization Quantization is the process of taking a continuous (analog) signal and breaking it up into a number of discrete steps. The binary code can be used to identify each of these discrete output steps. An A/D converter performs the operation of quantization.

Resolution The resolution of an A/D converter is the smallest change in its input (analog voltage) that can produce a change in its output (digital code). It is a function of the largest analog input it can convert (FSV) and the number of discrete steps the output can have. It is numerically equal to the voltage of the LSB and is given by

$$\text{Resolution (V)} = \begin{cases} \dfrac{\text{FSV}}{2^n - 1} & \text{for } n\text{-bit binary converter} \quad (10\text{-}15) \\ \dfrac{\text{FSV}}{10^d - 1} & \text{for } d\text{-digit BCD converter} \quad (10\text{-}16) \end{cases}$$

Quantizing Error The quantizing error is the difference between the analog input and the digital output. As the input to an A/D converter is varied, the quantizing error changes. An ideal A/D converter has a maximum quantizing error of one-half the LSB, which corresponds to one-half the resolution in volts. As an illustration consider a 4-bit converter with a resolution of 1 V. For an input of 0 V the output is 0000, which is zero error. As the input increases to 0.5 V, the output will jump to 0001; the error is now 0.5 V. As the input increases to 1.4 V, the output stays at 0001. During this time the error goes from 0.5 V to zero (when the input is 1 V) to −0.4 V. A plot of the quantizing error and output vs. input is shown in Fig. 10-10.

Offset Error This error represents the amount by which a best straight line is displaced from the origin along the analog axis (see Fig. 10-10). It can be adjusted to zero in some converters.

Gain Error Sometimes referred to as *scale-factor error*, the gain error is the difference in the slope of a best straight line from the ISF. It, too, is externally adjustable in many converters and is shown in Fig. 10-10.

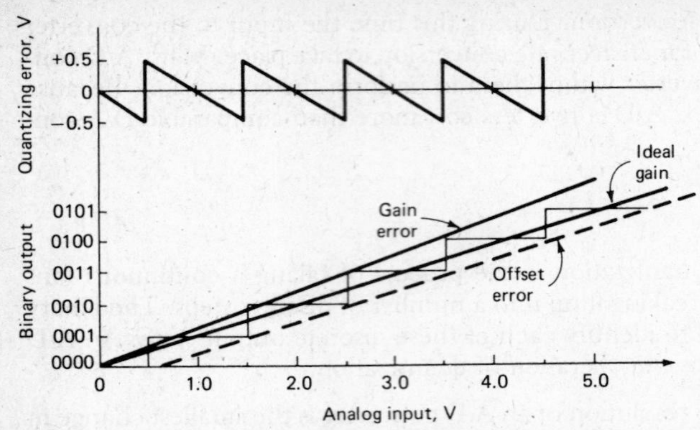

Fig. 10-10 Plots of output and quantizing error vs. input for an A/D converter.

Linearity As with the D/A converter, linearity refers to maximum deviation from a best straight line drawn through the output points of the characteristic. Standard linearity is a linearity of $\pm\frac{1}{2}$LSB.

Differential Linearity This represents the maximum deviation between any two successive steps. In other words it is the maximum deviation of the actual bit size from its theoretical value.

Monotonicity This has the same meaning as with the D/A converter; the output never decreases for an increase in input.

Missing Code This will occur in some A/D converters when the output skips a particular step. It happens if the differential linearity is ever greater than one LSB for any given bit.

Conversion Time This refers to the amount of time required for an A/D converter to produce a digital output corresponding to the analog input. In some converters it is a function of the input; therefore when specified it refers to the maximum conversion time. Some manufacturers specify the conversion time per bit. In this case the actual conversion time as defined here would be the number of bits of the converter times the conversion time per bit.

Conversion Rate This is a figure of merit for A/D converters and represents the number of conversions per second a particular converter can perform. It is numerically equal to the reciprocal of the conversion time. Since conversion times are sometimes a function of the input, the conversion rate (when specified) refers to the minimum conversion rate.

Example 10-6

The digital air-data computer of a jumbo jet aircraft senses barometric air pressure to determine the plane's altitude. To do this an A/D converter is used to interface the computer with a pressure transducer. The plane flies at altitudes ranging from sea level to 40,000 ft. The barometric air pressure thus varies from 14.7 lb/in^2 (at sea level) to 3.6 lb/in^2 (at 40,000 ft). If the gain of the pressure transducer is 600 mV/(lb/in^2) and the computer must sense changes in altitude of 5 ft, find

(a) The number of bits required for the A/D converter
(b) The FSV of the converter
(c) The resolution of the converter in volts

Solution

(a) Since the total dynamic range of altitude is 40,000 ft and the computer must sense changes in altitude of 5 ft, the A/D converter must have a minimum resolution of 8000 (40,000/5) codes (or steps). The question must be asked: How many bits will provide at least 8000 steps? If $n = 13$, $2^{13} = 8192$ steps, which is the requirement for this application. The actual resolution will be

$$\frac{40{,}000 \text{ ft}}{8192 \text{ steps}} = 4.88 \text{ ft/step}$$

The converter will be able to resolve this to ± 2.44 ft($\pm\frac{1}{2}$LSB). An A/D converter with 14 bits will be selected. The extra bit will be a sign bit and the binary code used could be two's complement. In this fashion altitudes below sea level can be sensed.

(b) If we assume for the purpose of this calculation that the maximum air pressure sensed will be that at sea level (14.7 lb/in^2), the FSV will be

$$\text{FSV} = [600 \text{ mV/(lb/in}^2)](14.7 \text{ lb/in}^2) = 8.82 \text{ V}$$

(c) Using Eq. (10-15), we find the resolution to be

$$\text{Resolution} = \frac{8.82}{2^{13} - 1} = \frac{8.82}{8191} = 1.08 \text{ mV} \approx 1 \text{ mV}$$

The pressure transducer must have a resolution at least as good as the required resolution of 5 ft. Pressure transducers are available that can sense a pressure change resulting from a change in altitude of $\frac{2}{10}$ ft.

10-2-2 Types of A/D Converters

Dual-Slope (or Dual-Ramp) Integrating Type. The many A/D converters of the integrating type all inherently convert the analog voltage by sensing the time required for an integrator to ramp up and down. The one discussed here is called a Dual-slope type. Its operation can best be described by referring to Figs. 10-11 and 10-12.

When the start-conversion pulse is received, the switch goes to position 1, causing the integrator to start ramping up. The moment the integrator output exceeds the comparator threshold, the control logic causes the counter to start counting up. When the count reaches a specified number N_1, it is reset and the switch is thrown to position 2. In position 2 a reference voltage E_R whose polarity is opposite to the input now causes the integrator to ramp down. At the same time that the switch was thrown, the counter started counting up again. When the integrator output (now decreasing) crosses the comparator threshold, the counter is stopped and its output (N_2 counts) is transferred to the storage-register output. The following points should be noted.

1. As long as the end-of-conversion (EOC) signal is high, the converter is busy and its input should be held constant.
2. When the data-valid signal is high, the storage-register output contains the results of the previous conversion. During the time the data-valid signal is low, the results of the conversion just completed are being transferred to the storage-register output. Therefore, during this interval the data is invalid (changing) and should be ignored.

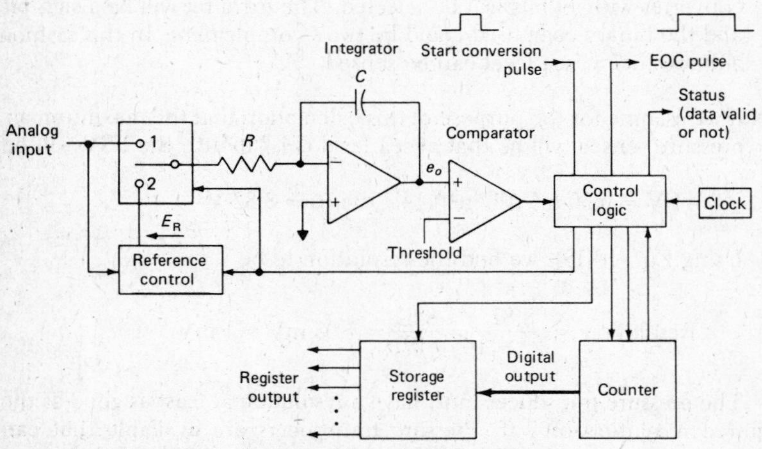

Fig. 10-11 Schematic of a dual-slope integrating A/D converter.

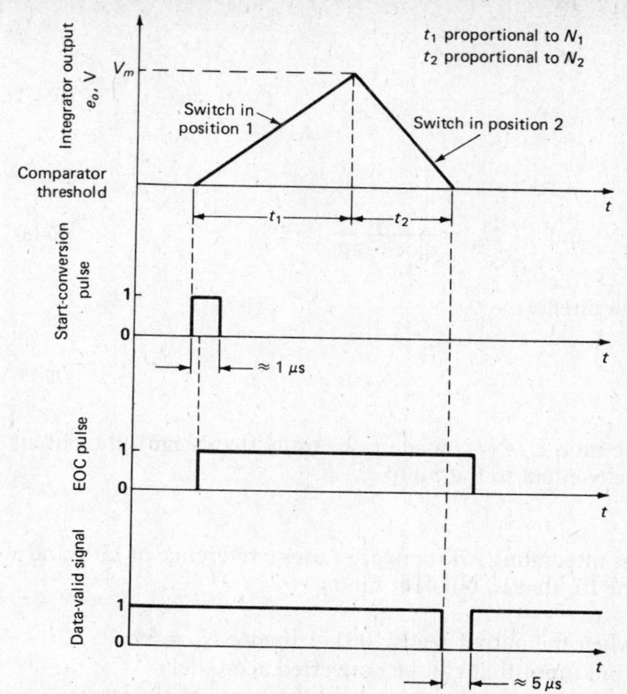

Fig. 10-12 Typical timing diagram of a dual-slope integrating A/D converter.

The following equations can be written for the integrator output.

For the interval t_1

$$e_o = V_m = e_i \frac{1}{RC} t_1 \qquad \text{(where } e_i \text{ is the analog input)}$$

For the interval t_2

$$e_o = V_m = E_R \frac{1}{RC} t_2$$

Equating the two above equations gives

$$e_i \frac{1}{RC} t_1 = E_R \frac{1}{RC} t_2$$

and solving leads to

$$e_i = E_R \frac{t_2}{t_1} \tag{10-17}$$

Since

$$N_1 = \frac{t_1}{\text{clock rate}} \qquad \text{and} \qquad N_2 = \frac{t_2}{\text{clock rate}} \tag{10-18}$$

Eq. (10-17) can be rewritten as

$$e_i = E_R \frac{N_2}{N_1} \tag{10-19}$$

Notice that if the ratio E_R/N_1 is made to be unity (by design), the output count N_2 is numerically equal to the input e_i.

Example 10-7

A 6-bit dual-slope integrating A/D converter uses a reference of 12 V and a fixed count N_1 of 10, that is, 001010. Find:

(a) The input when the output reads **100111** (hence $N_2 = 39$)
(b) The maximum input that can be converted accurately
(c) A technique that can be used to extend the range of the input

Solution

(a) From Eq. (10-19)

$$e_i = 12\left(\tfrac{39}{10}\right) = 46.8 \text{ V}$$

(b) Again using (10-19), we get

$$e_{i,\max} = E_R \frac{N_{2,\max}}{N_1} \tag{10-20}$$

Since $N_{2,\max}$ = **111111** = decimal 63,

$$e_{i,\max} = 12\left(\tfrac{63}{10}\right) = 75.6 \text{ V}$$

(c) If the fixed count N_1 is reduced to 5 as an illustration, then using Eq. (10-20), we have

$$e_{i,\max} = 12(\tfrac{63}{5}) = 151.2 \text{ V}$$

The reference can also be increased, but this is frequently impractical.

Example 10-8

For the converter of Example 10-7 find the conversion time and the conversion rate if the clock rate is 1 MHz.

Solution

Here the conversion time is a function of the input. The longest conversion time occurs for the largest input. Therefore $N_1 = 10$ and $N_2 = 63$. Using Eq. (10-18), we get

$$t_1 = \frac{10}{1 \text{ MHz}} = 10\ \mu s \qquad t_2 = \frac{63}{1 \text{ MHz}} = 63\ \mu s$$

The time required to transfer N_2 to the output register is about 5 μs. The total conversion time is the sum of the three times calculated, or

$$78\ \mu\text{s} \approx 80\ \mu\text{s}$$

The conversion rate is the reciprocal of the conversion time.

$$\text{Conversion rate} = \frac{1}{80\ \mu\text{s}} = 12{,}500 \text{ conversions/s}$$

This type of converter has the advantage of being relatively simple, inexpensive, very accurate, and having good noise rejection. Its accuracy results from the fact that its output is not affected by resistance and/or capacitance variations which occur over long periods of time. Its main disadvantage is its slow conversion time. For this reason it is popular for use in digital voltmeters or panelmeters, where faster conversion times are not necessary.

Counter-Type A/D Converter. A simplified diagram of a typical counter-type converter is shown in Fig. 10-13. It can be seen from the figure that this type of converter includes an internal D/A converter that determines the number of output bits (hence the resolution) and the full-scale input of the A/D converter.

When the start-conversion pulse is applied, the control logic starts the counter by gating the clock. At the same time the status signal (EOC pulse) goes high, indicating that a conversion has begun. As the count increases, the D/A converter produces an analog voltage proportional to the count. When the D/A output equals the analog input or exceeds it, the comparator trips. The

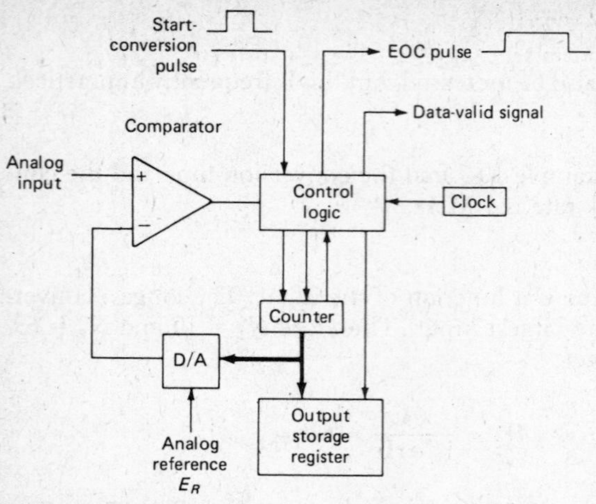

Fig. 10-13 Simplified diagram of a counter-type A/D converter.

control logic senses this, stops the counter, sets the data-valid signal low, and dumps the count onto the output register. The output register now contains a digital number equivalent to the analog input. At this point the EOC pulse goes low, the data-valid signal goes high, and the counter is reset. The analog input is now free to change. It must be noted that during the time when the EOC pulse is high (indicating a conversion in progress) the analog input must be held constant.

The largest analog input is determined by the D/A converter output. This in turn depends on the analog reference and the number of output bits. The longest conversion time would take place when the input was of full-scale value. In this case the counter would have to count up from 000 · · · to 111 · · · (all 1s). To do this, it requires $2^n - 1$ pulses from the clock. Therefore the total time required is

$$\text{Conversion time} = \frac{2^n - 1}{\text{clock rate}} \approx \frac{2^n}{\text{clock rate}} \tag{10-21}$$

Example 10-9

A 6-bit A/D converter of the counter type uses a 6-bit R-$2R$ ladder D/A converter with a reference of 10 V internally and a 1-MHz clock. Determine for the A/D converter:

(**a**) The full-scale input
(**b**) The resolution

(c) The conversion time
(d) The conversion rate

Solution

(a) The FSV of the A/D is determined by the FSV of the D/A used internally. Using Eq. (10-10), we have

$$\text{FSV} = 10\left(1 - \frac{1}{2^6}\right) = 10(0.9844) = 9.844 \text{ V}$$

(b) From Eq. (10-15) the resolution in volts is

$$\text{Resolution} = \frac{9.844}{2^6 - 1} = 0.156 \text{ V}$$

(c) From Eq. (10-21) the conversion time is

$$\text{Conversion time} = \frac{2^6}{1 \text{ MHz}} = \frac{64}{1 \text{ MHz}} = 64\ \mu\text{s}$$

(d) Since the conversion rate is the reciprocal of conversion time,

$$\text{Conversion rate} = \frac{1}{64\ \mu\text{s}} = 15{,}625 \text{ conversions/s}$$

This converter is somewhat faster than the dual-slope type, but it is limited in the magnitude of the input it can convert. Although relatively inexpensive and simple to build, it is a slow converter. In some applications it can be made to convert faster by making the counter an up-down counter. Here the counter is not reset after a conversion but is left in its previous state and either counts up or down depending on the magnitude of the next input. This is called a *servo-type* converter since feedback is used to direct the counter. Although on the average conversion times are shorter, the maximum conversion time is still the same as with the counter type.

Successive-Approximation Type. The successive-approximation type is the most widely used A/D converter. It is fast, having a fixed conversion time independent of the input. Because of its operation, data can be taken from it either in serial or parallel form. A simplified schematic is shown in Fig. 10-14 with a typical conversion shown in Fig. 10-15.

As soon as the start-conversion pulse is received (the leading edge), the MSB of the D/A is switched on, the EOC pulse goes high (indicating a busy

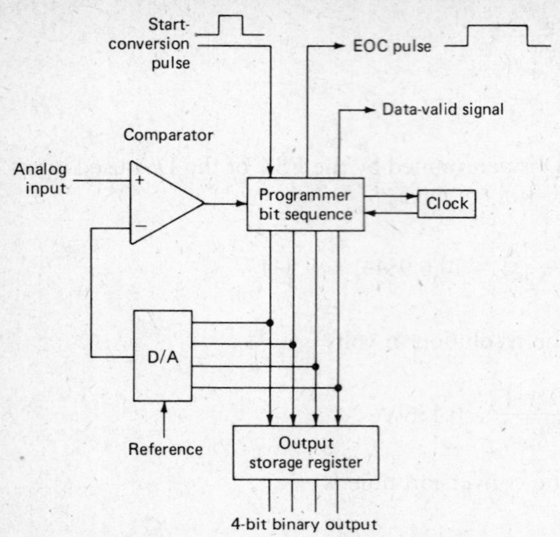

Fig. 10-14 Simplified schematic of a 4-bit successive-approximation A/D converter.

status), and the clock is gated to the bit sequencer. When the first clock pulse arrives after conversion has begun, two events happen simultaneously:

1. The second most significant bit is switched on.
2. The decision has been made whether or not to leave the MSB on. If the MSB was greater than the analog input, the decision is to turn it off. If the MSB was less than the analog input, the decision is to leave the MSB on. (The latter is the case in Fig. 10-15.)

When the second clock pulse arrives, two similar events happen simultaneously:

1. The third most significant bit is switched on.
2. The decision has been made whether or not to leave the second most significant bit on. In Fig. 10-15 the decision to turn it off was made since the digital output was greater than the analog input.

This procedure continues for each bit until the fourth clock pulse arrives. At this point there are no bits to turn on, and the decision is made whether or not to leave the LSB on or turn it off. Once this decision is made, the conversion is complete and it can be dumped onto a register.

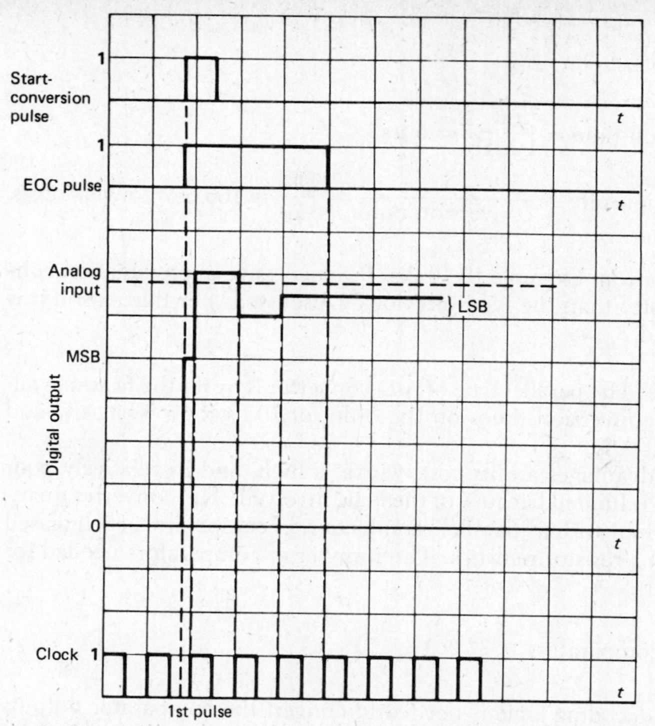

Fig. 10-15 Typical waveforms for a 4-bit successive-approximation A/D converter.

It is very important to note that when the data is to be taken serially, it is not to be accepted until it is deemed valid. This occurs at the leading edge of each clock pulse, i.e., MSB valid at leading edge of first pulse, second bit valid at leading edge of second pulse, etc.

It can be seen from the above discussion that the conversion time is approximately the same as the time required for four pulses for a 4-bit converter. In general, where n is the number of bits of the A/D converter

$$\text{Conversion time} = \frac{n}{\text{clock rate}} \tag{10-22}$$

Example 10-10

A 6-bit successive-approximation A/D converter uses a 1-MHz clock. Determine the conversion time and conversion rate.

Solution

Using Eq. (10-22), we have

$$\text{Conversion time} = \frac{6}{1\ \text{MHz}} = 6\ \mu\text{s}$$

$$\text{Conversion rate} = \frac{1}{\text{conversion time}} = \frac{1}{6\ \mu\text{s}} = 166{,}667\ \text{conversions/s}$$

It can be seen from Example 10-10 that the successive-approximation converter is much faster than the types previously discussed. For this reason it is generally preferred.

Parallel Type. The parallel type of A/D converter is by far the fastest available. It achieves conversion times on the order of 20 ns for a 3-bit A/D and 100 ns for an 8-bit A/D.

Its main disadvantages are its cost, which is high, and its relatively poor resolution, which is limited because of the logic involved. The converter quantizes the analog input with a "parallel" connection of comparators each biased one LSB apart by a resistor network. The number of comparators needed for an n-bit converter is

$$\text{Number of comparators} = 2^n - 1 \tag{10-23}$$

In addition, decoding logic is needed to convert the comparator outputs into binary or whatever digital output code is desired. The schematic of a 3-bit converter is shown in Fig. 10-16. When the analog input is applied, all comparators biased below the input will turn on (1). All comparators biased above the input will stay in the off (0) state. In Fig. 10-16 comparator 1 is biased at $(R/16R)E_R$ or $E_R/16$ V. Comparator 2 is biased at $(3R/16R)E_R$ or $3E_R/16$ V and so on up to comparator 7, which is biased at $13E_R/16$ V.

The resolution is given by the voltage corresponding to the LSB, which from Fig. 10-16 is

$$\text{Resolution (V)} = \frac{2R}{16R}E_R \quad \text{or} \quad \text{Resolution} = \frac{E_R}{8}$$

$$\text{or} \quad \text{Resolution} = \frac{E_R}{2^n} \tag{10-24}$$

Comparing Eq. (10-24) with (10-15), it is evident that the full-scale input for this converter is equal to the expression in Fig. 10-16.

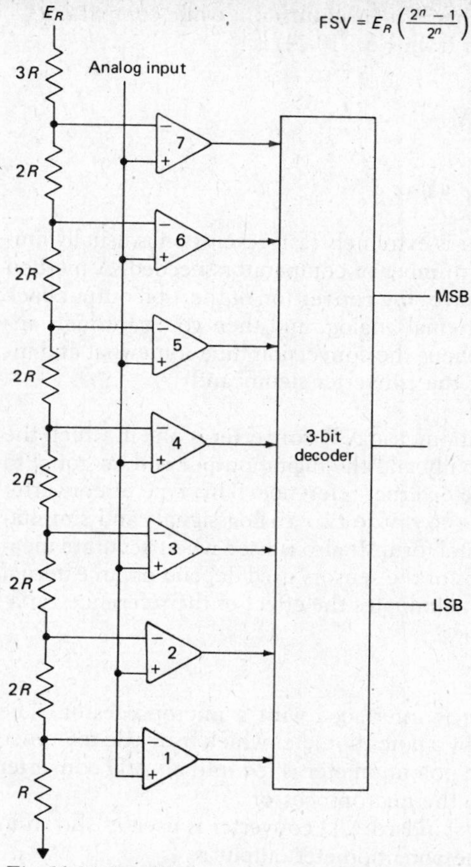

Fig. 10-16 Simplified schematic of a 3-bit parallel-type A/D converter.

Example 10-11

The converter of Fig. 10-16 uses a reference of 8 V. If an analog input of 4.2 V is applied, determine which comparators will be on and which will be off. What is the resolution in volts and the FSV input?

Solution

The following table lists each comparator and its bias voltage:

Comparator	1	2	3	4	5	6	7
Bias, V	0.5	1.5	2.5	3.5	4.5	5.5	6.5

If the input is 4.2 V, comparators 1 to 4 will turn on, while comparators 5 to 7 will be off. The resolution from Eq. (10-24) is

$$\text{Resolution} = \frac{8}{2^3} = \frac{8}{8} = 1 \text{ V}$$

The full-scale input is equal to $8(\frac{7}{8})$ or 7 V.

Although the parallel converter is extremely fast, resolution is usually limited to about 4 bits because of the number of comparators needed. A method for getting increased resolution involves the conversion of the 4-bit output back to analog, subtraction from the original analog, and then going through another parallel converter. This lengthens the conversion time somewhat and increases the cost and complexity of the converter significantly.

Ratiometric Conversion. A ratiometric A/D converter is one in which the reference can be connected externally and the digital output is then equal to the ratio of the analog input to the external reference. This type of converter finds great use when it is desirable to divide two analog signals and simultaneously convert the result into digital form. It also is used when accurate measurements are needed and the gains of the sensors used depend on an external reference. Ratiometric conversion eliminates the effect of the reference variation, as the following example shows.

Example 10-12

A positional servomechanism is interfaced with a microprocessor. The output shaft angle θ is sensed by a potentiometer which uses a dc reference voltage E_R. The output of the potentiometer is fed into an A/D converter whose output is connected to the microprocessor.

As a first consideration a standard A/D converter is used as shown in Fig. 10-17. From Eq. (2-2) the potentiometer output is

$$e_o = \frac{E_R}{\theta_{\max}} \theta \tag{2-2}$$

and e'_o will be the digital representation of e_o

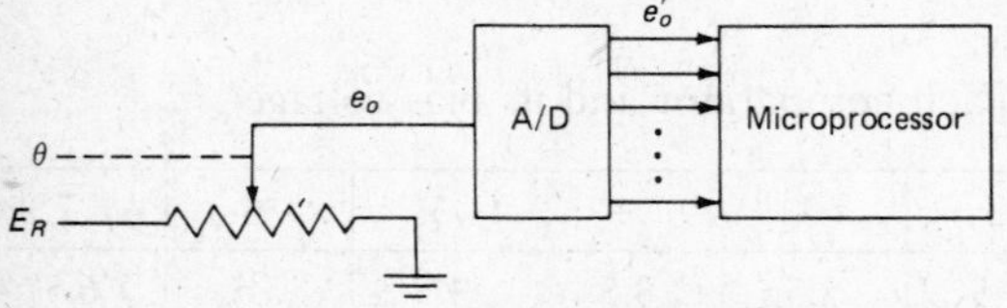

Fig. 10-17 Data conversion using standard A/D converter.

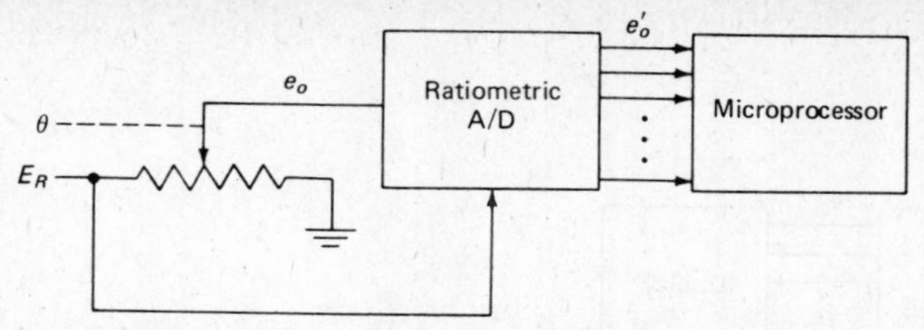

Fig. 10-18 Data conversion using ratiometric A/D converter.

$$e_o' = \frac{E_R}{\theta_{max}} \theta \qquad \text{in binary} \tag{10-25}$$

For a given angle θ, if the reference voltage E_R changes due to component tolerances and temperature effects, the digital output will also change. This is because e_o' is a function of E_R [see Eq. (10-25)]. This is undesirable since an error is now introduced in the measurements of θ; that is, the feedback gain has changed due to the variation in E_R.

As an alternate approach, a ratiometric A/D converter is used as shown in Fig. 10-18. As in the first case,

$$e_o = \frac{E_R}{\theta_{max}} \theta$$

but due to ratiometric conversion

$$e_o' = \frac{e_o}{E_R}$$

and $$e_o' = \frac{1}{\theta_{max}} \theta \tag{10-26}$$

From Eq. (10-26) it is apparent that the digital output e_o' is no longer a function of E_R. Therefore e_o' is not subject to reference-voltage variations.

Companding A/D Converters. As with D/A conversion, when it is necessary to transmit data which has a wide dynamic range, companding A/D converters are available. They make possible the conversion of large-magnitude signals with a good resolution using a minimal number of bits.

It should be mentioned that the output data registers shown in some of the A/D converters discussed are not necessarily included in A/D converters. It should also be noted that the converters presented here, although they may be the most popular, do not represent all that are available.

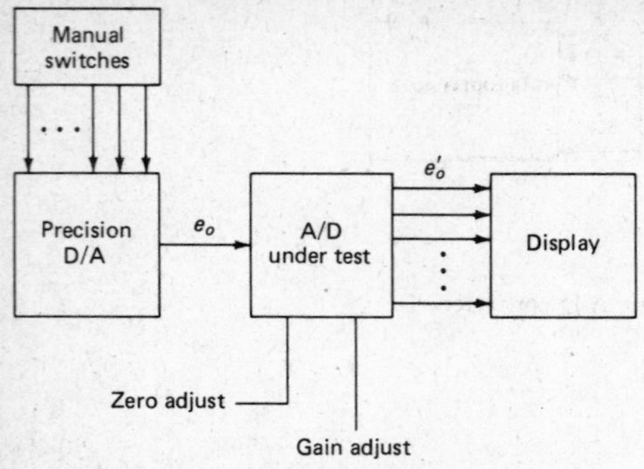

Fig. 10-19 A/D-converter test scheme.

10-2-3 Testing A/D Converters

A common test for A/D converters is a linearity test. It uses a precision D/A converter with the same scale factor as the A/D being tested and is shown in Fig. 10-19. With all the manual switches off, e_o will be zero and the zero adjust is varied to get e_o' equal to all 0s on the display. Next the switch controlling the MSB on the D/A is put on (MSB is now a **1**). The gain adjust is now varied to get only the MSB of e_o' to be a **1**. This is observed on the display. At this point the switches can be thrown in succession and the display observed. If the display does not correspond at any time to the switches thrown, the A/D converter is deemed nonlinear; in other words, it does not meet its linearity specification. There are other tests which can be performed on A/D converters, but they are too complex to be treated here.

10-3 SAMPLE-AND-HOLD CIRCUIT

The sample-and-hold circuit (S/H) is one which tracks (samples) an analog voltage, making it appear at its output terminal. Upon receipt of the proper command (a pulse) it freezes its output at the value it had when the pulse was applied and holds it at that value for the duration of the pulse. It is used in data conversion systems to hold the input of an A/D converter constant while the conversion is taking place. Sometimes the input to the A/D converter is of a slow enough frequency to be considered constant during the conversion time. Needless to say, when this happens, the S/H circuit is not needed (see Example 11-5). A simplified diagram of an S/H circuit is shown in Fig. 10-20. Typical waveforms are shown in Fig. 10-21.

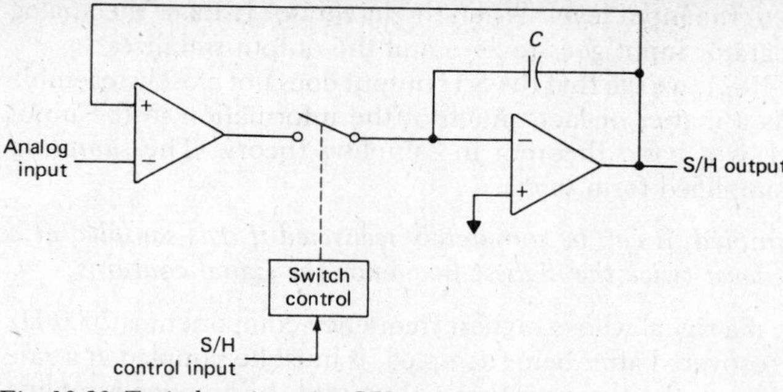

Fig. 10-20 Typical noninverting S/H circuit.

The switch in Fig. 10-20 is usually a field-effect transistor (FET) switch. Its impedance goes from a very high value (switch open) to a very low value (switch closed). The capacitor is called the *holding capacitor*. In this configuration the output amplifier is used as a very high gain integrator. Sometimes the capacitor is connected from the switch to ground, and when the switch is

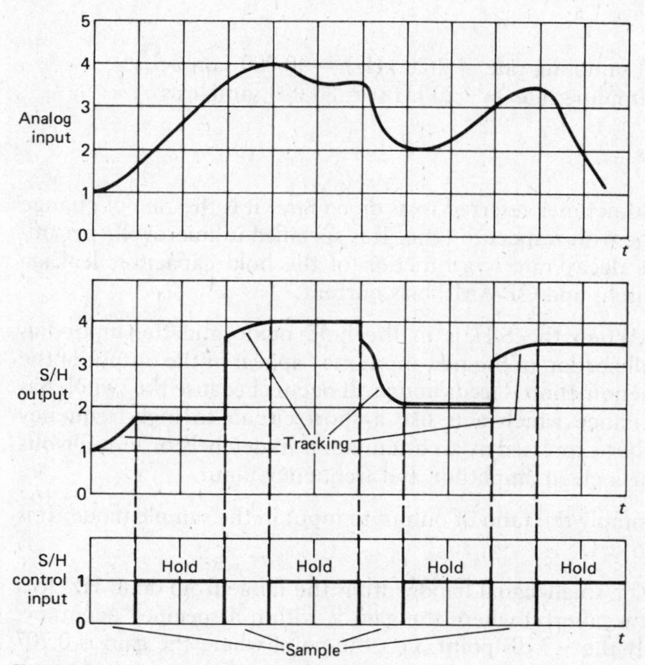

Fig. 10-21 Typical waveforms for a S/H circuit.

closed it charges up to the input level. When the integrator reaches the analog input level, the integrator input goes to zero and the output stabilizes.

Looking at Fig. 10-21, we see that the S/H output does not closely resemble the analog input. As a matter of fact, much of the information in the input signal is lost. There is a basic theorem in sampling theory (The *sampling theorem*) which in simplified form says:

> *If a signal is sampled, it can be completely recovered if it is sampled at a rate which is at least twice the highest frequency the signal contains.*

In other words, if a signal whose highest frequency component is 2000 Hz is to be completely recovered after being sampled, it must be sampled at a rate of 4000 samples/s. This is the theoretical sampling rate. In practice sampling rates of 2.5 to 4 times the highest frequency component are used.

Example 10-13

A S/H circuit is used to transmit data. The data contains signals from dc to 20 kHz. What are the theoretical and practical sampling rates in order to recover all the data?

Solution

$$\text{Theoretical sampling rate} = 2(20\text{ kHz}) = 40{,}000\text{ samples/s}$$
$$\text{Practical sampling rate} \approx 4(20\text{ kHz}) = 80{,}000\text{ samples/s}$$

10-3-1 Definitions

Decay Rate Sometimes referred to as *droop rate*, it is the rate of change of output voltage with respect to time. It is specified in microvolts per microsecond. The decay rate is a function of the hold capacitor, leakage through the switch, and OP-AMP bias current.

Feedthrough When the S/H is in the hold mode and the input has changed, a small portion of the new input may appear at the output of the circuit. This phenomenon is *feedthrough*. It occurs because the switch has inherent capacitance which acts like a short circuit to high-frequency signals. It is either expressed as an attenuation in decibels or in millivolts peak to peak for a given amplitude and frequency input.

Gain This is simply the ratio of output to input in the sample mode. It is usually equal to ± 1.

Bandwidth The small-signal bandwidth is the range from dc (0 Hz) to a frequency (in megahertz) where the gain is within a specified accuracy. This is generally the -3 dB point, i.e., the point where the gain is 0.707 times its midband value.

Aquisition Time This refers to the length of time between the command to sample and when the output permanently enters an error band around its final value (input times gain). It is similar to settling time.

Aperture Time This refers to the time it takes the switch to go from short to open once it has received a hold command.

Aperture Uncertainty Time (or Aperture Jitter) This refers to the maximum expected variation in aperture time.

Aperture Delay Time This is the time delay between the respective midpoints of the leading edge of the sample to hold pulse and the switch-resistance curve.

Some of the above definitions are shown graphically in Fig. 10-22. When used in conjunction with an A/D converter, the EOC pulse is used as the S/H command pulse. In this case the A/D converter controls the S/H circuit.

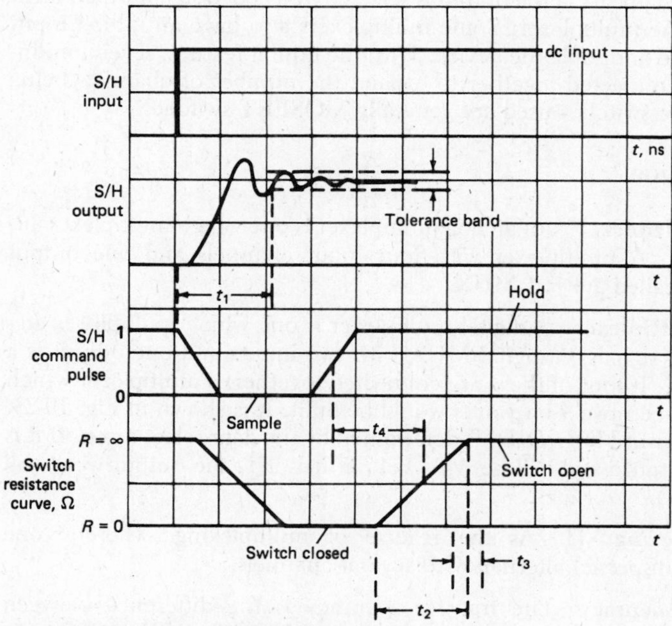

Fig. 10-22 Expanded plot of S/H waveforms:
t_1 = acquisition time ≈ 10 ns to 10 μs
t_2 = aperture time
t_3 = aperture uncertainty
t_4 = aperture delay time ≈ 10 to 100 ns

The holding capacitor has a great deal to do with the S/H specifications. It is generally in the range from 0.001 to 0.01 μF, although depending on what specification is most important other values are used. As the holding capacitance decreases, the acquisition time decreases and the bandwidth increases. As the holding capacitance increases, the droop rate, feedthrough, and the S/H errors are all decreased.

10-4 MULTIPLEXERS

The process in which a single path (or component) is shared alternately in time by several users is called *multiplexing*. A device which is used to accomplish this time sharing is called a *multiplexer* (MUX). A rotary switch whose output can be any one of its inputs depending on the switch position is in effect a multiplexer. Multiplexers have several input channels and one output channel. The input which appears at the output is selected by a coded signal which must be supplied to the multiplexer. Some multiplexers also have an inhibit input which can enable or disable the device. With the inhibit feature, several multiplexers can be connected together to expand the number of channels being multiplexed. The switches used are generally MOSFET switches.

10-4-1 Definitions

Analog Multiplexer An analog multiplexer is one which multiplexes analog signals. A multiplexer with four input channels and one output channel is called a 4 × 1 MUX.

Digital Multiplexer A digital multiplexer is one which multiplexes digital (0 or 1) signals. A digital MUX with two inputs and one output is a 2 × 1 MUX. If four of these are connected together, a multiplexer which selects between two 4-bit words would be built. It is shown in Fig. 10-23. The output word $D_1D_2D_3D_4$ is determined by the data-select input. If it is a 0, the output word will be $A_1A_2A_3A_4$; if it is a 1, the output word will be $B_1B_2B_3B_4$.

Demultiplexing This is the reverse of multiplexing, whereby one channel is dispersed alternately to several channels.

Transfer Accuracy The transfer accuracy is the difference between input and output expressed as a percentage of the input. It depends on the frequency of the input, source impedance, load impedance, and the switch resistance.

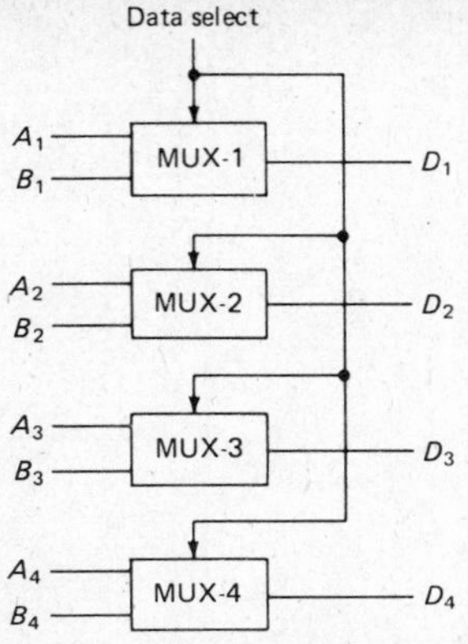

Fig. 10-23 A 4-bit 2 × 1 digital MUX.

Throughput Rate This is the fastest rate at which the MUX can switch between channels and maintain its specified accuracy. It depends on the settling time of the MUX.

Crosstalk The appearance of a signal at the output which is due to a signal present at an off input channel is crosstalk. The amount of this output as a percentage of the total input on all off channels is a measure of crosstalk. This percentage, however, is inverted and specified as an attenuation in decibels.

Input Leakage Current This represents the maximum current flowing in or out of an off channel due to switch leakage.

Example 10-14

A 4 × 1 MUX has 70 μV present at the output when the total input on the three off channels is 30 V. The on channel is grounded (0 input). Find the crosstalk in decibels.

Solution

$$\frac{70\ \mu V}{30\ V}(100\%) = 2.33 \times 10^{-4}\%$$

$$\frac{1}{2.33 \times 10^{-4}} = 4285.7$$

$$dB = 20 \log 4285.7 = 72.64\ dB$$

$$\text{Crosstalk} = -72.64\ dB$$

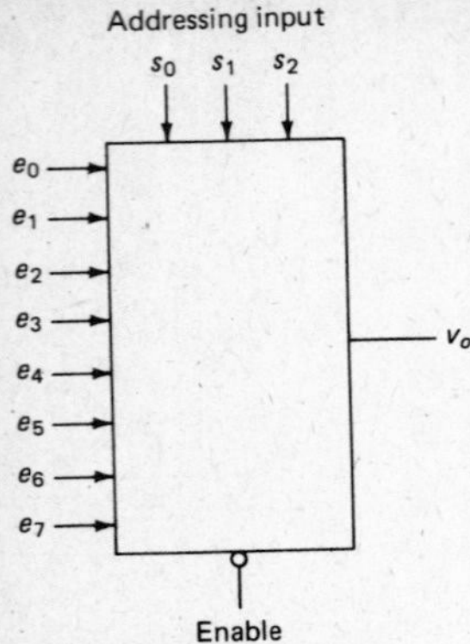

Fig. 10-24 Diagram of an 8 × 1 analog multiplexer.

10-4-2 Operation

An 8 × 1 analog MUX is shown in Fig. 10-24. When the enable is high **1**, the output e_o is held at ground. When the enable is grounded, the output is determined by the address code $S_1S_2S_3$. Table 10-1 is a truth table for the MUX shown in Fig. 10-24.

TABLE 10-1*

Enable	S_0	S_1	S_2	v_o
0	0	0	0	e_0
0	0	0	1	e_1
0	0	1	0	e_2
0	0	1	1	e_3
0	1	0	0	e_4
0	1	0	1	e_5
0	1	1	0	e_6
0	1	1	1	e_7
1	x	x	x	0

* x denotes that the code has no effect on the output.

Example 10-15

A 4 × 1 MUX is addressed by a 2-bit binary counter. The counter receives a pulse train with a frequency of 2 kHz. Find:

(a) The rate at which the MUX is switching
(b) The sampling rate per channel
(c) The theoretical maximum signal frequency permitted

(d) The time interval between samples per channel
(e) The time interval between samples at the MUX output

Solution

(a) The switching rate of the MUX is just the rate at which the addressing input is driven, which in this case is

2000 channels/s.

(b) $2000(\text{channels/s})\,\frac{1}{4\text{ channels}}$

or (500 samples/s)/channel

(c) $f_{max} = \frac{1}{2}(500) = 250$ Hz

(d) The time interval between samples per channel is

$$t = \frac{1}{500} = 2 \text{ ms}$$

(e) The time interval between samples at output is

$$t = \frac{1}{2000} = 0.5 \text{ ms}$$

10-5 FILTERING

In any analog data-acquisition system, there is noise which can cause significant errors. Noise is developed by the system components, picked up by electromagnetic radiation, and present in the data being transmitted. A low-pass filter is used to reduce the last of these sources of noise. Filters are also used to smooth the output of a D/A converter.

There are two basic types of filters, passive and active. Passive filters are circuits constructed only with resistors, capacitors, and inductors. Active filters are circuits made up of amplifiers, capacitors, and resistors. Inductors are rarely used in active filters.

Passive filters have the following advantages and disadvantages:

Advantages
1. They do not require power supplies.
2. They generate very little noise of their own.
3. They can be subjected to a wide range of voltage without fear of saturation.

Disadvantages
4. They attenuate signals rather than provide a gain of 1 or more.
5. It is hard to build them to have high input impedance and low output impedance.
6. They are more susceptible to magnetic pickup than active filters.

Active filters have the following advantages and disadvantages:

Advantages
1. They can provide large gain when needed.
2. Their output impedance can be made very low and the input impedance very high.
3. They can easily be adjusted with or without digital programming.

Disadvantages
4. They need a power supply.
5. They can contribute noise and offsets to a signal.

10-5-1 Definitions

Pole A pole is a root of the characteristic equation of a filter. The number of poles is numerically equal to the number of capacitors and inductors used in the filter.

Cutoff Frequency (f_{co}) This is the frequency at which the attenuation of the filter begins to increase sharply and continue to increase. It does not necessarily correspond to the 3-dB frequency defined previously.

Rolloff Rate This is the asymptotic slope of the frequency response in the region where the attenuation begins to increase sharply. It is usually expressed in decibels per octave. For each pole a filter contains the rolloff rate is 6 dB/octave. Therefore a four-pole filter would have a rolloff rate of 24 dB/octave.

Folding Frequency (f_d) In the discussion on sampling theory it was stated that to recover a signal completely it had to be sampled at a rate f_s twice the highest frequency component of the signal f_{max}. In reality, though, it was pointed out that f_s is usually about 3 or 4 times f_{max}. Since noise (containing frequencies greater than f_{max}) is generally in the signal, erroneous information (*aliases*) can be obtained. A filter must attenuate all such noise whose frequency is above the *folding frequency* f_d given by

$$f_d = f_s - f_{max} \tag{10-27}$$

Example 10-16

A signal being transmitted to a computer is being sampled at 3.5 times its maximum frequency component, which is 4 kHz. Noise which is present on the signal is to be attenuated to 2 percent of its nominal value at the folding frequency. Find the number of poles needed for the required filter. Assume that the filter is flat up to 4 kHz and that its cutoff is at 5 kHz.

Solution

$$f_{max} = 4 \text{ kHz}$$
$$f_s = 3.5 f_{max} = 14 \text{ kHz}$$
$$f_d = 14 \text{ kHz} - 4 \text{ kHz} = 10 \text{ kHz}$$

If noise is to be reduced at 10 kHz to 2 percent of its value at 4 kHz, it must attenuate by a factor of $\frac{1}{50}$, that is, $\frac{1}{50} = 2$ percent.

An attenuation of one-fiftieth is equivalent to

$$20 \log \tfrac{1}{50} = 20 \log 1 - 20 \log 50 = 0 - 20(1.7) = -34 \text{ dB}$$

The filter response shown in Fig. 10-25 is required. Since $f_{co} = 5$ kHz and $f_d = 10$ kHz, the rolloff is 34 dB in 1 octave. Therefore, a six-pole filter is needed. This will give a rolloff of 36 dB/octave. This is a complex

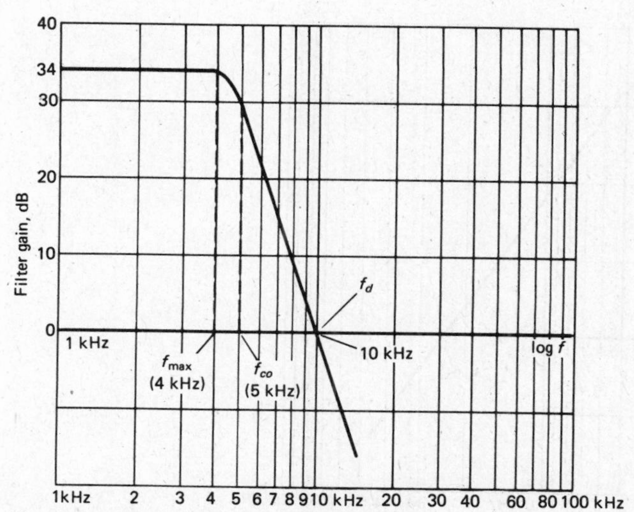

Fig. 10-25 Filter response for Example 10-16.

filter which is available. Example 10-17 shows a technique for easing the filter requirement.

Example 10-17

Solve the problem of Example 10-16 again if the sampling is done at twice the previous rate.

Solution

$$f_{max} = 4 \text{ kHz}$$
$$f_s = 7f_{max} = 28 \text{ kHz}$$
$$f_d = f_s - f_{max} = 24 \text{ kHz}$$

The filter response desired is shown in Fig. 10-26. The derivation will not be given here, but it can easily be shown that the decibels per octave is given by

$$\text{dB/octave} = \text{attenuation} \times \frac{\log 2}{\log (f_d/f_{co})} \tag{10-28}$$

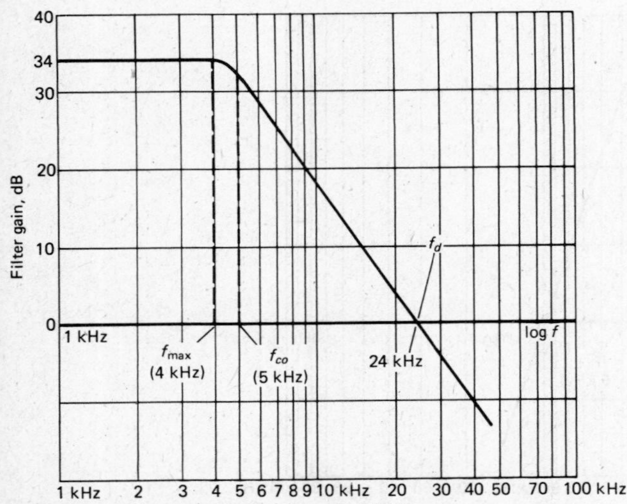

Fig. 10-26 Filter response for Example 10-17.

Using Eq. (10-28), we get

$$\text{dB/octave} = \frac{34 \log 2}{\log \frac{24}{5}} = \frac{34(0.3)}{0.68} = 14.97 \approx 15 \text{ dB/octave}$$

In this case, a filter of three poles (18 dB/octave) will do the job.

It should be pointed out that sampling rates are not just arbitrarily increased. Considering just the time element, the faster sampling rate dictates the need for an A/D converter with a faster (shorter) conversion time. Finally, a cost trade-off must always be made.

10-5-2 Types of Filters

There are many types of filters available. Among the popular ones are Butterworth, Bessel, Chebyshev, elliptic, and *RC* filters. Some differences in filters are the flatness of the passbands, the sharpness of the cutoffs, and the type of phase shift introduced. The phase shift introduces a time delay. For a linear-phase filter, the delay is given by

$$\text{Delay} = \frac{0.125}{f_{co}} \text{ s/pole} \tag{10-29}$$

at the cutoff frequency. Because of the complexity of the various filter designs, they will not be analyzed or presented in detail. Their brief presentation here is for the user and not the filter designer.

10-6 AMPLIFIERS

In the design of data-acquisition systems amplifiers are sometimes needed to raise signals to a full-scale level. For example, an accelerometer is used to sense a body's vibration or acceleration, and the output never exceeds 2 V. Much better resolution can be obtained if the 2-V signal is amplified by 5 so that the signal now varies from 0 to 10 V. If an active filter is used, it can be used to provide the necessary gain.

OP-AMPs are designed for general-purpose and special applications. General-purpose OP-AMPs can usually be used to meet most requirements. Low-voltage-drift OP-AMPs are chosen when it is not easy to adjust the offsets externally in a particular application. These would be used in test equipment and as summers in control systems. In applications where high output voltage and/or current is needed, a high-output OP-AMP is selected.

In data-acquisition and transmission systems, OP-AMPs with a wide bandwidth, fast slew rate, and fast settling time are desired. The selection must be properly matched to the other system components. As an illustration, consider an OP-AMP used on a signal which is being sampled. The highest frequency component of the signal is 20 kHz. From noise considerations a sampling rate of 100 kHz is chosen. This in turn means that the signal is being sampled every 10 μs (1/100 kHz). The OP-AMP used must then be able to respond to a full-scale change and settle to within a specified accuracy within 10 μs. In other words, the specified settling time of the OP-AMP should be less than 10 μs.

Figure 10-27 illustrates the importance of OP-AMP settling time. As the curves show, the S/H output indicates that the original signal is 40 percent greater than it actually is. When fed into a computer this error in transmission could cause corrective action which in reality is not warranted. This in turn could cause unwanted oscillations in the overall system. If, on the other hand, the OP-AMP had settled before the sample at 20 μs, this error would not have occurred.

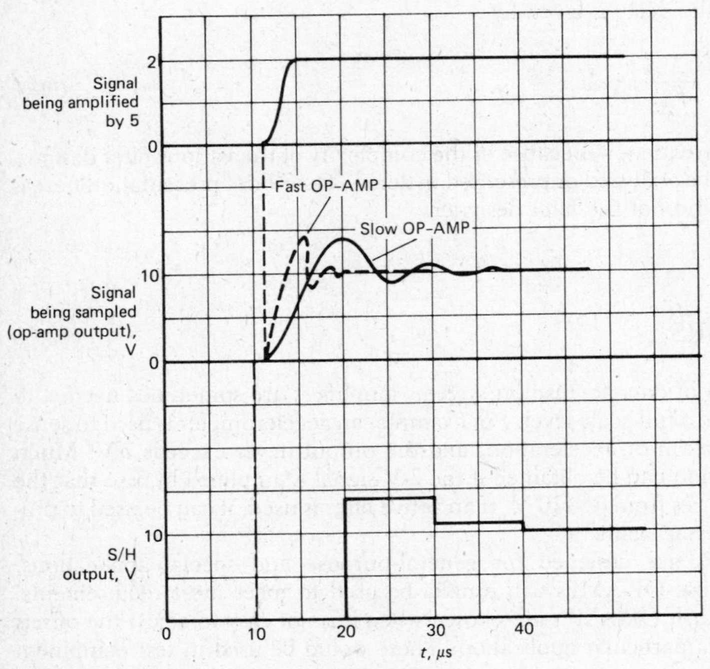

Fig. 10-27 Sampled data using a slow OP-AMP.

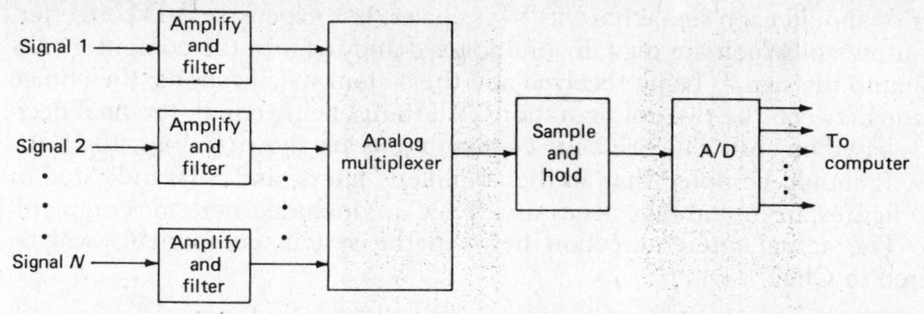

Fig. 10-28 *N*-signal data-acquisition system using an analog multiplexer and one A/D converter.

Care should be taken not to consider just the OP-AMP slew rate. As a case in point, an ultrafast hybrid OP-AMP is available with a slew rate of 1000 V/μs. This implies that a 10-V change in output can occur in 10 ns; that is,

$$\frac{10\text{ V}}{1000\text{ V}/\mu\text{s}} = 10\text{ ns}$$

The OP-AMP settling time, however, is specified as 100 ns for a 10-V change to 0.1 percent of the final value. The difference in the times obtained here is due to the oscillatory nature (low damping) of the OP-AMP, as shown in Fig. 10-27.

10-7 THE COMPLETE DATA-ACQUISITION SYSTEM

When a computer is used to receive data from a control system, a decision must be made. Should several analog signals be multiplexed and fed into a single, very fast, expensive A/D converter whose output is fed directly into the com-

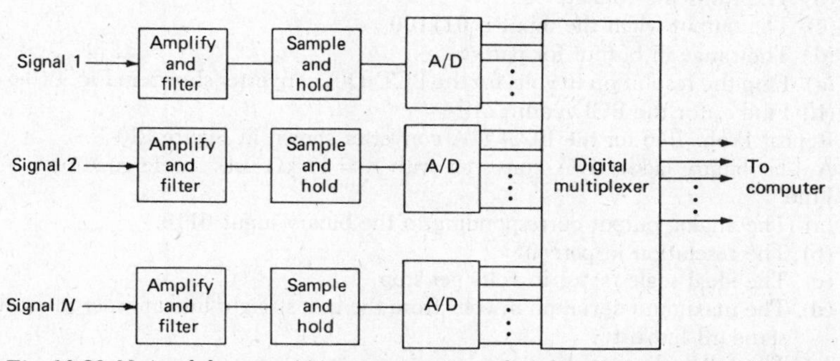

Fig. 10-29 *N*-signal data-acquisition system using a digital multiplexer and *N* A/D converters.

puter or should each signal have its own slower, less expensive A/D converter, the outputs of which are digitally multiplexed and fed into the computer? According to the signals being received and the system specifications, the choice is made between the two configurations. All things being equal, the final decision is based on cost. The two basic configurations are shown in Figs. 10-28 and 10-29. It should be noted that all the amplifiers, filters, and S/Hs indicated in these figures are not always necessary. They are included here for completeness. The actual interconnection between the system components will be covered in Chap. 11.

PROBLEMS

10-1. Define the following D/A converter terms: resolution, monotonicity, linearity, scale factor, and glitch.

10-2. What is two-quadrant multiplication? Four-quadrant multiplication?

10-3. An 8-bit R-$2R$ binary ladder DAC has $R = 10\ \text{k}\Omega$ and $E_R = 10$ V. Find:
- (a) The resolution in percent
- (b) The resolution in volts
- (c) The output when the input is 00010001
- (d) The input when the output is 7.5 V
- (e) The full-scale output

10-4. A two-digit BCD D/A converter is a weighted-resistor type with $E_R = 1$ V (refer to Fig. 10-7), $R = 1\ \text{M}\Omega$, and $R_f = 10\ \text{k}\Omega$. Find:
- (a) The resolution in percent
- (b) The resolution in volts
- (c) The output for the decimal input 82
- (d) The BCD number when the output is 0.37 V
- (e) The full-scale output

10-5. A 6-bit R-$2R$ ladder D/A converter has a reference voltage of 6.5 V. It meets standard linearity. Find:
- (a) The resolution in percent and volts
- (b) The full-scale voltage
- (c) The output when the input is 011100
- (d) The range in output for part (*c*)

10-6. (a) Find the resolution in volts for the BCD D/A converter shown in Fig. P10-6.
(b) Find e_o for the BCD count of 38.

10-7. Repeat Prob. 10-6 for the BCD D/A converter shown in Figure P10-7.

10-8. A 4-bit binary ladder D/A converter with $R = 10\ \text{k}\Omega$ uses a reference of 5 V. Find:
- (a) The analog output corresponding to the binary input 0110.
- (b) The resolution in percent
- (c) The ideal scale factor in volts per step
- (d) The maximum deviation in volts from the best straight line in order to meet standard linearity
- (e) The full-scale output

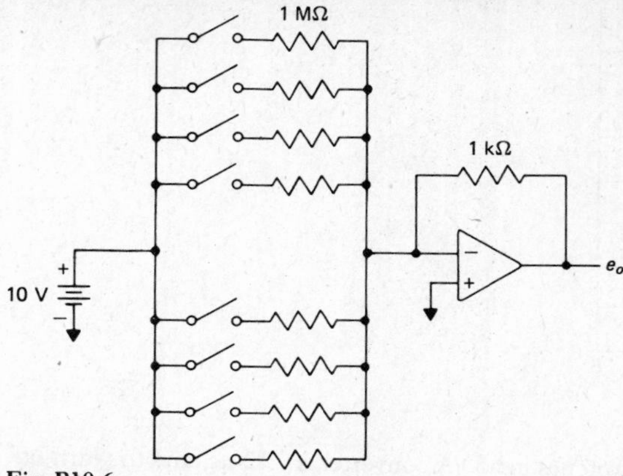

Fig. P10-6

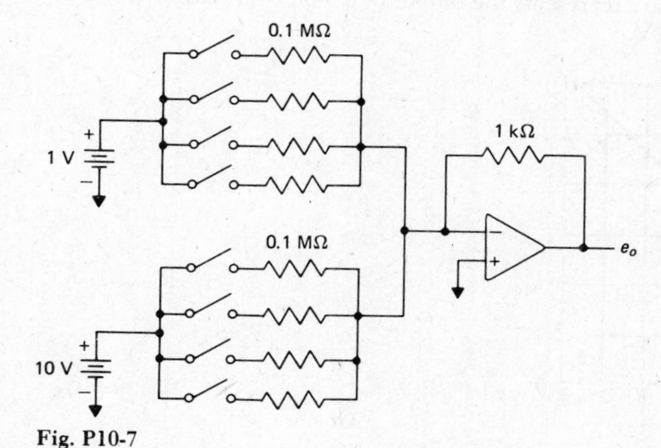

Fig. P10-7

10-9. A two-digit BCD D/A converter utilizes two 4-bit binary ladders, each using R = 10 kΩ and a reference of 8 V for the MSD and 0.8 V for the LSD. Find:
(a) The resolution of the converter in percent
(b) The ideal scale factor in volts per step
(c) The analog output for a BCD count of 72
(d) The full-scale output

10-10. Data collected for a 3-bit ladder D/A converter, using a reference of 4 V is shown in Fig. P10-10.
(a) Is the DAC monotonic?
(b) Does the DAC meet standard linearity?
(c) What is the actual scale factor?

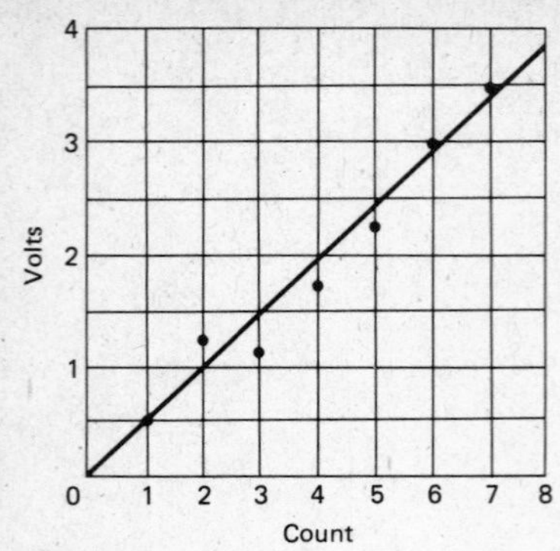

Fig. P10-10

10-11. The turn-on time of the switches in a D/A converter is 0.12 μs, and the turn-off time is 0.11 μs. If the input to the D/A converter changes from a binary count of 4 to a count of 1, sketch the glitch that results. Show relative magnitudes.

10-12. The plot in Fig. P10-12 represents the output of a 3-bit R-$2R$ ladder D/A converter with $E_R = 1.6$ V.

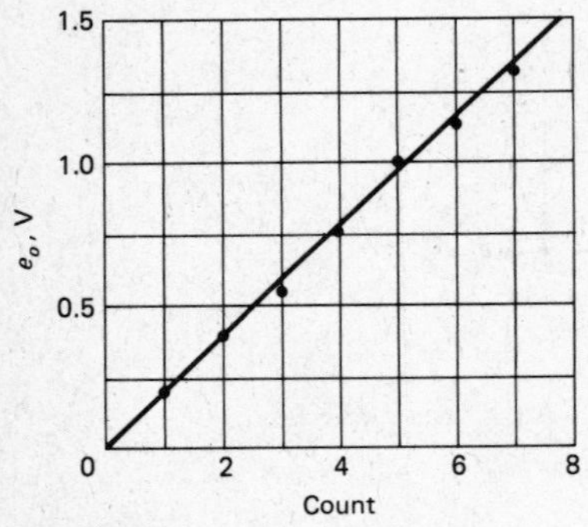

Fig. P10-12

(a) Find the theoretical resolution in volts.
(b) Find the actual scale factor.
(c) Is the D/A converter linear? Explain.
(d) What is the ideal FSV?

10-13. The switches in a D/A converter have a turn-on time of 0.08 μs and a turn-off time of 0.06 μs. The DAC is 5-bit binary and loaded with 00101. If the number is changed to 01100, sketch the output (using a scale factor of 1) showing the old value, the glitch (show size), and the new value.

10-14. In the weighted-current-source D/A converter shown in Fig. 10-3, if $V_{cc} = 12$ V and R = 1 kΩ, find:

(a) The resolution in milliamperes (assume the transistor emitter-base voltage is 0.7 V.)
(b) The full-scale current output
(c) The range in output when the input is **1011** (assume standard linearity)

10-15. A 4-bit D/A converter with E_R = 10 V is tested. It is a weighted-resistor type with R = 100 kΩ and R_f = 1 kΩ. The test results for each bit are

0001 – 0.15 V

0010 – 0.25 V

0100 – 0.35 V

1000 – 0.80 V

Is the converter monotonic? Does it meet standard linearity? Explain your answers.

10-16. Define the following A/D converter terms: quantization, resolution, quantizing error, offset error, gain error, conversion time, conversion rate, and missing code.

10-17. A temperature transducer is used to sense the outside temperature of a space shuttle. An A/D converter is used to interface the transducer output (volts) to an on-board computer. The temperature can vary from −55 to 125°C. The computer must sense changes in temperature of 0.2°C. How many bits are required for the A/D converter?

10-18. A 6-bit dual slope A/D converter uses a reference of −3 V and a 1-MHz clock. It uses a fixed count of 40(**101000**). Find:
(a) The input if the output register contains **100111** at the end of a conversion
(b) The maximum conversion time
(c) The minimum conversion rate
(d) The maximum input that can be converted
(e) The fixed count that should be used so the output register will represent the input

10-19. Repeat Prob. 10-18 with a reference of −10 V.

10-20. A 3-bit dual-slope A/D converter uses a 100-kHz clock, −1 V reference, and a fixed count of **111**. Find:
(a) The input when the output register reads **010**
(b) The maximum conversion time
(c) The minimum conversion time
(d) The conversion rate

10-21. Repeat part (*a*) of Prob. 10-20 if the A/D converter develops BCD and the number in the registers is 73. The converter now has 8 bits.

10-22. A counter type A/D converter contains a 4-bit binary ladder and a counter driven by a 2-MHz clock.
(a) Find the conversion time and conversion rate.
(b) If the ladder reference is 2 V, determine the minimum comparator resolution required.

10-23. Repeat Prob. 10-22 if the ladder is replaced by a two-digit BCD-to-analog converter with a scale factor of 0.1 V per step (the counter is also changed to BCD).

10-24. Find the conversion time of a successive approximation A/D converter which uses a 2-MHz clock and a 5-bit binary ladder containing an 8-V reference. What is the conversion rate?

10-25. A servo type A/D converter contains a 4-bit binary ladder with a 2-V reference and uses a 2-MHz clock.
(a) Find the conversion time.
(b) What is the minimum conversion time?
(c) What is the conversion rate?

10-26. An 8-bit successive approximation A/D converter uses a 2-MHz clock.
(a) What is the conversion time?
(b) What is the conversion rate?

10-27. A 4-bit parallel type A/D converter uses a 6-V reference.
(a) How many comparators are required?
(b) What is the resolution in volts?

10-28. What is a ratiometric A/D converter? When is one used?

10-29. A S/H circuit is used to sample a signal whose highest frequency component is 75 Hz. In order to be able to recover the signal, what sampling rate should be used (a) theoretically and (b) practically?

10-30. Define the following S/H terms: droop rate, feedthrough, and acquisition time.

10-31. In a S/H circuit, state the effect of the holding capacitor on:
(a) acquisition time
(b) bandwidth
(c) droop rate
(d) feedthrough

10-32. A S/H circuit has for its input the signal shown in Fig. P10-32. Sketch the output if it is sampling at a 2-kHz rate. The first sample is made at $t = 0$.

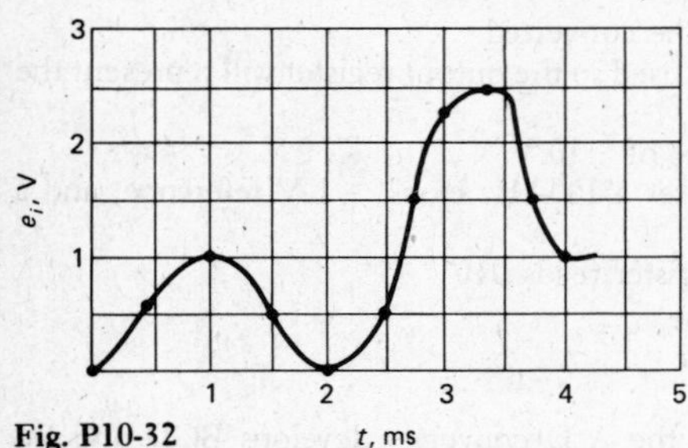

Fig. P10-32

10-33. Repeat Prob. 10-32 if the sampling rate is 1 kHz.

10-34. Define the following multiplexer terms: crosstalk, throughput rate, and transfer accuracy.

10-35. The inputs to a four-channel MUX have the following bandwidths:

Channel 1 50 Hz

Channel 2 200 Hz

Channel 3 75 Hz

Channel 4 90 Hz

Find the theoretical minimum sampling rate of the MUX.

10-36. A five-channel MUX is operating at 1000 channels per second.
(a) What is the sampling rate per channel?
(b) What is the theoretical maximum signal frequency allowed?
(c) What is the interval between samples per channel?
(d) What is the interval between samples at the MUX output?

10-37. Determine the values of A and B to enable input 3 to pass through the digital multiplexer shown in Fig. 10-37.

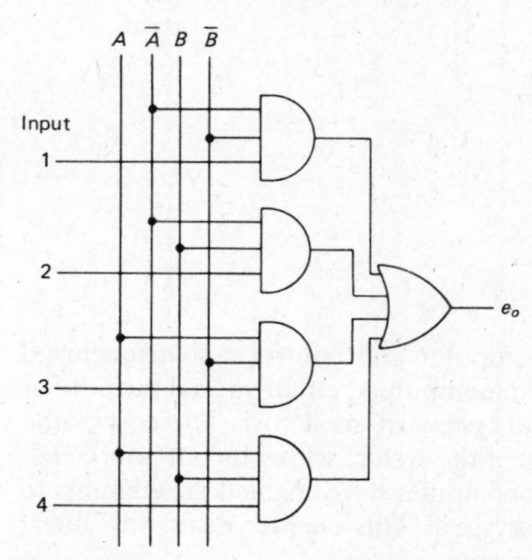

Fig. P10-37

10-38. An 8 × 1 MUX has 150 μV present at its output when the ON channel is grounded, and the total input on the seven off channels is 56 V. What is the crosstalk in decibels?

10-39. A multiplexer is sampling channels whose signals are slower than 100 Hz. The maximum data rate into the D/A being fed by the MUX is 10 kHz. What is the maximum number of MUX channels permitted?

10-40. What is a passive filter? An active filter?

10-41. Define the following filter terms: pole, cutoff frequency, and rolloff rate.

10-42. The maximum frequency content of a signal is 400 Hz. It is being sampled at 1.2 kHz. The noise present on the signal is to be attenuated to 3 percent of its nominal value at the folding frequency. Assuming the filter is flat up to 400 Hz and its cutoff is at 500 Hz, find the number of poles needed in the filter.

10-43. Repeat Prob. 10-42 if the sampling rate used is 2 kHz.

chapter 11

Interfacing The Computer and The Control System

Until recently, interfacing between computer and control system occurred only for large-scale control systems. Minicomputers, either special-purpose or general-purpose, were interfaced with a system at great cost. The cost of the computer quite often exceeded the cost of the system with which it interfaced. The low cost and small size of the microcomputer have enabled the engineer to use it in almost every type of control system. This chapter deals with interfacing microcomputers with control systems.

11-1 INTRODUCTION

In Chap. 8 the computer was shown in two basic configurations involving control systems, off-line and on-line. When the computer is connected off-line, data from the system is fed into it by an operator. The computer can take this data and use it to provide a solution to a complex problem. From the solution the operator can supply inputs to the system to alter its behavior or take corrective action. The advantages of the computer in this case are high speed and accuracy. If it has been properly programmed, the computer can solve the problem in a few seconds quite accurately. If an engineer were to solve the problem, it would take him minutes, possibly hours, to do the same. In addition, being human, engineers are susceptible to error. An example of this application is the task of sending a rocket to the moon. During the flight, corrective action due to unexpected events must be taken. A computer on earth is used off-line to solve the problem quickly and accurately. The results are then used to supply commands to the system from a transmitting station on earth.

When the computer is connected on-line, it replaces hardware (electronics). The cost of the materials and the labor needed to build the circuitry is reduced tremendously. In addition the maintenance costs are reduced because hardware, which occasionally fails, has been replaced with software (a computer program), which is far more reliable. The system's accuracy is also improved because hardware gains, which depend on resistors and capacitors (whose values change with age), are replaced with software. Software does not change with age; it remains exactly as it was when originally created. There is a final advantage of a computerized control system over a 100 percent hardware system. Sometimes a system has been operating in a certain fashion for some time. When changes in its operation are desired, extensive labor may be involved to make hardware changes. If a computer is used on-line, the changes can be made by simply changing an instruction in the program. This principle is finding great use today in complex telephone systems. Telephone numbers at a large facility are controlled by a microprocessor. When extension numbers are changed, wiring changes are replaced by microprocessor program changes.

11-2 PARALLEL- VS. SERIAL-DATA TRANSFER

Whenever a computer is interfaced with a control system, data must be transferred from one to the other. Since the computer operates only on parallel data, conversion to this form must ultimately be made.

When the computer is located close to the system, the data is transferred in parallel. For high-frequency signals, which require fast sampling rates, the conversion from analog to digital should take place as close to the computer as possible. The A/D converter output should then be in parallel.

When the system is located at some distance from the computer, it can be connected to the computer by a cable, which should be a twisted pair to reduce the effects of transients picked up from the surrounding area. The signal can be either serial digital data or analog data with the conversion taking place at the computer location. Analog data is preferred when high sampling rates are involved.

At very large distances a cable connection is impractical or impossible. A tracking station which controls the path of a satellite is a perfect example. In this case, data is usually in serial digital form and transmitted by radio waves using pulse-code modulation (PCM) (see Sec. 11-3-1).

11-3 SERIAL-DATA INTERFACING

In many cases digital data is received in serial form. If it is received over a telephone line, it most likely will be serial binary data.

When a telephone line is used to connect an input device such as a teletype to a computer, a device called a MODEM (modulator-demodulator) is used. The teletype is connected to the phone line by an *originate MODEM*. The serial data fed into the MODEM from the teletype is converted into analog signals. If the data is a 1, the MODEM sends a signal of 1270 Hz. If the data is a 0, the MODEM sends a signal of 1070 Hz. At the computer end of the phone line is another MODEM which is called the *answer MODEM*. The answer MODEM receives the analog signals and reconstructs the serial data to be fed into the computer. When the computer sends data to be printed by the teletype, it sends serial data to the answer MODEM. The answer MODEM converts the data into analog form by converting a 1 into a signal of 2225 Hz and a 0 into a signal of 2025 Hz. These signals are then received by the originate MODEM, which converts them back into serial binary. The serial data now gets fed into the teletype, and the message is printed. However, when the data is transmitted over radio waves, it will probably be a PCM (pulse-code-modulated) signal.

11-3-1 Pulse-Code Modulation

Pulse-code modulation (PCM) refers to the technique of taking a discrete-level signal (quantized signal) and converting it into a binary code for transmission. Figure 11-1 illustrates the process. The signal $d(t)$ is a quantized signal representing some system parameter, e.g., air pressure in a satellite orbiting the earth. The signal is first encoded; in this case it is encoded to RZ and NRZ unipolar binary. Unipolar binary is a binary code in which a 1 is represented by a voltage level, that is, 5 or 10 V, and a 0 is represented as 0 V. Bipolar binary is a binary code in which a 1 is represented by a positive voltage level (+5 V) and a 0 is represented by a negative voltage level (−5 V). The RZ means that the code returns to zero between bits. The standard form is NRZ, which means that the code does not return to zero between bits. In this manner the only time the PCM signal changes is when the data actually changes.

After encoding, the PCM signal is transmitted serially. Usually the LSB is transmitted first for each successive level of the quantized signal. In Fig. 11-1, the order of the bits transmitted is 010 110 100 001 110. Upon reception, the PCM signal is decoded, generally with a sampler which must be synchronized to the bit transmission. It is then converted back into its original quantized

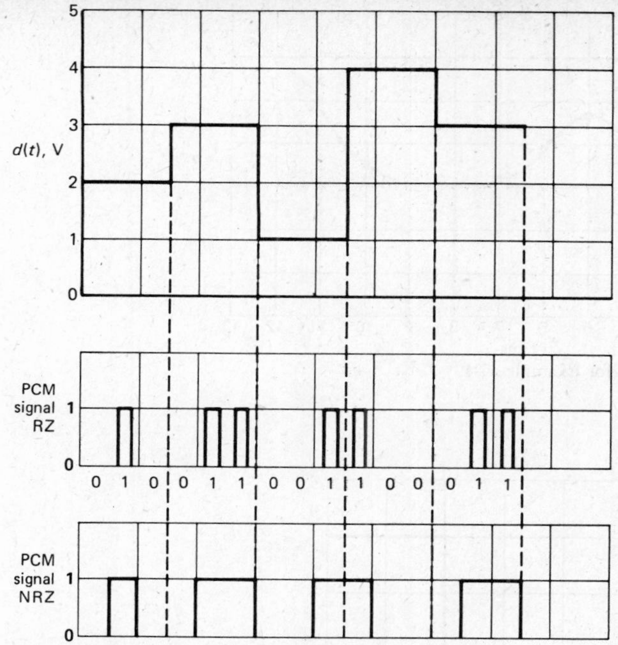

Fig. 11-1 PCM of a quantized signal $d(t)$.

state. The description presented here is by no means complete; it is presented merely to acquaint the reader with PCM.

Example 11-1

The signal shown in Fig. 11-2 is sampled at 0.25 kHz starting at $t = 0$. It is converted into 4-bit PCM (NRZ bipolar binary). Sketch the PCM signal for the first three sampling periods.

Solution

Since the signal is sampled at 0.25 kHz, a sample is taken every

$$\frac{1}{0.25\ \text{kHz}} = 4\ \text{ms}$$

starting at $t = 0$. Figure 11-3 shows the sampled signal and the PCM signal in NRZ unipolar binary.

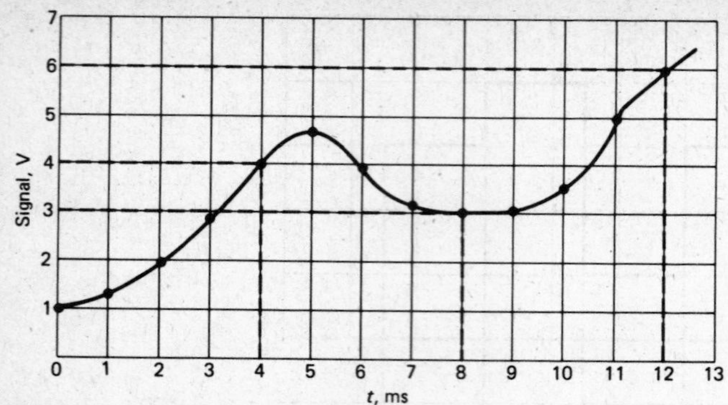

Fig. 11-2 Signal waveform for Example 11-1.

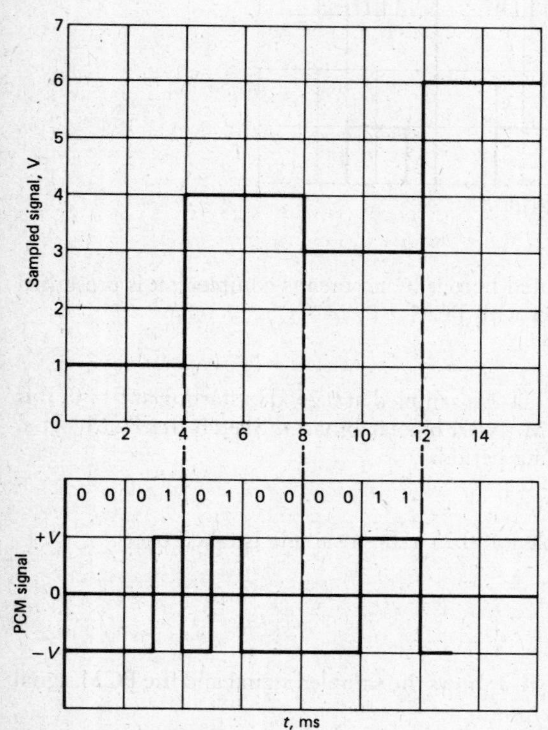

Fig. 11-3 Solution to Example 11-1.

11-3-2 Hardware Considerations*

When serial transmission of data is used, the data cannot be fed directly into the microprocessor unit (MPU). An interface device must be used which can receive serial data, convert it into parallel data, and then feed it into the MPU. In addition, the device must be able to accept parallel data from the MPU and convert it into the proper serial form for transmission back to the system.

Intel makes a device which performs the above operations and connects directly to the 8080A, called a universal synchronous-asynchronous receiver-transmitter (USART) 8251. Motorola makes a similar device which connects directly to the M6800, called an asynchronous communications interface adapter (ACIA) MC6850. It can be used only for asynchronous data. The operation of both units is programmed by their respective MPUs. In addition, they both have two serial-data lines on the peripheral side. This enables them to transmit and receive serial data simultaneously.

11-4 PARALLEL-DATA INTERFACING

Since the MPU is built to operate internally with parallel data, it is desirable to interface in parallel whenever possible. When time is a critical parameter, as is generally the case when a computer is used, it is particularly important to interface in parallel. It takes 8 times longer to enter an 8-bit serial word than an 8-bit parallel word. As with serial interfacing, the input cannot be fed directly to the MPU.

11-4-1 Hardware Considerations

A special device for interfacing parallel data with a microprocessor must be used. Intel offers two options; the first is an 8-bit input-output port 8212, and the second is a programmable peripheral interface (PPI) 8255. The PPI is by far the more flexible. The input-output port is merely a buffer, which can be used to send data to the MPU data bus or receive data from the data bus. To accomplish both, two input-output ports are needed. The PPI, like Motorola's peripheral interface adapter (PIA) MC6820, can do both. They can receive parallel data from a peripheral and send it to the data bus and vice versa. Both the PPI and the PIA contain three-state bidirectional buffers, which are controlled by the MPU either to transmit data to or receive data from the data bus. The buffers can assume three states. They can be high (1) or low (0) for the transmission of binary data. In addition they can assume a third, high-impedance state

* There are several microprocessors available today. Reference here will be made only to two of the most popular, the Motorola M6800 and the Intel 8080A.

which in effect is like disconnecting them from the data bus. This third state is essential because the data bus is shared by many devices, not all of which can be connected to it at the same time.

11-5 SOFTWARE: THE ALTERNATIVE TO HARDWARE

Whenever a MPU is used in a system there is an increasing trend toward replacing some of the digital or analog hardware with software. As an example, when an analog control system is interfaced with a computer, it is necessary to perform an A/D conversion. As an alternative to using an A/D converter, the cost of which for the application might be excessive, one can reduce the hardware expense by combining some software with a minimum of hardware. Figure 10-14 shows a successive-approximation A/D converter which has five basic components. When a MPU is used, all that is necessary to perform the A/D conversion is the comparator and the D/A converter.

Figure 11-4 shows a typical configuration. It is a general diagram and by no means complete. The offset and calibration adjustments, as well as the reference voltages, have all been omitted for simplicity. The complicated portions of the A/D converter, the bit sequencer-programmer and storage register, have been replaced by a software program (subroutine) and the MPU memory, respectively. The program performs the operation of turning the IDAC bits on or off starting with the MSB. When the conversion is complete, the program

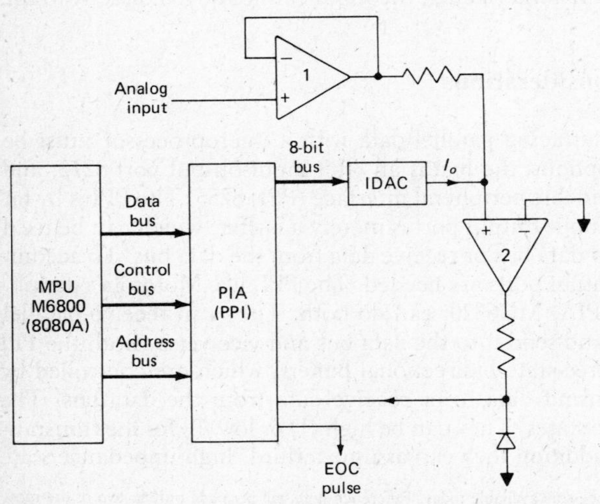

Fig. 11-4 Typical software-hardware successive-approximation A/D converter.

indicates the fact by producing a 1 on one of the PIA data-bus bits (EOC pulse). The result of the conversion is stored at a memory address specified by the program. It is then accessible for use by the main program of the system.

The only disadvantages of this technique are the increase in MPU memory used and the conversion time. For an 8-bit A/D converter interfaced with a MPU approximately 40 to 50 memory locations are required for the control of the A/D and PIA. If the software technique described above is used, more than 100 memory locations would be needed. The conversion time using the software technique is about 10 times that of the hardware technique. If a computer is being used in a large system, it would be unwise to have it stop whatever it is doing and spend 500 μs or more to perform a conversion, but in some instances it might be the preferred technique.

11-6 THE COMPLETE SYSTEM

Having examined the control system and those components used for interfacing it with a computer, we can now construct a picture of the overall system.

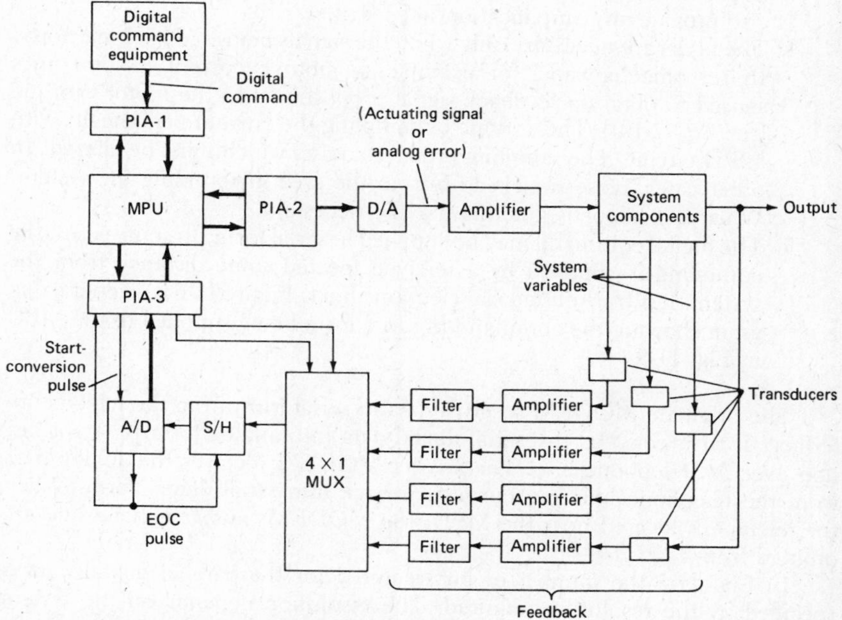

Fig. 11-5 A complete on-line computer control system.

When the computer and the control system are located together or reasonably close, a configuration like that in Fig. 11-5 can be used. The following points should be made with reference to the figure.

1. Even though three PIAs are shown, only one may be necessary. The MC6820 PIA has two 8-bit input-output (I/O) lines on the peripheral side. This is a total of 16 bits. If the system specifications are such that only 4-bit resolution is needed, only one PIA is required. Four bits each would be connected to the D/A, A/D, and the digital command equipment. In addition, 2 bits are needed to address the MUX since it has four inputs. The last 2 available bits are needed to transmit the start-conversion pulse and receive the EOC pulse.
2. Not all the transducers indicated may be necessary. In some instances, system variables which are sensed and fed back for control are voltages, eliminating the need for transducers.
3. The amplifiers indicated are needed to scale the feedback signals up to obtain better resolution when they are converted into digital form. In many cases, proper selection of transducers will eliminate the need for this scaling. Any gain needed for dynamic system behavior can usually be provided by the software. In addition, if an active filter is used, it can also provide any amplification needed.
4. The filters are necessary only when the signals being fed back are noisy. In servomechanisms, for example, a motor's current is sometimes sensed to obtain a feedback signal proportional to the motor's torque [see Eq. (2-10)]. This is done by sampling the current periodically with a S/H circuit. The sampling produces noise, which must be filtered. In other cases, noise is picked up from the area surrounding the system. When this happens, filtering may be necessary.
5. The digital command may be supplied in serial form. In some cases, the command is supplied by a teletype located some distance from the system. It is transmitted via telephone lines. Figure 11-6 is a general diagram showing the command in serial form being supplied to the MPU in Fig. 11-5.

The originate MODEM receives data in serial form from the teletype interface. It then converts this serial digital data into analog form for transmission over the telephone line. The answer MODEM receives the analog data from the telephone line and converts it back into serial digital form. When the teletype is located near the MPU, the MODEMs and telephone line are omitted from Fig. 11-6.

In Fig. 11-5 the number of bits required for the parallel digital data is specified by the resolution required. The resolution depends on the system specifications, especially the steady-state error or, equivalently, the accuracy.

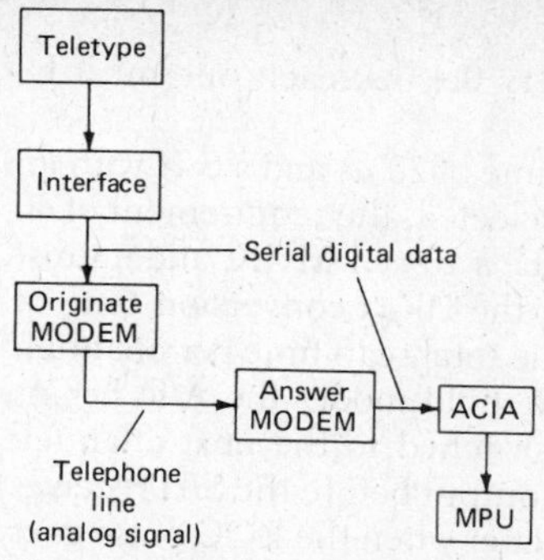

Fig. 11-6 Remote serial command for system of Fig. 11-5.

Example 11-2

A microprocessor is used to position a gun turret on board a battleship. It is used on line in a positional servomechanism. The turret must position itself to ±0.1 percent of full scale in response to step commands. Determine the number of bits required for the A/D converter used.

Solution

Nonlinearities which can introduce steady-state errors will be ignored to simplify the problem. In addition, the servomechanism has a steady-state error of zero (see the discussion on errors in Chap. 5). Therefore the only error considered will be the quantization error. In Chap. 10 it was shown that the quantization error is $\pm\frac{1}{2}$LSB. Relating this to the specified error (±0.1 percent), we see that the LSB must be 0.2 percent of full scale. The resolution is therefore 0.2 percent. From Eq. (10-5)

$$\text{Resolution}(\%) = \frac{100}{2^n}$$

$$0.2 = \frac{100}{2^n}$$

$$2^n = \frac{100}{0.2} = 500$$

Since $2^8 = 512$, 8 bits are required.

Example 11-3

The example just considered is further complicated. The MPU must now control 10 gun turrets sequentially in 0.4 ms. On a time basis, specify the requirements of the A/D converter, S/H circuit, and the multiplexer used.

Solution

If 10 turrets are sequentially controlled every 0.4 ms, each one must be scanned in 0.04 ms (0.4 ms/10), or 40 μs.

If an A/D converter with a conversion time of 20 μs and a S/H with an acquisition time (settling time) of 10 μs are selected, the requirement of 40 μs will be safely met. The multiplexer must be a 10×1 MUX, and it must be able to switch between channels during the 20-μs conversion time of the A/D. This time does not contribute to the total scan time per channel. In other words, once the S/H goes into the hold mode, the A/D begins converting. Simultaneously, the MUX is switched to the next channel. The next channel must appear at the MUX output before the S/H receives its next command to sample, which is 20 μs later when the EOC pulse goes low.

Example 11-4

Examine the errors introduced by the S/H and MUX in Example 11-3. The following information is known about the hardware used. A full scale of 10 V is used.

MUX	S/H
Leakage current = 50 nA Switch's on resistance = 2 kΩ Output resistance = 15 Ω	Input resistance = 10^8 Ω Gain error <0.01% Leakage current = 10 nA
OP-AMP output resistance = 5 Ω	

Solution

The S/H introduces a gain error no greater than 0.01 percent. In addition there is an offset error because the S/H leakage current causes a voltage drop across the MUX switch when it is on. This offset is

$$(10 \text{ na})(2.0 \text{ k}\Omega) = 20 \ \mu\text{V}$$

$$\frac{20 \ \mu\text{V}}{10 \text{ VFS}}(100\%) = 0.0002\%$$

The MUX has a transfer error due to loading by the S/H, given by voltage division

$$\text{S/H input} = \text{MUX output} \frac{\text{switch resistance}}{\text{input resistance} + \text{switch resistance}}$$

$$\frac{\text{S/H input}}{\text{MUX output}} = \frac{2 \text{ k}\Omega}{2 \text{ k}\Omega + 10^8} \approx 2 \times 10^{-5}$$

or an error of

$$2 \times 10^{-5}(100\%) = 2 \times 10^{-3}\% = 0.002\%$$

In addition the MUX leakage current causes an offset voltage at its input by creating a voltage drop across the OP-AMP output resistance

$$\text{Offset} = (50 \text{ na})(5\ \Omega) = 0.25\ \mu\text{V}$$

or an error of

$$\frac{0.25\mu}{10 \text{ VFS}}(100\%) = 2.5 \times 10^{-6}\%$$

It can be seen that all these errors are indeed negligible compared with the quantizing error specified in Example 11-2. There are also errors due to temperature variations, but these are also usually very small when compared with the quantizing error.

Example 11-5

In Example 11-3, under what conditions could the S/H be eliminated from the system?

Solution

The purpose of the S/H in this application is to hold the input to the A/D constant during the conversion time while the MUX is switching to the next channel. If the MUX does not switch channels until after the conversion, the only consideration is the frequency of the signal being converted. If during the conversion time the input to the A/D does not change by one LSB (the resolution in volts), the S/H is not necessary. The relationship between conversion time t_c and maximum signal frequency f_{max} will now be derived. Figure 11-7 shows a sine wave with an amplitude equal to the FSV of the converter. The maximum rate of change of the input occurs at the zero crossing and is obtained by differentiating the input and letting $t = 0$.

$$\text{Max. rate} = R_{max} = 2\pi f(\text{FSV}) \cos \omega t|_{t=0} = 2\pi f(\text{FSV}) \qquad \text{V/s}$$

The maximum change in voltage Δv during the conversion time is

$$\Delta v = R_{max} t_c = 2\pi f(\text{FSV}) t_c$$

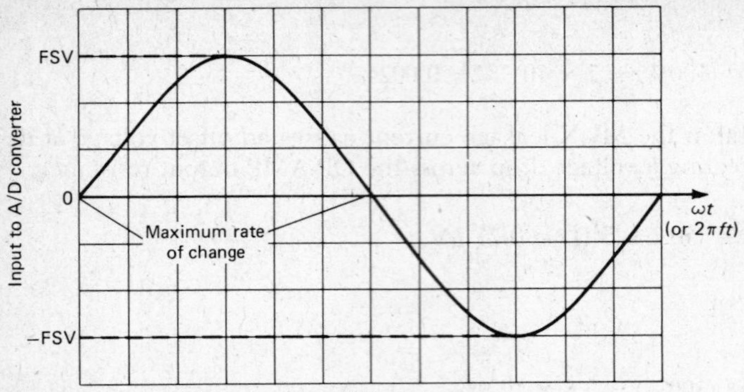

Fig. 11-7 Input to A/D converter: FSV sin $2\pi ft$

To eliminate the S/H, Δv must be less than one LSB, which is the resolution in volts. From the equation for an n-bit converter

$$\text{Resolution} = \frac{\text{FSV}}{2^n - 1} \approx \frac{\text{FSV}}{2^n} \qquad \text{(10-15)}$$

$$\Delta v < \text{resolution} \qquad \text{and} \qquad 2\pi f(\text{FSV})t_c < \frac{\text{FSV}}{2^n}$$

$$\text{Then} \quad f < \frac{1}{2\pi t_c 2^n} \qquad \text{(11-1)}$$

$$\text{or} \quad t_c < \frac{1}{2\pi f 2^n} \qquad \text{(11-2)}$$

In Example 11-3, $t_c = 20\ \mu s$ and $n = 8$; therefore

$$f < \frac{1}{2(3.14)(20\ \mu)2^8} = 31 \text{ Hz}$$

If the maximum frequency fed back to the computer is less than 31 Hz, the S/H is not necessary. However, it is important to note that the MUX cannot switch to the next channel now until the conversion is over. The MUX settling time ($\approx 1\ \mu s$) must be added to the 20-μs conversion time of the A/D to find the total conversion time ($\approx 21\ \mu s$). This is still within the requirement of 40 μs stated in Example 11-3.

PROBLEMS

11-1. What are the advantages of using computers on-line? Off-line?

11-2. When should data be transferred from control system to computer in parallel? In serial?

11-3. What is a MODEM? What does it do?

11-4. What frequencies does the originate MODEM convert a 1 and 0 to? What frequencies does the answer MODEM convert them to?

11-5. What is PCM? When is it used?

11-6. Define unipolar binary, bipolar binary, RZ, and NRZ.

11-7. The signal shown in Fig. P11-7 is sampled at 0.5 kHz starting at $t = 0$ and converted into 3-bit bipolar NRZ. Sketch the PCM signal over the first four sampling periods.

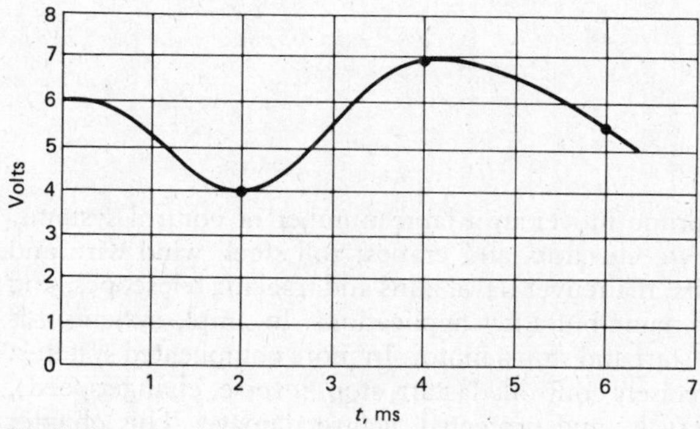

Fig. P11-7

11-8. Repeat Prob. 11-7 for a 4-bit unipolar NRZ PCM code.

11-9. Repeat Prob. 11-7 for a 3-bit unipolar RZ PCM code.

11-10. What interface device is used when transferring serial data to and from a microprocessor?

11-11. What interface device is used when transferring parallel data to and from a microprocessor?

11-12. The M6800 is used on-line in a radar-antenna control system. The antenna rotates 360°. The MPU must sense changes in antenna position of 0.2°. How many MC6820 PIAs are necessary to interface the MPU with the system?

11-13. Repeat Prob. 11-12 if the MPU must sense (a) a 1° change, (b) a 1.5° change.

11-14. A microprocessor is used on-line with 16 different systems. The highest frequency signal of the 16 systems is 100 Hz. If one A/D is used with a S/H and a 16 × 1 MUX, specify the requirements of the A/D conversion time and S/H acquisition time. Base your results on the theoretical sampling rate required.

11-15. Repeat Prob. 11-14 if the highest frequency is (a) 20 Hz, (b) 500 Hz.

11-16. Repeat Prob. 11-15 specifying the A/D conversion time if the S/H is eliminated from the system. Assume the MUX switching time between channels is 1 μs and that the resolution is 8 bits.

chapter 12

Motor Controls

The motor is the basic prime mover in a large number of control systems. Motors are used to operate elevators and cranes, roll steel, wind wire and paper, move conveyer belts, maneuver naval guns and tracking telescopes, and operate machine tools, to name but a few applications. In simple systems it is enough to use a switch to start and stop a motor. In more complicated systems, motors may have to be precisely controlled (start, stop, reverse, change speed), operated for specific intervals, and protected against damage. This chapter deals with the specific techniques for controlling the operation of motors.

12-1 FUNDAMENTALS OF MOTOR CONTROL

Motor controls are designed to meet the operating requirements of the motor under specified load conditions and are either manual or automatic. Manual controllers are hand-operated systems which generally perform simple operations. The three- and four-point starters used with dc shunt motors are examples of manual controllers. The operator merely moves a handle slowly until the motor comes up to full speed. Manual controllers are generally used to operate a simple machine and are usually located near the machine. With automatic control, an operator is required to start the process, but the actual control requires no human intervention. Automatic motor starters, for example, reproduce the features of manual starters but only require an operator to activate a momentary push-button switch. The switch can even be located remotely and used to start more than one machine.

Most controllers, whether manual or automatic, contain one or more protective devices for the motor. These include overload protection, low-voltage protection, and low-voltage release.

Manual controllers are basically open-loop devices. Feedback has to be supplied by the operator. For example, with a manual speed controller, the operator sets the control for a specific speed. When the speed changes due to variation in the load, the operator must readjust the controller to bring the speed back to the desired value. Automatic controllers may be open-loop or closed-loop. In the open-loop (no-feedback) system, the operator starts the controller, which then proceeds through a fixed sequence of events. Any change in the sequence due to power or load variations requires human intervention to reestablish the sequence.

With a closed-loop automatic controller, feedback is used to maintain the desired operating conditions. For example, an automatic speed controller will maintain speed at the set value regardless of changes in line voltage and load (within design limits) without human intervention. This means that a feedback device such as a tachometer must be included with the controller.

While the most common motor controllers are used for starting, stopping, and speed variation, others are used for positioning. A jogging controller is used to start the motor repeatedly for short periods of time to achieve a desired degree of rotation. Some controllers include timing devices to sequence one or more motors through various maneuvers, e.g., run motor 1 for 1 min, then start motor 2 only if motor 1 has stopped. The equipment used for sequencing, timing, and logic until recently was exclusively electromechanical (relays). Today, more and more control is being achieved with static (electronic) devices. With increasing use of the microprocessor, complex control functions can be achieved, and this implies increasing use of static control.

12-2 CONTROL COMPONENTS

This section is designed to introduce the various components required to start, stop, protect, and control the operation of motors. These include switches, circuit breakers, relays, and monitoring (pilot) devices.

12-2-1 Switches and Circuit Breakers

A switch is used to make, break, or change connections in an electrical system. Safety switches are manually operated devices used to connect or disconnect power to a circuit. The rating of the switch contacts must exceed the rating of the device to which power is being applied. It is obvious that a 1-A switch cannot be used to feed a motor which draws 5 A.

The number of positions on a switch is designated as the *throw*, while the number of switches which operate simultaneously is designated as the number of *poles*. Figure 12-1 illustrates a single-pole single-throw (SPST) switch. Single-throw implies that there is only one make position, the closed position. Single-pole means that there is only one switch. The single-pole double-throw switch (Fig. 12-2) has two make positions labelled NO (normally open) and NC (normally closed). The contactor arm or common terminal C normally makes contact with NC. When the switch is activated, C makes contact with NO while NC breaks. Figure 12-3 illustrates a double-pole single-throw (DPST) switch. This is actually two electrically isolated switches operated simultaneously by the same mechanism. To perform multiple functions, switches are also available with three, four, or more poles.

A circuit breaker is a device used to sense overload and interrupt the power to a system. Unlike fuses, which must be replaced after an overload condition, breakers can be moved back into the original closed position. Most breakers are designed for short-circuit protection and can handle the heavy current or arc which is created when a circuit is interrupted. Some low-voltage breakers employ a bimetallic strip, which flexes due to increases in temperature in the vicinity of the motor caused by overloads. The bimetallic type tend to be slow-acting except for short-circuit conditions when heat is generated rapidly. Magnetic breakers use a coil of wire wrapped around an armature to open the protective contacts due to sustained overload conditions.

12-2-2 Relays

A relay is an electromagnetic switch which can be used as a protective device, an amplifier, or a logic element. The basic components of the relay are the *coil*, which is wrapped around an iron core, and one or more sets of *contacts* as illustrated in Fig. 12-4. The relay is energized by applying rated current or voltage to the coil. The current creates a magnetic field which moves the armature to open or close the contacts.

Typically, a relay uses low voltage at the coil to control larger voltages through the contacts. In a home heating system, the thermostat is used to apply 24 V to the coil of a relay, which switches 115 or 230 V to operate the fan motor, as shown in Fig. 12-5.

Relays can also be used to provide a *latch*. In many motor starters, the operator momentarily pushes a start button which activates a relay to apply power to the starter. In Fig. 12-6 if switch 1 is closed momentarily, the relay will energize and close its own contacts, labeled A_1. The relay keeps itself energized through these contacts even after switch 1 is opened to provide power to the starter. If switch 2, the stop button, is opened momentarily, the relay deenergizes and removes power from the starter.

Fig. 12-1 Single-pole single-throw switch.

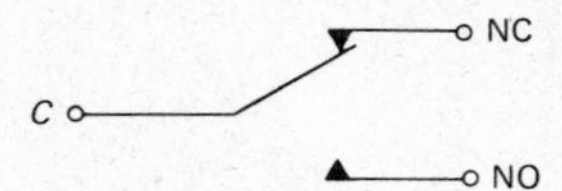

Fig. 12-2 SPDT switch.

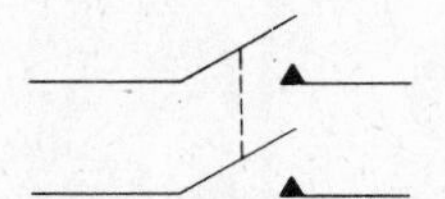

Fig. 12-3 DPST switch.

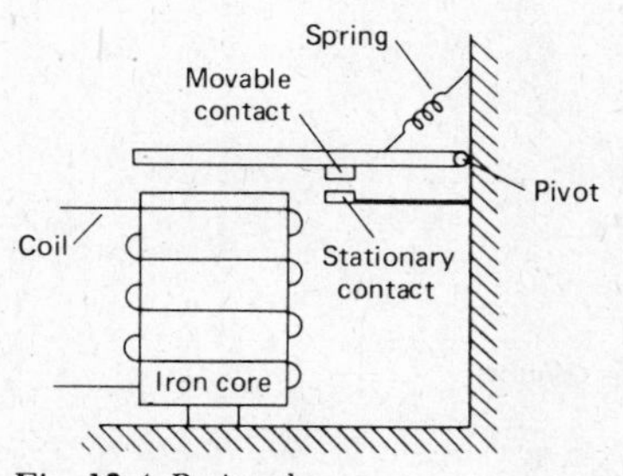

Fig. 12-4 Basic relay.

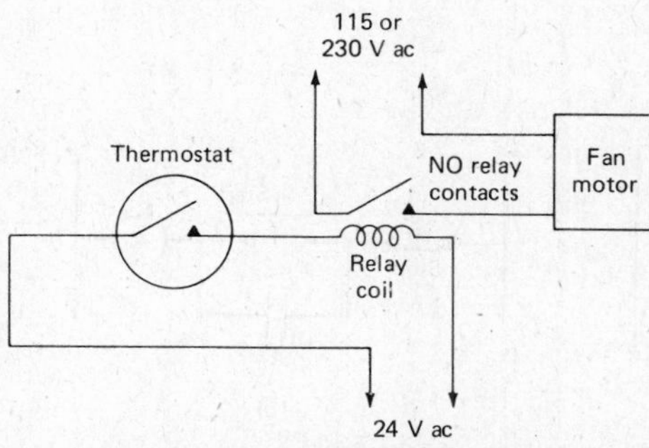

Fig. 12-5 Using a relay to switch voltage to a motor.

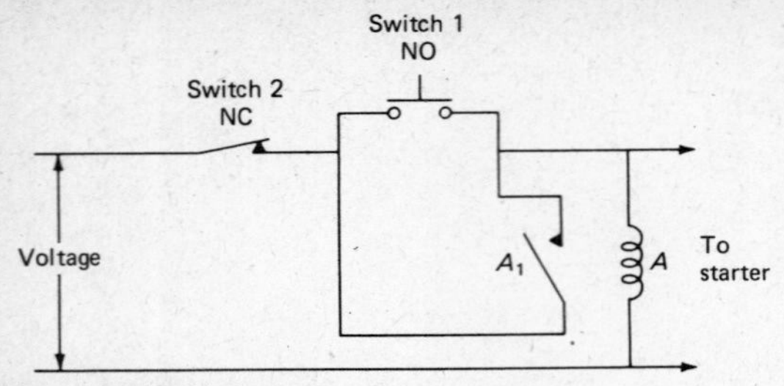

Fig. 12-6 Relay latching circuit.

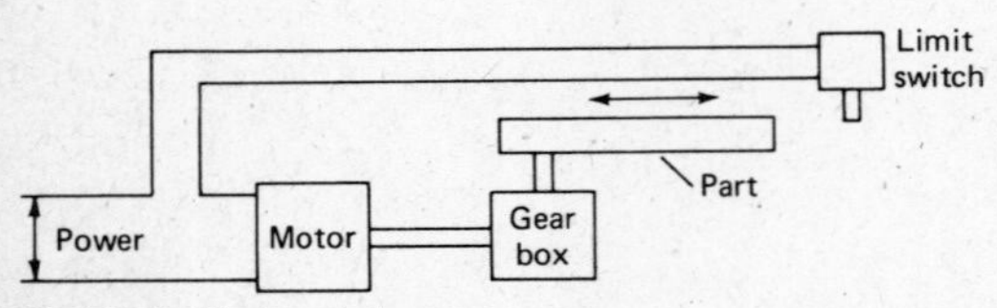

Fig. 12-7 Limit switch.

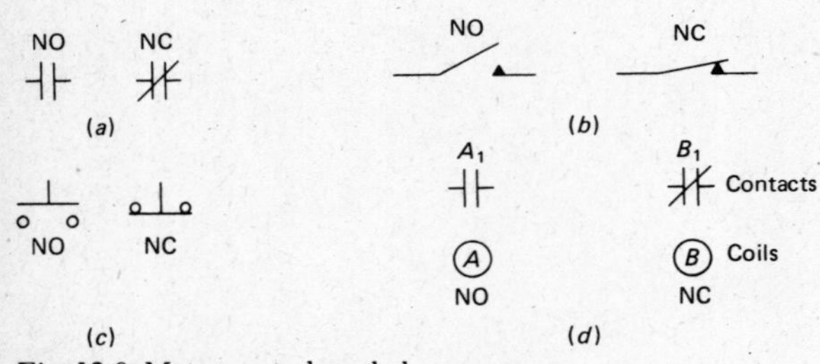

Fig. 12-8 Motor control symbols.

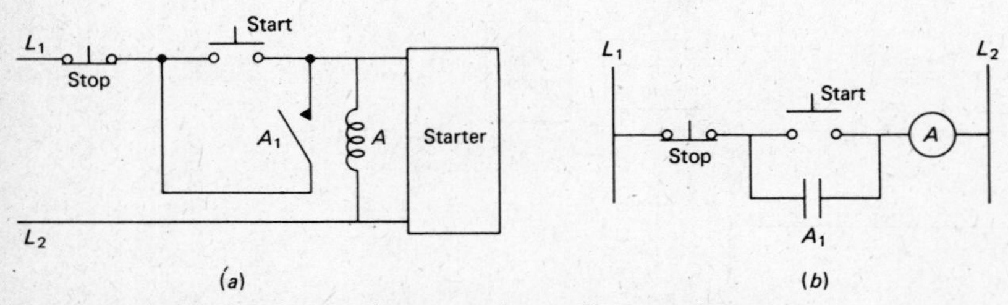

Fig. 12-9 Comparison of wiring diagrams and control-circuit schematics.

Time-delay relays are used to provide time sequencing in control circuits or to prevent sudden overloads from disconnecting power to a control circuit. The armature in these relays moves through a fluid damper similar to the shock absorber in an automobile. A sudden inrush of current to the relay coil cannot overcome the friction in the damper. Sustained current of the proper magnitude will cause armature motion after a specified interval.

12-2-3 Sensing Devices

Sensing devices are used to provide feedback in control circuits. For example, a float switch can be used to stop a motor operating a pump delivering liquid to a tank. When the level reaches a set value in the tank, the float-switch contacts open and stop the pump.

Limit switches are used to prevent a motor from turning or moving beyond a certain point. In Fig. 12-7 a motor is used to position a part to the right through a gear box. At the extreme right-hand position, the part hits the limit-switch arm, causing the switch contacts to open and stop the motor.

A *plugging switch* is used to bring a motor to an abrupt stop. The contacts in the plugging switch are centrifugally operated and connected to reverse polarity to the motor when the stop button is pressed. The sudden reversal causes quick deceleration, which opens the contacts and disconnects power.

Many other sensing devices provide control functions of such variables as pressure and temperature. The reader is referred to textbooks on motor controls and instrumentation for further details.

12-3 CONTROL CIRCUITS

The design of control circuits is carried out by preparing circuit diagrams which use symbolic representations for the various components. The interconnection between these components is actually the "logic" required to perform specified functions.

12-3-1 Symbols

Figure 12-8 illustrates a typical set of symbols used for motor control circuits. The automatically operated contacts in Fig. 12-8*a* are shown in their deenergized state. Contacts of this type are part of a relay, starter, or some other automatic switch. Figure 12-8*b* represents manual toggle-type switches. The momentary push-button switches of Fig. 12-8*c* are used extensively in motor starters. Relay coils and their associated contacts are shown in Fig. 12-8*d*.

12-3-2 Control Circuit Schematics

Figure 12-9 demonstrates the difference between a wiring diagram and a control-circuit schematic. Figure 12-9*a* demonstrates the latching circuit used to start a motor automatically. Notice the simplicity of the schematic in Fig. 12-9*b*. In complex control-circuit design, the schematic makes it easier for the engineer or technician to develop the logic required to perform a specified sequence of functions. Notice that Fig. 12-9*b* does not include the starter. It merely demonstrates the circuit necessary to latch the relay after the start button is pressed.

Suppose we want to design a circuit to start two different motors but the second motor can be started only after the first motor has been started. Figure 12-10 shows the control circuit necessary to meet these specifications. When switch 1 is depressed, relay A energizes and starts motor 1. Relay contacts A_1 are used to latch relay A, and contacts A_2 allow motor 2 to start when switch 2 is depressed. If the stop switch for motor 2 is pressed, only motor 2 will stop. If the stop switch for motor 1 is pressed, motor 1 stops but relay contacts A_2 also open and stop motor 2.

12-4 MOTOR STARTERS

A starter is a control component used to start a motor and bring it up to speed. Starters not only deliver power to the motor but also provide the necessary current limiting to protect the motor during the start-up period. In manual starting, the operator usually moves a lever through a sequence of positions. Automatic starters are generally controlled by remote momentary push buttons, which activate a logic circuit to carry out the desired starting sequence.

12-4-1 Manual DC Motor Starters

When a dc motor is started, the back emf is so low that the line voltage would cause a large inrush of current to the armature. A series resistor will reduce the current but also prevent rated current flow when the motor comes up to speed. For proper operation, this series resistor must be reduced as the motor speed increases, finally dropping to zero at rated speed.

Figure 12-11 illustrates a manual commercial starter for a dc shunt motor. As the lever is rotated clockwise, the series resistance is gradually reduced while the motor comes up to speed. At the extreme right-hand position, the resistance is zero. The starter also has connections to provide power for a holding coil and the shunt field. The holding magnet energized by the coil is used to

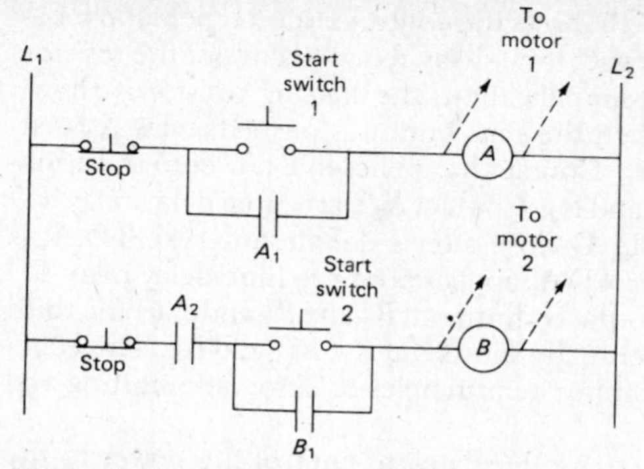

Fig. 12-10 Control circuit to start two motors.

hold the lever in the RUN position. The lever is spring-loaded and in the event that power fails or line voltage drops, the magnet releases the lever and disconnects power to the motor.

12-4-2 Automatic DC Motor Starters

Automatic starters can be open-loop or closed-loop devices. In an open-loop starter, the power to the motor is controlled in a definite sequence independent

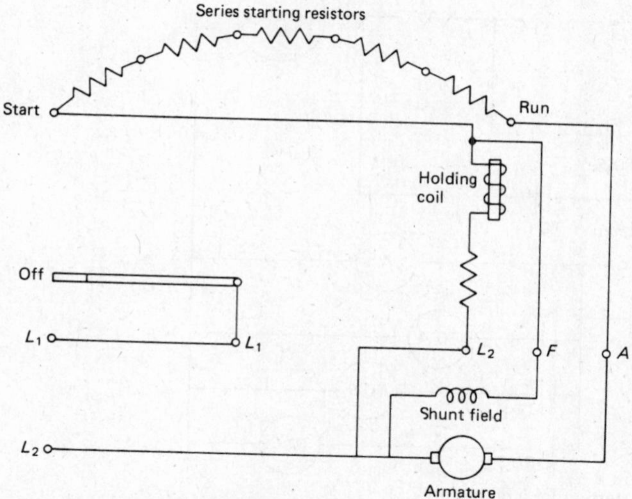

Fig. 12-11 Commercial starter for a dc motor.

of the motor operation. Figure 12-12 shows the control-circuit schematic for an open-loop dc motor starter which uses time-delay relays to cut out the resistors in series with the armature. For simplification, the starting resistor is shown being cut out in three stages. When the start button is pushed, relay A is energized and latched by contacts A_1. Contact A_2 applies line power to the armature circuit in series with R_1, R_2, and R_3. Contact A_3 starts time delay relay TD A, whose contacts TD A_1 and TD A_2 close after a definite interval. TD A_1 is used to shunt out resistor R_1 while TD A_2 applies power to time-delay relay TD B. When this relay activates, its contacts shunt out R_1 and R_2 and start the third time-delay relay TD C, which eventually shorts out R_1, R_2, and R_3. This schematic is only designed to demonstrate the principles of automatic starting and is by no means complete.

Closed-loop automatic starters use feedback to control the power to the motor in response to the operation of the motor. Figure 12-13 illustrates the use of relays connected across the armature which energize at specific values of back emf developed by the armature. When the motor is started, the emf across the armature is too small to energize any of the relays, so that R_1, R_2, and R_3 remain in series to limit the armature current. As the motor accelerates, relay B is energized, causing contact B_1 to close and shunt out R_1. As the speed increases, relays C and D energize in sequence, causing R_2 and R_3 to be shunted. Relay A is used to provide the power latch.

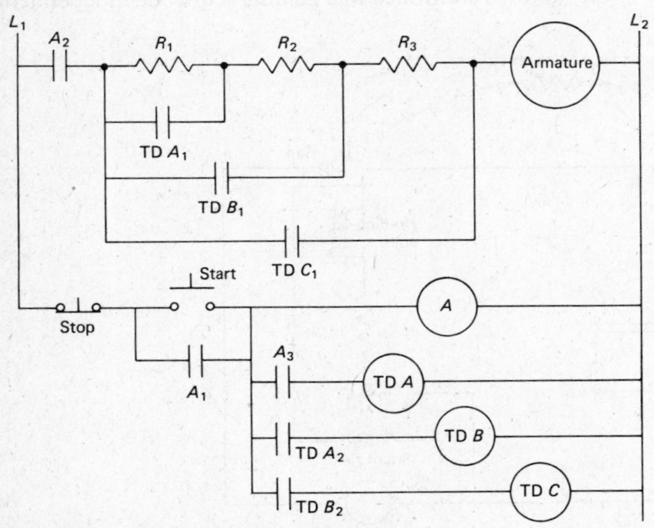

Fig. 12-12 Automatic starting for a dc motor.

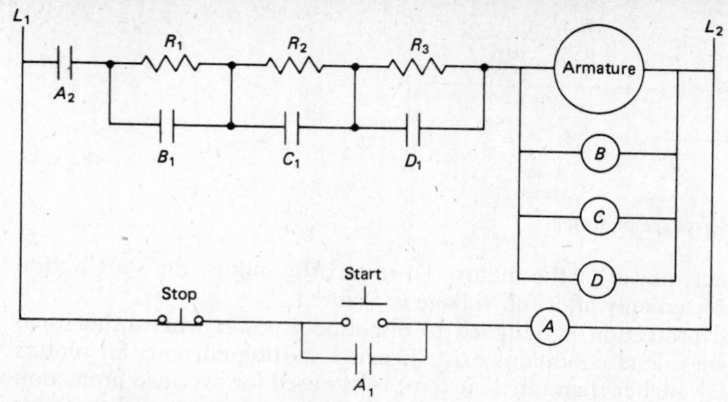

Fig. 12-13 Automatic starter with feedback.

12-5 MOTOR-PROTECTION DEVICES

There are four basic forms of protection for motors: undervoltage release, undervoltage protection, overload protection, and short-circuit protection. Undervoltage release is a technique used to disconnect power to a motor when voltage drops below a specified value but restore it when the voltage returns to normal. Figure 12-14 illustrates a circuit employing an undervoltage relay. When the line voltage drops, the contacts A_1 and A_2 of relay A open and disconnect power to the motor control circuit. When the voltage returns to normal, the relay energizes and supplies power to the circuit.

With undervoltage protection, a motor must be restarted manually after the line voltage drops below a specified value. The same undervoltage relay used in Fig. 12-14 can be wired as shown in Fig. 12-15 to provide undervoltage protection. If relay A in Fig. 12-15 senses a drop in voltage, contact A_1 opens

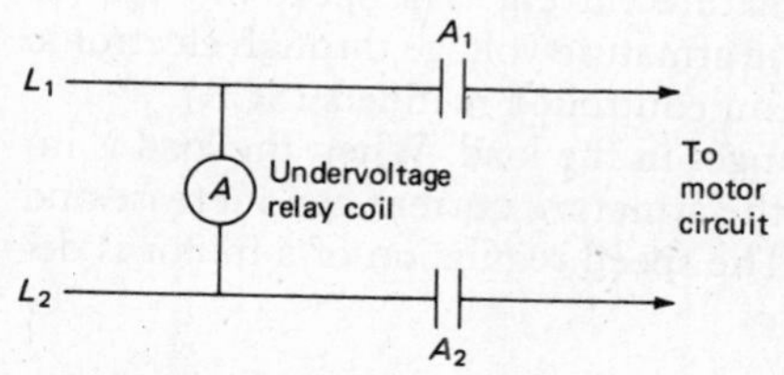

Fig. 12-14 Undervoltage release.

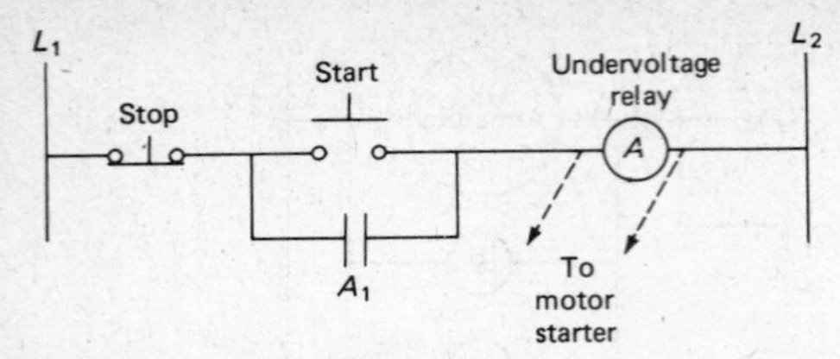

Fig. 12-15 Undervoltage protection.

and disconnects power to the motor. To restart the motor, the start button must be depressed only after full voltage is restored.

Overload protection is designed to disconnect power when other than temporary excess-load conditions exist. Because starting currents for motors are often much higher than rated current, relays used for overload protection must be capable of handling these currents without causing a disconnect during start-up. This is easily accomplished because of the slow response time of most overload relays. For short-circuit protection, fuses and magnetic or thermal breakers with much faster response times are used. The current ratings of these devices are much higher than the overload relays used in the same circuit.

12-6 SPEED CONTROL

When the field flux in a dc motor is reduced, the back emf drops, causing an increase in the armature current. The resulting increase in the developed torque, at a fixed load, increases the speed to a point where the torque delivered is reduced sufficiently to drive the load at the higher speed. Similarly, an increase in flux causes a reduction in speed. A common method of controlling the speed of shunt and compound motors is to change the resistance in the field circuit, thereby changing the field flux.

In servomechanisms employing small permanent-magnet dc motors, speed is automatically varied by changing the armature voltage. The changes in armature voltage cause changes in the armature current. The speed of larger dc motors can also be adjusted by varying the armature voltage through electronic controls employing devices such as silicon controlled rectifiers (SCR).

Motor speed is also affected by changes in the load. When the load is increased, the motor slows down, so that the armature current can increase and supply the additional torque required. The speed regulation of a motor is defined by

$$\text{Regulation}(\%) = \frac{n_{nl} - n_{fl}}{n_{fl}} \qquad \text{(12-1)}$$

where n_{nl} is no-load speed and n_{fl} is full (rated) load speed, both in revolutions per minute. Clearly, a large percent regulation implies poor regulation.

Example 12-1

A 120-V dc shunt motor whose armature resistance is 0.2 Ω draws an armature current of 20 A at 1000 r/min. When the field flux is reduced 2 percent, the armature current eventually drops to 5 A. Find the new speed of the motor.

Solution

Using Fig. 12-16, we can determine the back emf as

$$V_b = V_t - I_a r_a$$

Use the subscript 1 to denote values at 1000 r/min and the subscript 2 to indicate the new speed; then

$$V_{b1} = 120 - 20(0.2) = 116 \qquad V_{b2} = 120 - 5(0.2) = 119$$

Because the flux dropped 2 percent,

$$\Phi_2 = 0.98\Phi_1$$

The back emf V_b is proportional to flux and speed,

$$V_b \propto \Phi n$$

Then $$\frac{V_{b2}}{V_{b1}} = \frac{\Phi_2 n_2}{\Phi_1 n_1}$$

or $$n_2 = \frac{\Phi_1 n_1 V_{b2}}{\Phi_2 V_{b1}} = \frac{\Phi_1}{\Phi_2}\frac{V_{b2}}{V_{b1}} n_1 = \frac{1}{0.98}\frac{119}{116}(1000) = 1047 \text{ r/min}$$

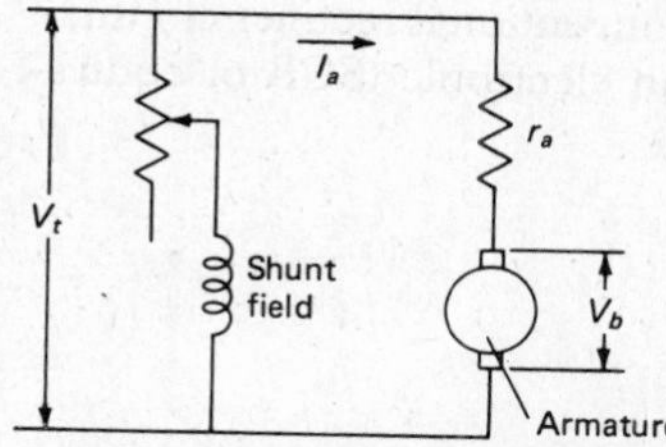

Fig. 12-16 Schematic for Example 12-1.

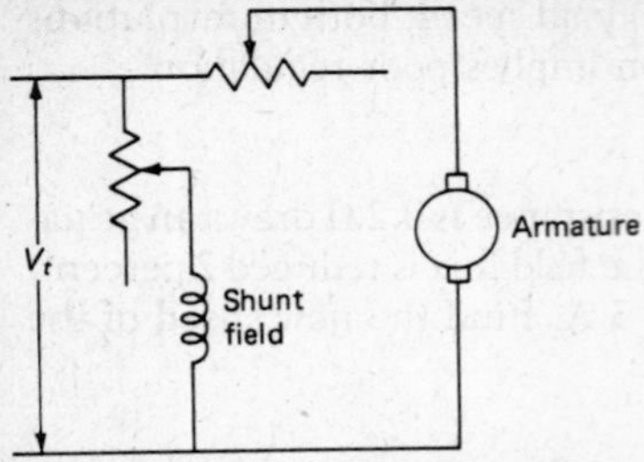

Fig. 12-17 Varying speed with field and armature resistance control.

Example 12-2

At no load, the speed of a motor is 1050 r/min. At rated load the speed drops to 1000 r/min. Determine the speed regulation.

Solution

$$\%\text{ regulation} = \frac{n_{nl} - n_{fl}}{n_{fl}} \times 100 = \frac{1050 - 1000}{1000} \times 100 = 5\%$$

12-6-1 Methods of Speed Control

Figure 12-17 shows how field and armature resistance can be varied to change the speed of a dc motor. As the field resistance is increased, flux is reduced and the speed increases. While this technique has little effect on speed regulation due to load variation, if the field flux is reduced considerably, extreme increases in speed can occur. When the resistance in series with the armature is increased, the armature current drops and produces a reduction in speed. However, this technique results in poor speed regulation as well as producing considerable power (heat) loss in the series resistor.

The speed of permanent-magnet or separately excited dc motors can be adjusted by varying the dc voltage applied to the armature, as shown in Fig. 12-18. This method provides a wide range of speed variation, relatively low power loss, and good speed regulation. The variable dc voltage can be supplied from a dc generator (Ward-Leonard system), a conventional rectifier circuit, a power amplifier (servo-motor applications), or an electronic (SCR or equivalent) control.

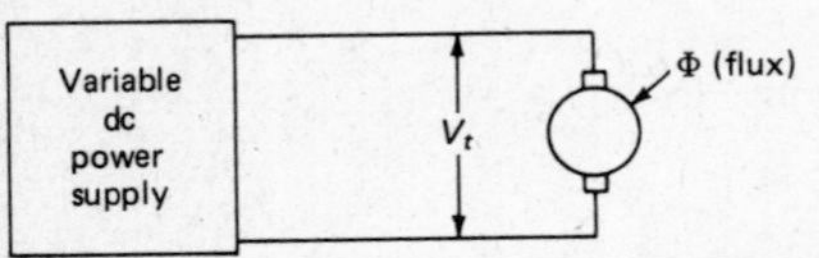

Fig. 12-18 Speed control using armature voltage.

12-6-2 Electronic Motor Control

The SCR provides a convenient and efficient method for controlling the dc armature voltage for a motor from an ac power source. When an ac voltage is applied between the anode and cathode of the SCR shown in Fig. 12-19, current can flow only during the positive half cycle of the sine wave. Furthermore, the duration of this dc flow can be controlled by the application of a positive voltage pulse to the gate. By controlling the time at which the gate pulse is applied, the time at which the SCR is triggered can be changed. For example, if the gate pulse is applied after one-third of the positive half cycle, the SCR will only conduct for the remaining two-thirds of the positive half cycle. The variation of the average direct current through the armature due to controlled application of the gate voltage results in control of speed over wide ranges.

To reduce ripple due to half-wave rectification, full-wave configurations such as SCR bridge circuits or three-phase circuits are generally used. In some cases, 12 SCRs are used with three-phase ac power to provide bidirectional speed control. Speed control of small permanent-magnet dc motors can also be accomplished using pulse-width modulation. An integrated circuit is used to change the duty cycle ("ON" time) of a series of short constant-voltage pulses applied to the armature. Since this, in effect, changes the average dc level in the armature, the speed will vary with the on time of the pulses.

12-6-3 Maintaining Constant Speed with Feedback

Two basic feedback methods are used to maintain speed as the load on the motor changes. In the first method, a tachometer is used to sense speed and vary the voltage applied to the motor in a standard closed-loop system (Fig. 12-20). To eliminate the ripple and loading effects of mechanically coupled dc tachometers, optically coupled ac tachometers can be used when a high degree

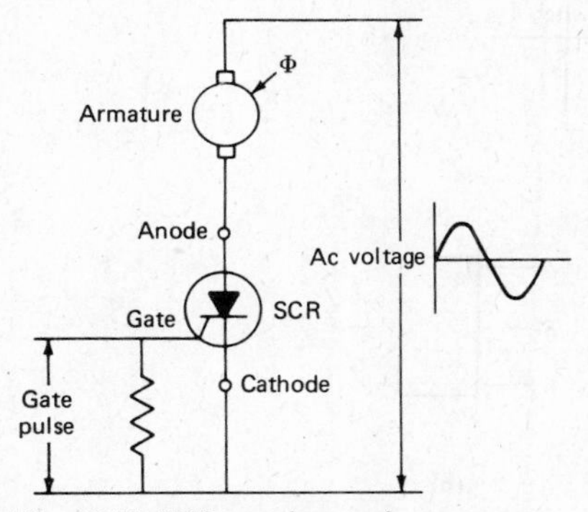

Fig. 12-19 SCR speed control.

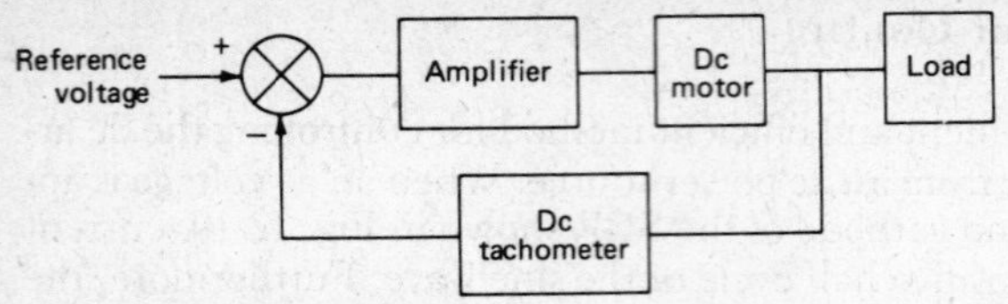

Fig. 12-20 Tachometer feedback.

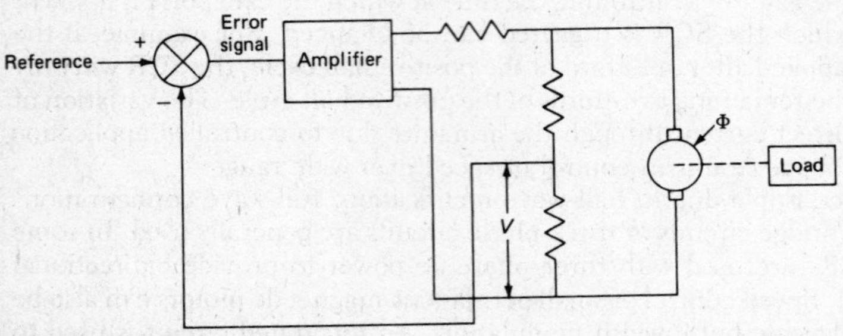

Fig. 12-21 Back-emf feedback.

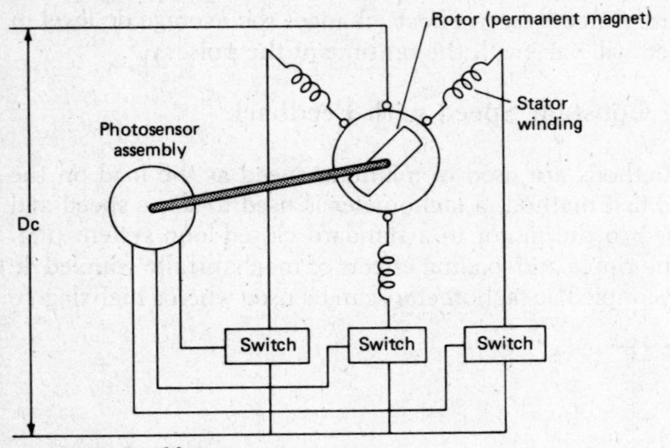

Fig. 12-22 Brushless motor.

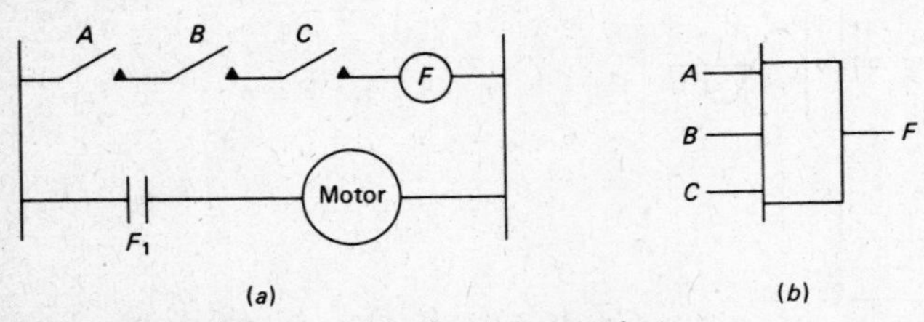

Fig. 12-23 AND gates: (*a*) relay logic; (*b*) static logic.

of speed control is desired. These systems employ a crystal oscillator to provide a frequency and phase-lock loop.

The second method for speed control uses the back emf of the motor to provide the feedback signal. Figure 12-21 shows a voltage divider connected across the armature of the motor. As the load on the motor increases, the back emf is reduced and V decreases. The drop in feedback voltage causes the error signal to rise and increase the armature voltage applied to the motor. This tends to bring the motor back up to speed.

12-6-4 Speed Control of Brushless Motors

A brushless motor usually contains a permanent-magnet rotor, a wound stator, and a method for sensing rotor position. Electronic switches operated by the rotor position sensors act like commutators to energize the proper stator winding for continuous torquing in one direction. Brushless motors require less maintenance than conventional motors, eliminate brush arcing, are more efficient than small servomotors, and permit use of small signals to control speed. Figure 12-22 shows a three-winding brushless motor which uses an optical method to sense rotor position. As the rotor turns, the optical sensors switch the proper windings on and off. Because the power circuitry is included in the brushless motor, low-power techniques can be used for speed control. Pulse-width modulation can be used to control the time the switches remain on after they are switched by the photosensors. The electronic switching signals can also be used to provide digital signals proportional to speed. The digital signal can be integrated into a closed-loop servo.

12-7 STATIC CONTROL

Historically, the automatic control of electrical machines was accomplished with relay logic and relay switching circuits. The requirement for low power, high speeds, and complex computer-compatible logic circuits resulted in the development of static control. Static control replaces electromechanical devices (relays) with electronic components.

The decision making in a static-control circuit is performed by elements known as *logic gates*, *memory*, and *delay*. Actual switching of the machinery can be performed with solid-state relays.

12-7-1 Static Control Elements

The logic gate is designed to develop a signal based on the presence or absence of other signals. Static control recognizes only two signals called ON and OFF, 1 and 0, or high and low. Figure 12-23 compares a relay logic circuit with a static AND circuit. In the relay circuit, if all three switches are closed, the relay coil F will energize and close contact F_1. The static AND circuit shows that F will be ON only if inputs A, B, and C are all ON. While the relay circuit shows both logic and switching, the static control circuit merely illustrates the logic.

Figure 12-24 illustrates an OR gate. In this case, if any one or more switches are closed, the relay is energized. The static OR gate will provide an ON signal at F if input A or B or C or any combination of inputs is ON.

The relay NOT function in Fig. 12-25 causes the normally closed relay contact F_2 to open when switch A is closed. In the static NOT circuit, if A is ON, F is OFF and vice versa. Some static control systems employ NAND and NOR logic instead of AND and OR. While the circuit configuration is different with NAND and NOR, the basic principles are the same and will not be covered here.

The relay memory is a latching circuit described earlier for motor starters. When momentary button A in Fig. 12-26 is depressed, relay F is latched by contact F_1. Contact F_2 maintains power to the motor. When contact B is opened, the latch is released. In the static memory, if A is ON momentarily, F is ON and remains ON. When B is turned ON momentarily, F returns to the OFF state.

Static delays illustrated in Fig. 12-27 provide the same functions as time-delay relays. In the on-delay type, F will turn ON some time after A is turned ON. With the off-delay type, F is ON when A is OFF. When A is turned ON, F turns OFF after the delay period.

12-7-2 Static-Control Output Signals

The output of static control circuits cannot be used to drive motors, relays, or other switches directly. Instead, these signals are amplified before application to any external load. However, for many applications, static switches can be connected to the static control circuit. The solid-state relay is a static switch used to reproduce the action of an electromechanical relay. The logic signal can be used to operate a light-emitting diode (LED), which optically couples the signal to a power-switching circuit.

12-7-3 Development of a Static Control Circuit

This section demonstrates the development of a simple static control circuit, which is to turn on a small mixing motor only when a level switch in the mixing tank is on; 10 s after the motor starts, it is to turn on a valve which feeds raw material into the tank. The circuit is shown in Fig. 12-28. When the start signal is turned ON, the memory latches signal A ON. As soon as the level switch produces an ON signal at B, the AND turns signal F on, which activates the

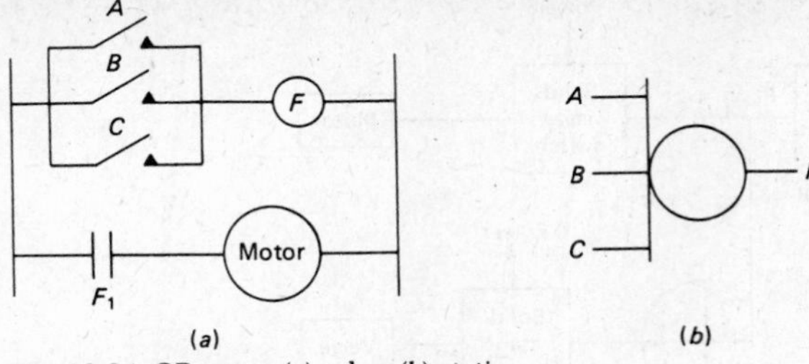

Fig. 12-24 OR gates: (*a*) relay; (*b*) static.

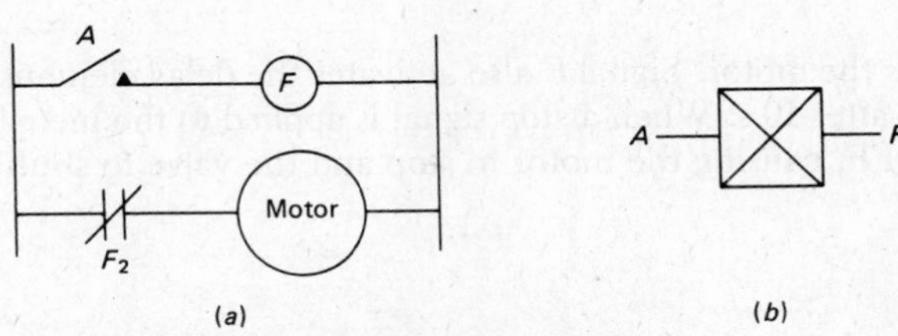

Fig. 12-25 NOT circuits: (*a*) relay; (*b*) static.

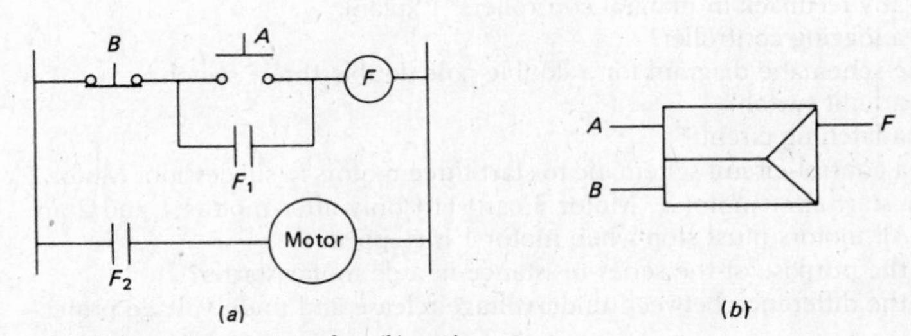

Fig. 12-26 Memory: (*a*) relay; (*b*) static.

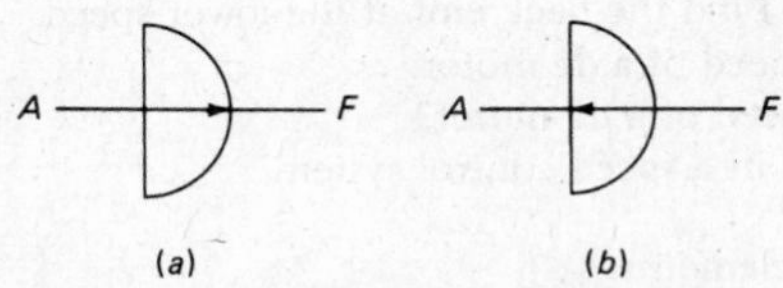

Fig. 12-27 Static delays: (*a*) ON delay; (*b*) OFF delay.

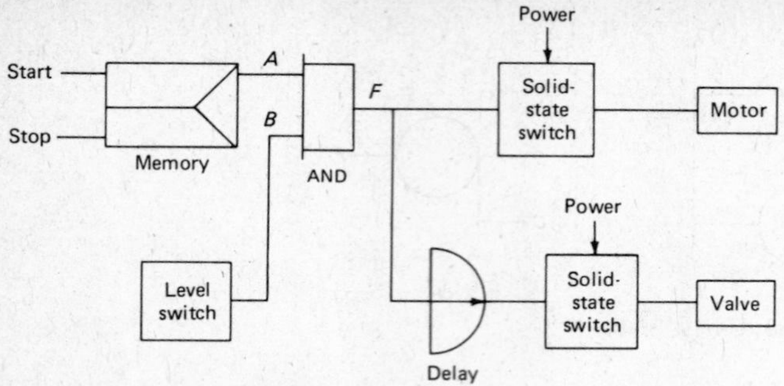

Fig. 12-28 Static control circuit.

solid-state relay and starts the motor. Signal *F* also activates the delay element which operates the valve after 10 s. When a stop signal is applied to the memory, signal *A* is turned OFF, causing the motor to stop and the valve to shut.

PROBLEMS

12-1. What is the main difference between a manual and automatic motor controller?
12-2. Name three types of protective devices normally included with motor controllers.
12-3. Is there any feedback in manual controllers? Explain.
12-4. What is a jogging controller?
12-5. Draw the schematic diagram for a double-pole double-throw switch.
12-6. What is a limit switch?
12-7. What is a latching circuit?
12-8. Prepare a control-circuit schematic to start three motors in succession. Motor 2 can only start after motor 1. Motor 3 can start only after motors 1 and 2 are started. All motors must stop when motor 1 is stopped.
12-9. What is the purpose of the series resistance in a dc motor starter?
12-10. What is the difference between undervoltage release and undervoltage protection?
12-11. Determine the no-load speed of a motor whose speed at rated load is 900 r/min and whose speed regulation is 3 percent.
12-12. At 1000 r/min, the back emf of a dc motor is 115 V. When the flux is increased 4 percent, the speed drops to 980 r/min. Find the back emf at the lower speed.
12-13. Name three methods for varying the speed of a dc motor.
12-14. How is the SCR used to control the speed of a dc motor?
12-15. What is the purpose of the tachometer in a speed control system?
12-16. What is a brushless motor?
12-17. Repeat Prob. 12-8 using static control elements.

chapter 13

Analog Computers and Simulation

An analog computer is a device which simulates the variables of a system in terms of similar physical quantities easy to generate and control continuously. Today, all analog computers are electronic, so that the quantity used is voltage. As an example, consider the differential equation

$$\ddot{y} + a\dot{y} + by = c \sin \omega t$$

This equation is merely a mathematical expression and is not necessarily related to electrical quantities. Suppose it is assumed that y is a voltage; then $\dot{y}$, $\ddot{y}$ and $c \sin \omega t$ would also be voltages. In reality, the equation may represent a chemical process, but there is no way of "seeing inside" this process. However, the process can be simulated electronically on a computer in order to observe the dynamics of the system.

The advantage of simulation via an analog computer is that the results are usually presented in the form of a continuous graph of the required variables so that system performance is easily observable. However, analog computers are inherently inaccurate, so that they are never used as a substitute for a digital computer when detailed mathematical information is desired.

13-1 BLOCK DIAGRAMS

The required building blocks of an analog computer are

1. Integrator
2. Differentiator
3. Constant multiplier
4. Summer

These building blocks operate on electrical signals in such a way as to produce the desired changes at their output. For example, see Fig. 13-1. Since the input is an electrical quantity, the output is also an electrical quantity. To show how these building blocks are used, again consider the equation

$$\ddot{y} + a\dot{y} + by = c \sin \omega t \tag{13-1}$$

Solving for $\ddot{y}$, we get

$$\ddot{y} = c \sin \omega t - a\dot{y} - by \tag{13-2}$$

Equation (13-2) is represented in Fig. 13-2. Starting with $\ddot{y}$, Fig. 13-3 shows how to generate y. But $\ddot{y}$ can be simulated as shown in Fig. 13-2. Then the interconnection suggested by Figs. 13-2 and 13-3 constitutes an analog simulation, which is shown in Fig. 13-4.

Once the system is started by application of the signal $c \sin \omega t$ and the proper initial conditions (discussed later) are inserted, the variables $\ddot{y}$, $\dot{y}$, and y will appear as continuous voltages which can be recorded.

To simplify the construction of diagrams like the one in Fig. 13-4, the following symbols will be used:

1. Integration

   $$y = \int \dot{y}\, dt = \frac{1}{s}\dot{y}$$

   See Fig. 13-5.
2. Differentiation

   $$y = \frac{dy}{dt} = sy$$

   See Fig. 13-6.

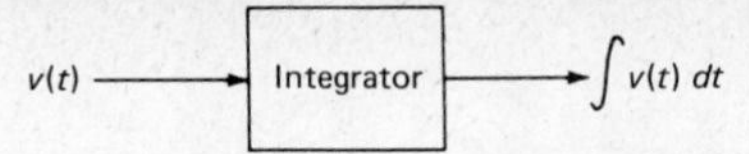

Fig. 13-1 Block diagram of an integrator.

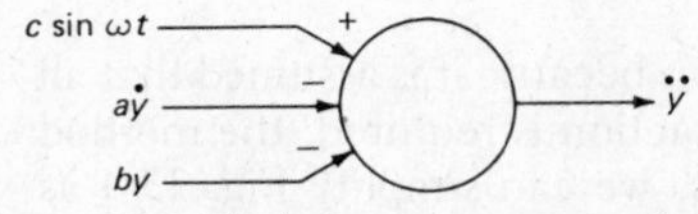

Fig. 13-2 Block-diagram representation of Eq. (13-2).

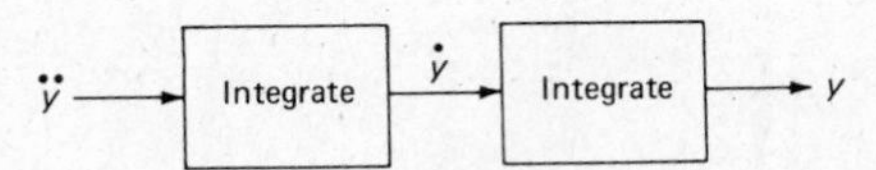

Fig. 13-3 Generating *y* by integration.

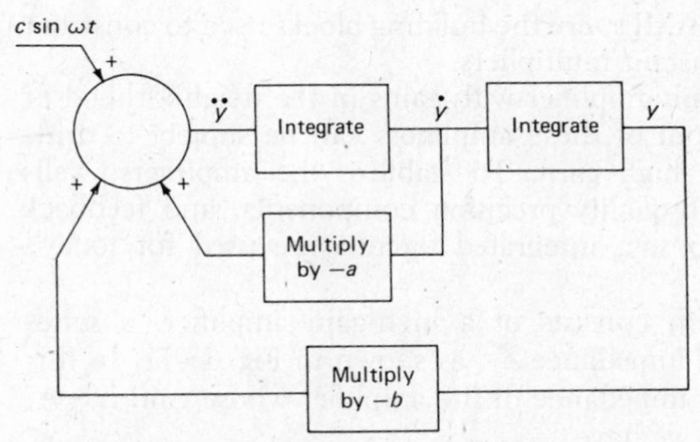

Fig. 13-4 Simulation of $\ddot{y} + a\dot{y} + by = c \sin \omega t$.

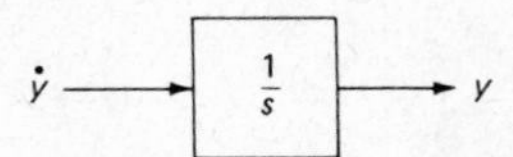

Fig. 13-5 Block diagram for integration.

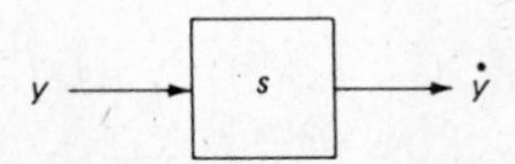

Fig. 13-6 Block diagram for differentiation.

3. Multiplication by a constant

$$y = ax \qquad \text{or} \qquad y = -ax$$

See Fig. 13-7.
4. Summation, see Fig. 13-8.

Notice that no plus or minus signs are shown because it is assumed that all values entering the junction are added. If subtraction is required, the method of Fig. 13-9 can be used. Using these symbols, we can simplify Fig. 13-4 as shown in Fig. 13-10.

13-2 OP-AMPS

Today's electronic analog computers are composed mainly of electronic amplifiers, resistors, and capacitors. Inductance is avoided because of the presence of appreciable amounts of resistance and capacitance in inductors. The electronic amplifiers used, the OP-AMPs, are the building blocks used to construct integrators, summers, and constant multipliers.

The OP-AMP is a high-gain amplifier with gains in the neighborhood of 5×10^4 to 30×10^6. The output of these amplifiers can be subject to drift, which is accentuated by the high gain. To stabilize the amplifiers, well-regulated power supplies, high-quality precision components, and feedback circuits are used. Needless to say, integrated circuits are used for today's OP-AMPs.

An OP-AMP configuration consists of a high-gain amplifier, a series impedance Z_i, and a feedback impedance Z_f, as shown in Fig. 13-11. In this figure, Z_g represents the input impedance of the amplifier whose gain is $-A$. Using node and loop analysis, we have

$$E_i = I_i Z_i + E_g \tag{13-3}$$

$$I_i + I_f = \frac{E_g}{Z_g} \tag{13-4}$$

$$I_f = \frac{E_o - E_g}{Z_f} \tag{13-5}$$

but $$E_o = -AE_g \tag{13-6}$$

Now let $Z_g \rightarrow \infty$ and $A \rightarrow \infty$

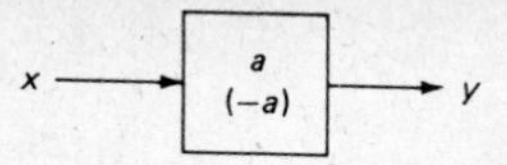

Fig. 13-7 Block diagram for multiplication by a constant.

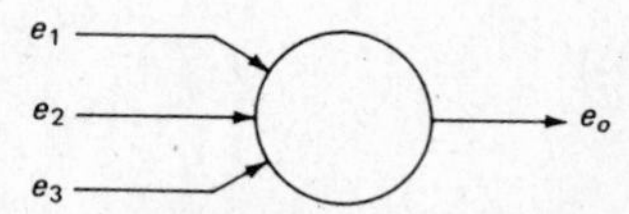

Fig. 13-8 Block diagram for summation.

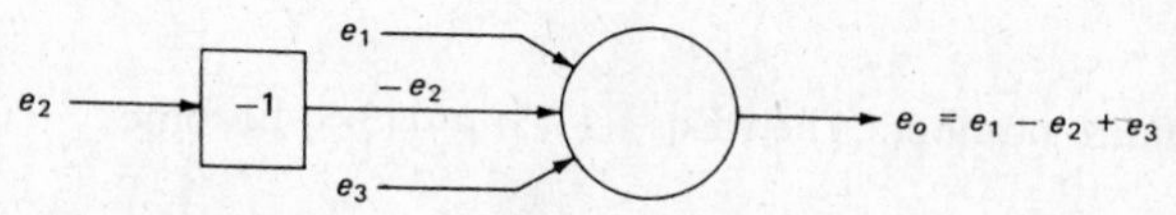

Fig. 13-9 Modifying Fig. 13-8 to include subtraction.

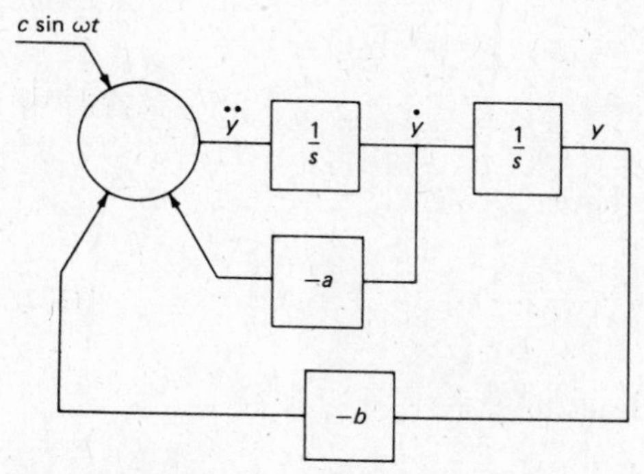

Fig. 13-10 Simplification of Fig. 13-4.

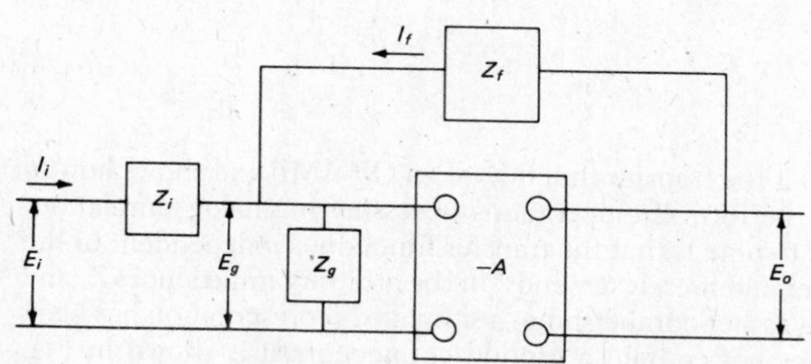

Fig. 13-11 OP-AMP configuration.

Then Eq. (13-4) becomes

$$I_i + I_f = 0 \tag{13-7}$$

and rewriting (13-6) as

$$E_g = -\frac{E_o}{A} \tag{13-8}$$

gives

$$E_g \approx 0$$

since $A \approx \infty$ and E_o is a finite number. Then Eqs. (13-3) to (13-5) become

$$E_i = I_i Z_i \tag{13-9}$$

$$I_i + I_f = 0 \qquad \text{or} \qquad I_i = -I_f \tag{13-10}$$

$$I_f = \frac{E_o}{Z_f} \tag{13-11}$$

Substituting (13-11) into (13-10), we have

$$I_i = -\frac{E_o}{Z_f} \tag{13-12}$$

and substituting (13-12) into (13-9) leads to

$$E_i = -\frac{E_o}{Z_f} Z_i \tag{13-13}$$

$$\text{or} \quad \frac{E_o}{E_i} = -\frac{Z_f}{Z_i} \tag{13-14}$$

Equation (13-14) is the transfer function of an OP-AMP and shows how to utilize this device to perform the operations necessary in analog simulation. The important thing to note is that the transfer function is independent of the gain A of the amplifier and merely depends on the arbitrary impedances Z_i and Z_f. Since Fig, 13-11 is rather cumbersome, a simplified representation has been developed which does not contain a ground connection; it is shown in Fig. 13-12. Remember that when the actual computer circuits are made, the ground

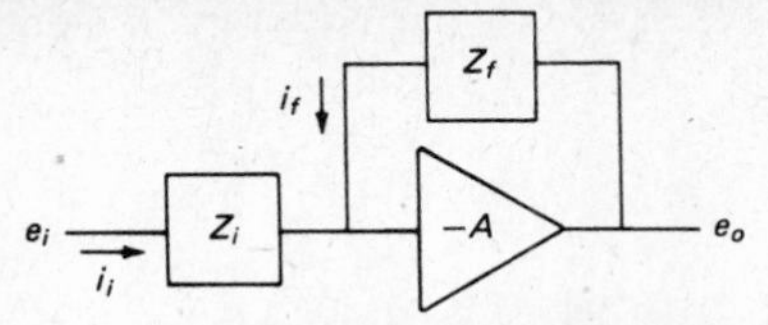

Fig. 13-12 Simplified configuration for an OP-AMP.

connection must be included. In Fig. 13-12, lowercase letters are used to represent voltage, since subsequent examples are carried out in the time domain.

Practical amplifiers are linear for only certain ranges of the input. Common OP-AMPs are capable of producing outputs from 5 to 30 V. If the input increases to a point beyond which the output attempts to exceed these limits, the OP-AMP tends to saturate. Because the gains are so high, the inputs are limited to very small values. As a result of the output limitation, OP-AMPs are not used to perform differentiation because the output of the amplifier would be proportional to the rate of change of the input. Even if the input were kept very small, a further restriction would have to be placed on the frequency content to avoid fast changes. Differentiation is not performed for another reason; every signal has some noise associated with it, and differentiation tends to increase the noise level. This is just the opposite of integration, which is a smoothing process (averaging).

13-3 OP-AMP CONFIGURATIONS

Multiplication by a constant

$$\frac{e_o}{e_i} = -\frac{Z_f}{Z_i} = -\frac{R_f}{R_i} = -a \qquad \text{where} \qquad a = \frac{R_f}{R_i}$$

is shown in Fig. 13-13. The simplified schematic is shown in Fig. 13-14.

Integration is shown in Fig. 13-15:

$$Z_i = R_1 \qquad Z_f = \frac{1}{sC}$$

$$\frac{E_o}{E_i} = -\frac{Z_f}{Z_i} = -\frac{1}{RCs} \qquad \text{or} \qquad E_o = -\frac{E_i}{RCs}$$

$$\text{and} \quad e_o = -\frac{1}{RC}\int e_i\,dt = -b\int e_i\,dt \qquad \text{where } b = \frac{1}{RC}$$

The simplified schematic is given in Fig. 13-16.

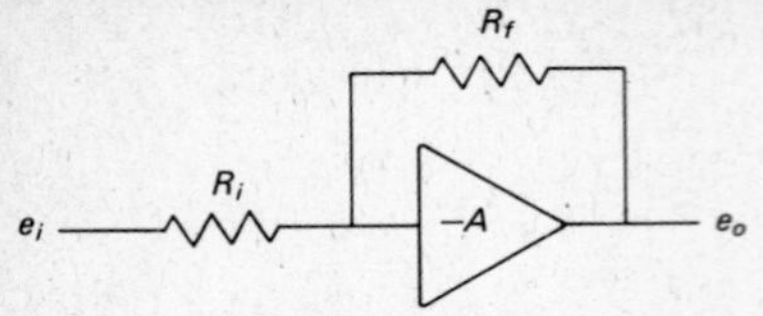

Fig. 13-13 OP-AMP configuration for multiplying by a constant.

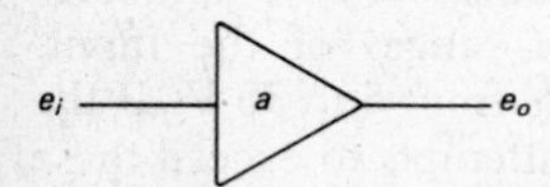

Fig. 13-14 Simplified schematic of Fig. 13-13.

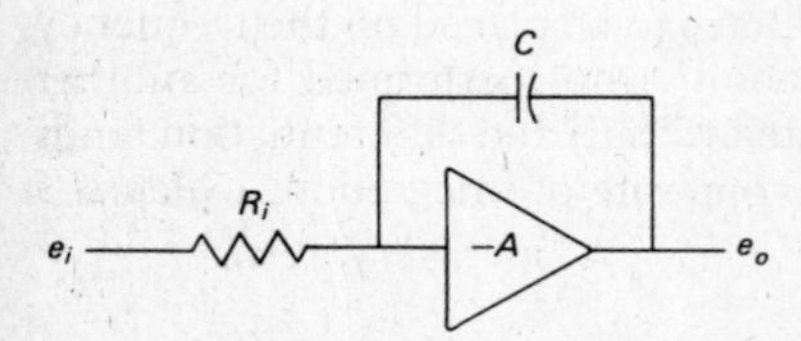

Fig. 13-15 OP-AMP configuration for integration.

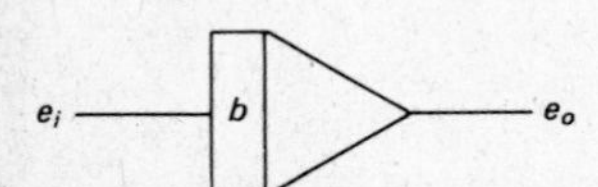

Fig. 13-16 Simplified schematic of Fig. 13-15.

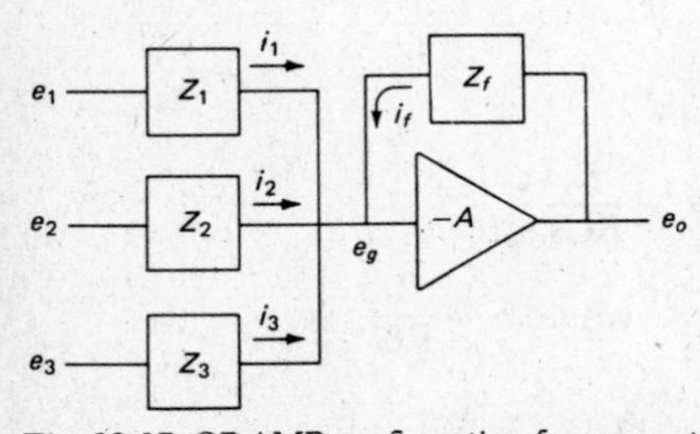

Fig. 13-17 OP-AMP configuration for summation.

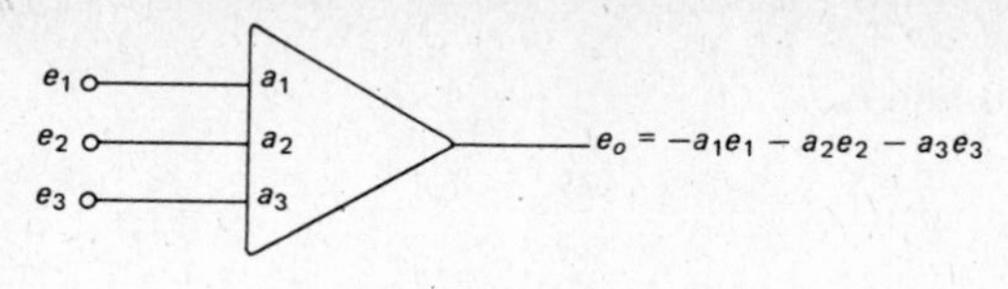

e_1
e_2
e_3
a_1
a_2
a_3
$e_o = -a_1e_1 - a_2e_2 - a_3e_3$
$a_1 = \frac{R_f}{R_1}$ $a_2 = \frac{R_f}{R_2}$ $a_3 = \frac{R_f}{R_3}$

Fig. 13-18 Simplified schematic of a summing amplifier.

Summation is shown in Fig. 13-17. Since A $\rightarrow \infty$ and the input impedance of the amplifier $Z_g \rightarrow \infty$, it was shown that

$$e_g \approx 0 \tag{13-15}$$

and $$i_i = -i_f \tag{13-16}$$

But in this case

$$i_i = i_1 + i_2 + i_3 \tag{13-17}$$

Since $$i_1 = \frac{e_1}{Z_1} \qquad i_2 = \frac{e_2}{Z_2} \qquad i_3 = \frac{e_3}{Z_3} \tag{13-18}$$

we have $$\frac{e_1}{Z_1} + \frac{e_2}{Z_2} + \frac{e_3}{Z_3} = -i_f = -\frac{e_o}{Z_f} \tag{13-19}$$

and $$e_o = -\frac{Z_f}{Z_1}e_1 - \frac{Z_f}{Z_2}e_2 - \frac{Z_f}{Z_3}e_3 \tag{13-20}$$

For addition, let $Z_1 = R_1$, $Z_2 = R_2$, $Z_3 = R_3$, and $Z_f = R_f$, so that the schematic becomes that shown in Fig. 13-18, and for a summing integrator, let $Z_1 = R_1$, $Z_2 = R_2$, $Z_3 = R_3$, and $Z_f = 1/sC$, so that the representation is as shown in Fig. 13-19.

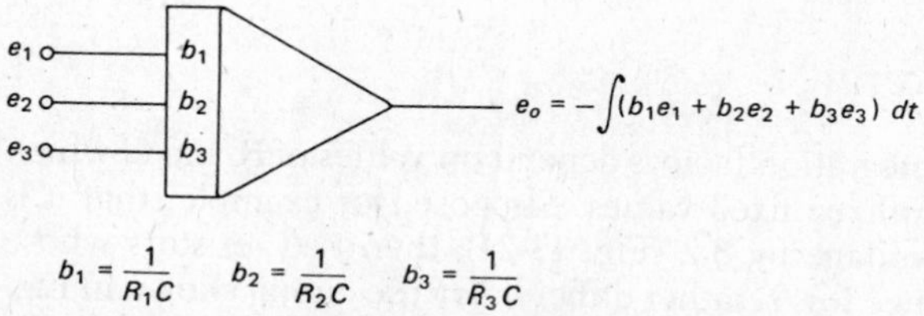

e_1
e_2
e_3
b_1
b_2
b_3
$e_o = -\int (b_1e_1 + b_2e_2 + b_3e_3)\, dt$
$b_1 = \frac{1}{R_1C}$ $b_2 = \frac{1}{R_2C}$ $b_3 = \frac{1}{R_3C}$

Fig. 13-19 Simplified schematic of a summing integrator.

Example 13-1

For the circuit in Fig. 13-20 find e_o.

Solution

$$a_1 = \frac{R_f}{R_1} = \frac{1\ \text{M}\Omega}{1\ \text{M}\Omega} = 1$$

$$a_2 = \frac{R_f}{R_2} = \frac{1\ \text{M}\Omega}{0.5\ \text{M}\Omega} = 2$$

$$a_3 = \frac{R_f}{R_3} = \frac{1\ \text{M}\Omega}{3\ \text{M}\Omega} = \frac{1}{3}$$

The simplified schematic is given in Fig. 13-21. Then

$$e_o = -(e_1 + 2e_2 + \tfrac{1}{3}e_3)$$

Example 13-2

For the circuit in Fig. 13-22, find e_o.

Solution

$$b_1 = \frac{1}{R_1C} = \frac{1}{(1 \times 10^6)(1 \times 10^{-6})} = 1$$

$$b_2 = \frac{1}{R_2C} = \frac{1}{(0.5 \times 10^6)(1 \times 10^{-6})} = 2$$

$$b_3 = \frac{1}{R_3C} = \frac{1}{(3 \times 10^6) \times 10^{-6}} = \frac{1}{3}$$

The simplified schematic is given in Fig. 13-23. Then

$$e_o = -\int_0^t (e_1 + 2e_3 + \tfrac{1}{3}e_3)\, dt$$

13-4 POTENTIOMETERS

The multiplication and integration factors depend on values of R and C which generally come in standardized fixed values. Suppose, for example, that it is necessary to multiply a voltage by 8.2 (Fig. 13-24); then two resistors whose ratio is 8.2 are needed. Since this is rather difficult, try the circuit shown in Fig. 13-25. Schematically this becomes Fig. 13-26. It is *important* to note that the

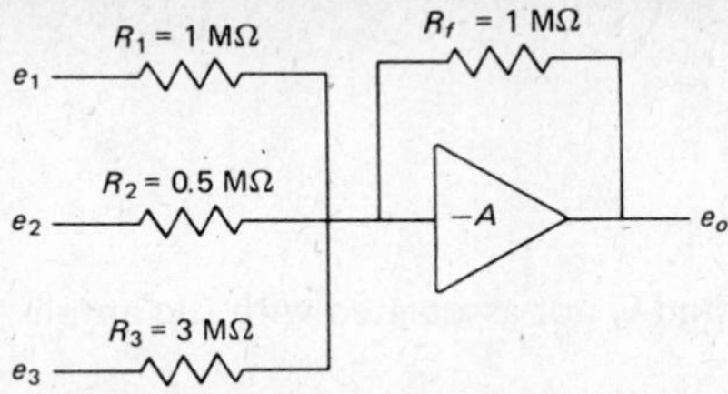

Fig. 13-20 Circuit for Example 13-1.

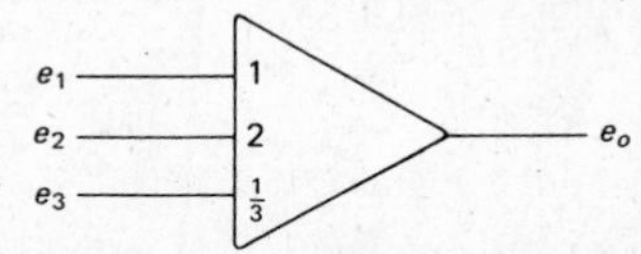

Fig. 13-21 Simplified schematic for Example 13-1.

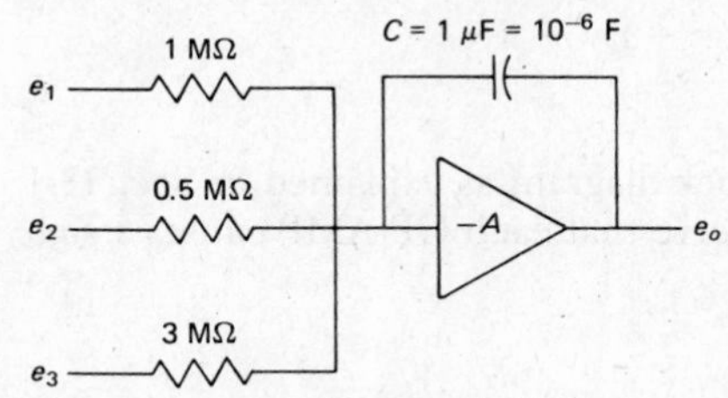

Fig. 13-22 Circuit for Example 13-2.

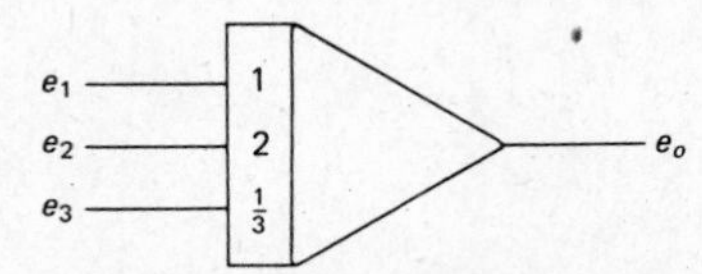

Fig. 13-23 Simplified schematic for Example 13-2.

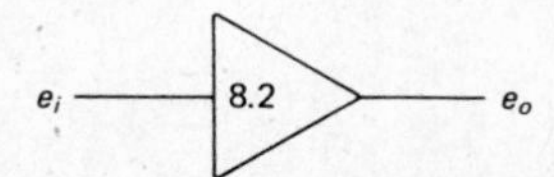

Fig. 13-24 Multiplication by a constant.

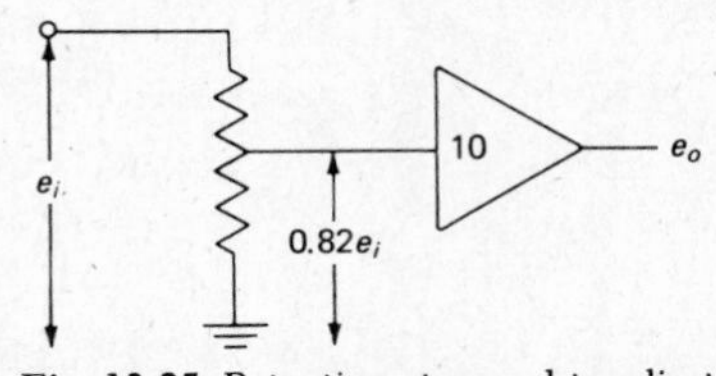

Fig. 13-25 Potentiometer used to adjust multiplication constant.

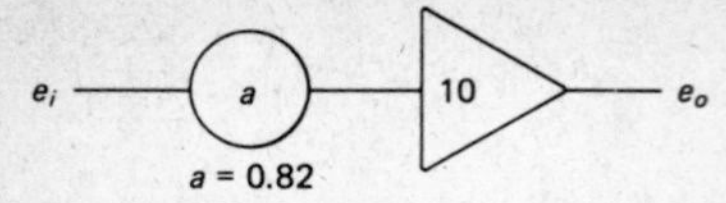

Fig. 13-26 Schematic representation of Fig. 13-25.

potentiometer is used as a voltage divider and is not associated with the amplifier input resistor.

13-5 SIMULATION

Example 13-3

Derive the analog simulation for the differential equation

$$\ddot{y} + a_1\dot{y} + a_2 y = 0 \qquad \text{or} \qquad \ddot{y} = -a_2 y - a_1\dot{y}$$

Solution

The first approach is to set up a block diagram as explained in Sec. 13-1 (Fig. 13-27.) Before continuing, observe that each OP-AMP causes a sign change, i.e.,

$$\frac{e_o}{e_i} = -\frac{Z_f}{Z_i}$$

The analog circuit is shown in Fig. 13-28.

At summing amplifier 1

$$\ddot{y} = \left(-\frac{R_1}{R_2}\right)\dot{y} + \left(-\frac{R_1}{R_3}\right)y$$

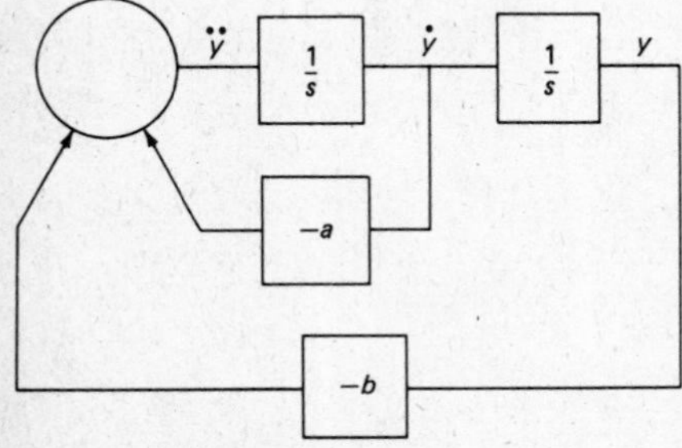

Fig. 13-27 Circuit for Example 13-3.

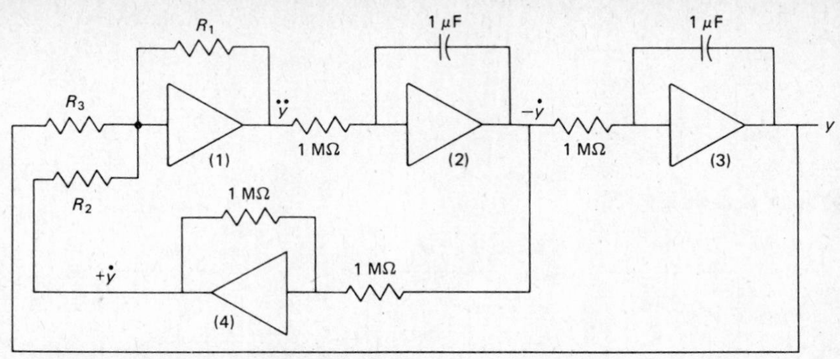

Fig. 13-28 Analog circuit for Example 13-3.

$$\text{Let} \quad \frac{R_1}{R_2} = a_1 \qquad \text{and} \qquad \frac{R_1}{R_3} = a_2$$

Since amplifier 2 not only integrated but also changed the sign, amplifier 4, which is called a *sign changer*, is necessary to generate $+\dot{y}$.

Four OP-AMPs are required for the simulation of this second-order system. As the system order increases, the number of amplifiers required can be very large, so that every attempt is made to reduce this number. One approach is to ask whether the highest derivative is a required output. In general, the answer is no. Let us again look at our equation:

$$\ddot{y} = -a_2 y - a_1 \dot{y}$$

Integrating both sides gives

$$\dot{y} = \int (-a_1 \dot{y} - a_2 y)\, dt = -\int (a_1 \dot{y} + a_2 y)\, dt$$

This output is easily generated using the summing integrator shown in Fig. 13-29, but if $\dot{y}$ is available, y can be generated by integration. Then the new simulation becomes that shown in Fig. 13-30. A sign changer is still required in this configuration, but the number of amplifiers has been reduced to three.

Bear in mind, however, that $\ddot{y}$ is no longer available. Do *not* be fooled into thinking that the input to amplifier 1 is $\ddot{y}$ (prove this to yourself; *hint:* the input to an OP-AMP is practically at ground potential). Notice also that no provision was made for initial conditions; in this case, assume that the initial conditions are zero.

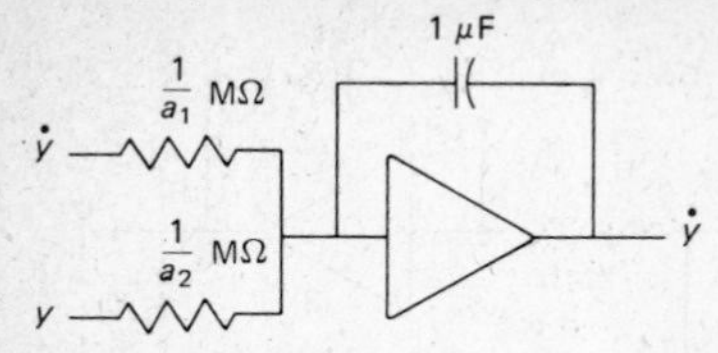

Fig. 13-29 Generating $\dot{y}$.

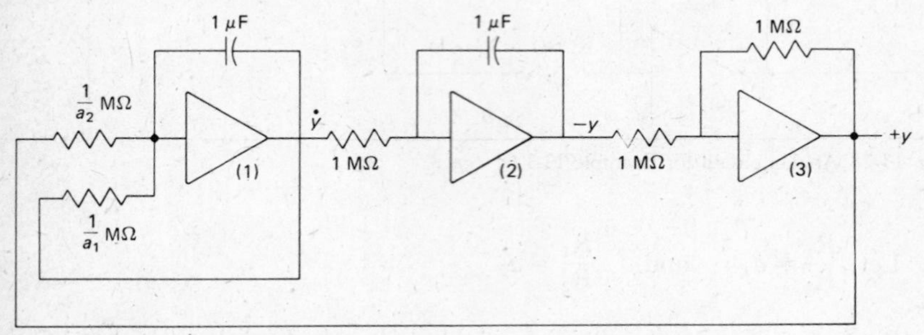

Fig. 13-30 Simulation for Example 13-3.

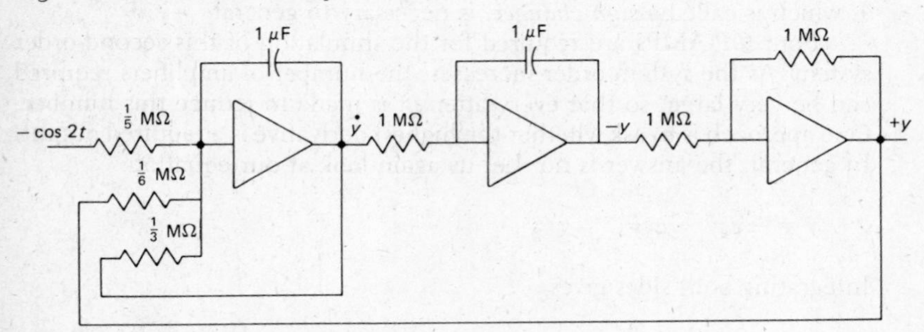

Fig. 13-31 Circuit for Example 13-4.

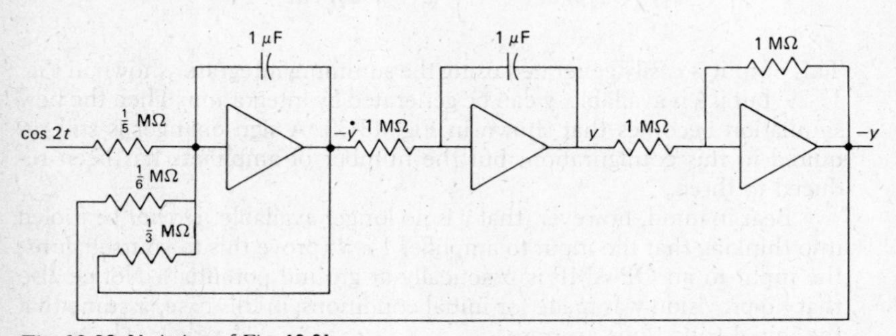

Fig. 13-32 Variation of Fig. 13-31.

Example 13-4

Find the analog simulation for the differential equation

$$\ddot{y} + 3\dot{y} + 6y = 5\cos 2t$$

Solution

$$\ddot{y} = 5\cos 2t - 3\dot{y} - 6y$$

or $$\dot{y} = \int (5\cos 2t - 3\dot{y} - 6y)\,dt = -\int (3\dot{y} + 6y - 5\cos 2t)\,dt$$

See Fig. 13-31.

A slight variation of this setup which is generally used is to force the output of the first integrator to be $-\dot{y}$. Then the new circuit is as shown in Fig. 13-32.

The rules for setting up an analog-computer circuit are as follows:

1. Solve the differential equation for the highest-order derivative.
2. Set up a summing integrator whose output is the negative of the next-to-highest-order derivative.
3. Generate the required integrals and insert sign changers where necessary.

Example 13-5

Simulate the differential equation

$$\frac{d^3y}{dt^3} + \frac{5dy}{dt} + 4y = 3x(t)$$

where $x(t)$ is some specified function.

Solution

$$\frac{d^3y}{dt^3} = 3x(t) - 4y - \frac{5dy}{dt}$$

and $$\frac{d^2y}{dt^2} = \ddot{y} = \int [3x(t) - 4y - 5\dot{y}]\,dt$$

Because of the sign change in an integrator

$$-\ddot{y} = -\int [3x(t) - 4y - 5\dot{y}]\,dt$$

so that the simplified schematic becomes that in Fig. 13-33. The simulation is shown in Fig. 13-34.

In a practical example the limitations of the analog computer must be considered. In this case, only gains of 1 or 10 can be obtained, so that potentiometers must be used.

Example 13-6

Find the simulation for

$$\ddot{q} + 7\dot{q} + 0.83q = 9.3 \cos t$$

Solution

$$-\dot{q} = -\int (9.3 \cos t - 7\dot{q} - 0.83q)\, dt$$

See Fig. 13-35. With the addition of potentiometers this becomes Fig. 13-36, and the simulation is as shown in Fig. 13-37.

13-5-1 Initial Conditions

An nth-order differential equation requires n initial conditions in order to specify the transient response. In the analog simulation of differential equations exactly n integrators were required. It is true that one of these integrations was part of a summation, but it is nonetheless an integration. Each integrator requires a feedback capacitor capable of storing an initial charge. Therefore, the initial conditions can be inserted as initial voltages across these capacitors.

In Fig. 13-38 the switch is kept closed until $t = 0$. The initial voltage across the capacitor is therefore E_o. After the switch is opened,

$$-y = \left(\frac{-1}{1}\int_0^t y\, dt\right) - E_o \qquad \text{or} \qquad y = \int_0^t y\, dt + E_o$$

Thus, $E_o = y(0)$.

Initial conditions can be illustrated as follows:

$$y(0) = 0 \qquad \text{then} \qquad E_o = 0 \text{ (short circuit)}$$

(see Fig. 13-39).

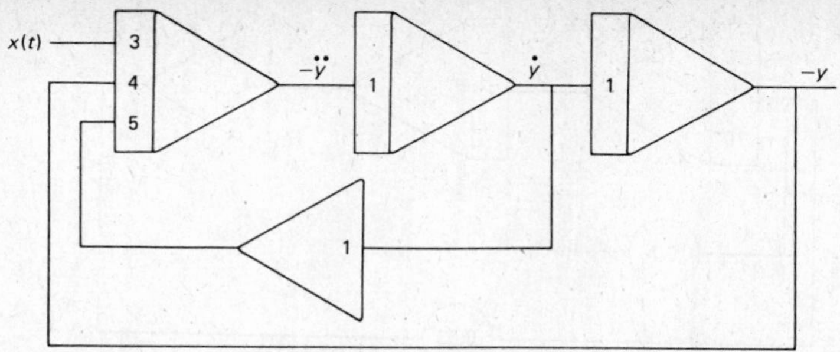

Fig. 13-33 Schematic circuit for Example 13-5.

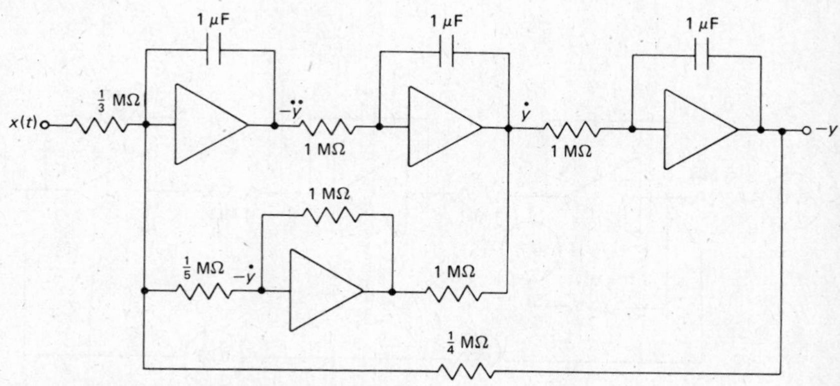

Fig. 13-34 Final simulation for Example 13-5.

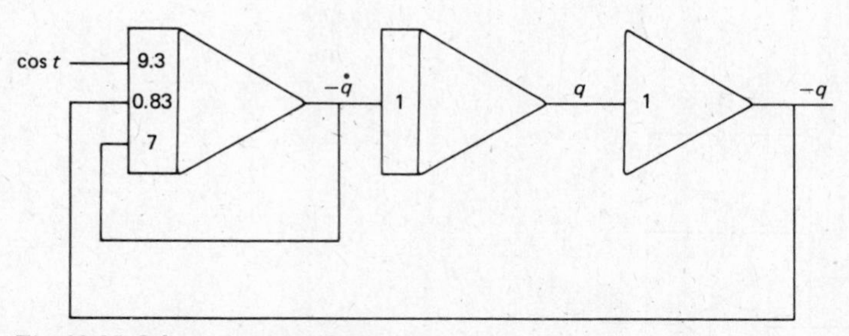

Fig. 13-35 Schematic circuit for Example 13-6.

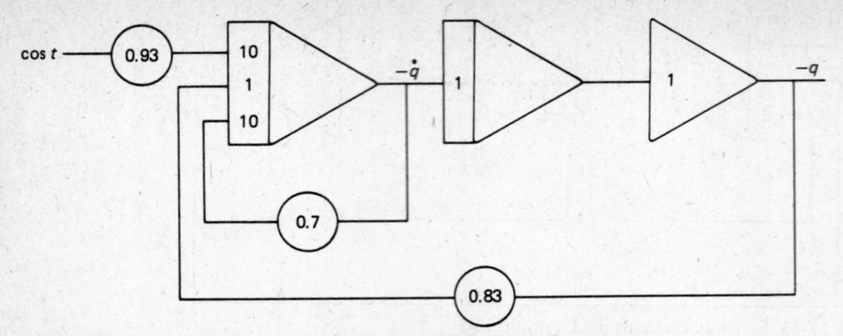

Fig. 13-36 The circuit of Fig. 13-35 with the addition of potentiometers.

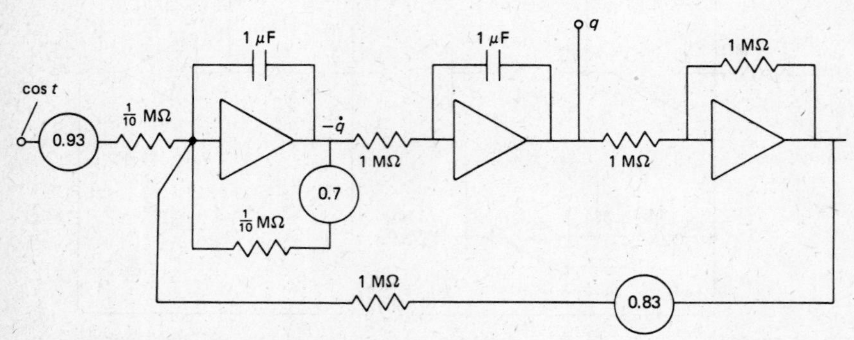

Fig. 13-37 Final simulation for Example 13-6.

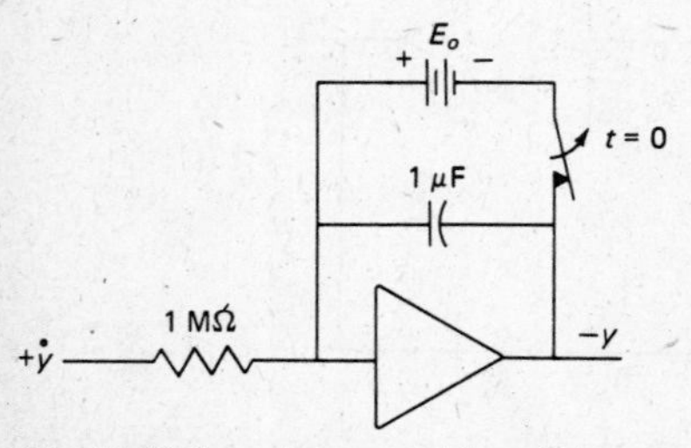

Fig. 13-38 Inserting an initial condition.

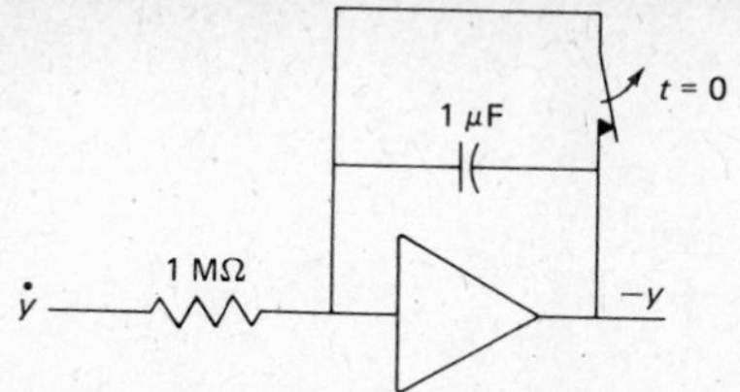

Fig. 13-39 Initial condition = 0.

For

$$y(0) = +5 \qquad E_o = 5$$

see Fig. 13-40.
For

$$y(0) = -5 \qquad E_o = -5$$

see Fig. 13-41.

Example 13-7
Find the simulation of

$$\ddot{y} + 3\dot{y} + 0.6y = 5\cos 2t \qquad \text{for } \begin{matrix} y(0) = -7 \\ \dot{y}(0) = 3 \end{matrix}$$

Solution

$$\begin{aligned} \ddot{y} &= 5\cos 2t - 3\dot{y} - 0.6y \\ &= 0.5(10\cos 2t) - 0.3(10y) - 0.6y \end{aligned}$$

$$\text{or} \quad -\dot{y} = -\int [0.5(10\cos 2t) - 0.3(10y) - 0.6y]\,dt$$

The simplified schematic simulation is given in Fig. 13-42. IC_1 refers to the first integrator, whose output is $-\dot{y}$, but

$$\dot{y}(0) = 3 \qquad \text{or} \qquad -\dot{y}(0) = -3 \qquad \text{and} \qquad IC_1 = -3$$

IC_2 refers to $y(0)$, but

$$y(0) = -7 \qquad \text{or} \qquad IC_2 = -7$$

The final simulation is shown in Fig. 13-43.

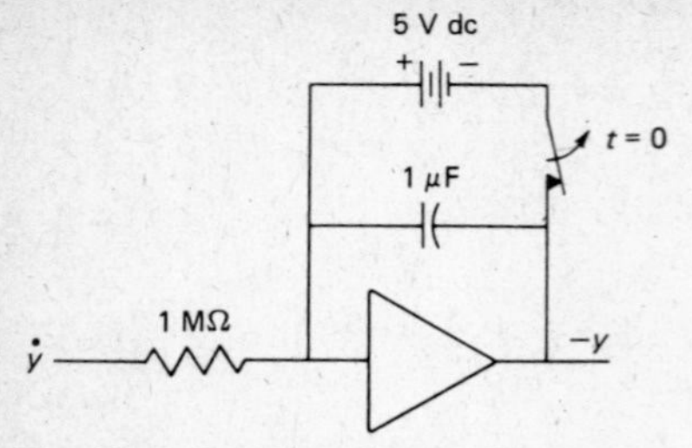

Fig. 13-40 Initial condition = 5.

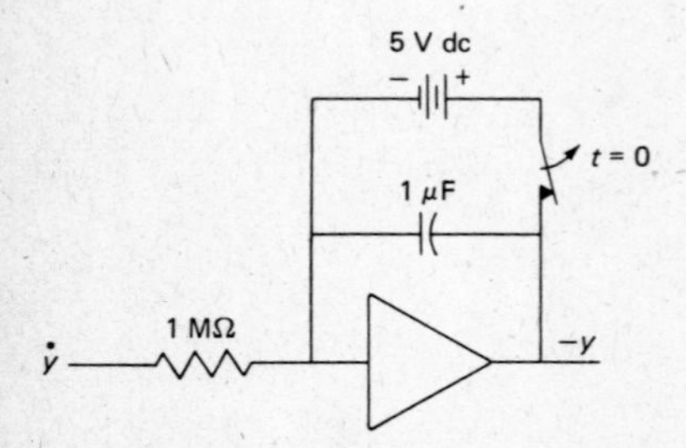

Fig. 13-41 Initial condition = −5.

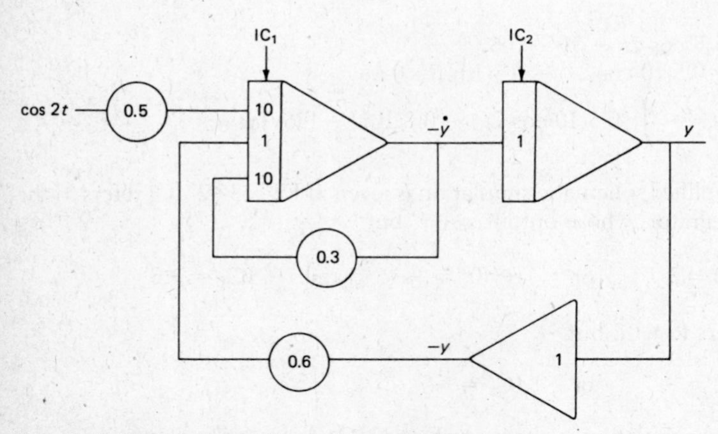

Fig. 13-42 Simplified simulation for Example 13-7.

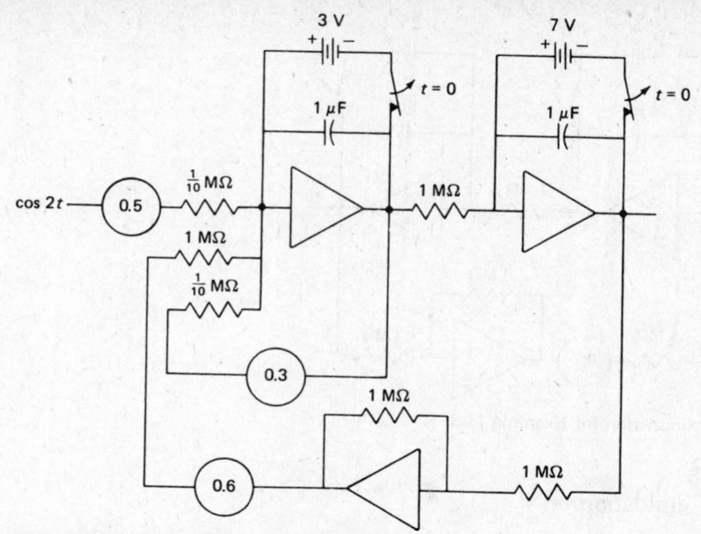

Fig. 13-43 Final simulation for Example 13-7.

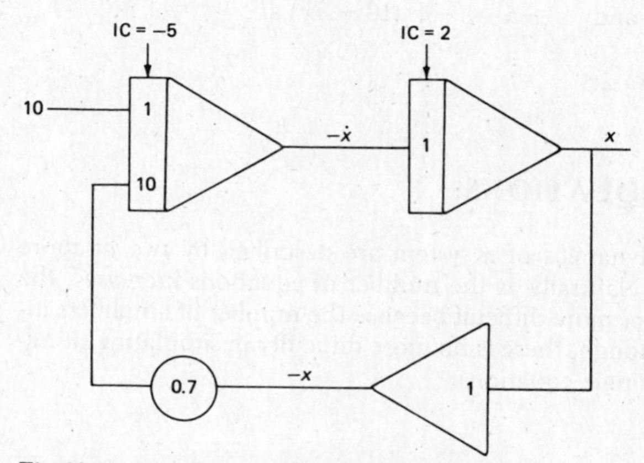

Fig. 13-44 Simplified simulation for Example 13-8.

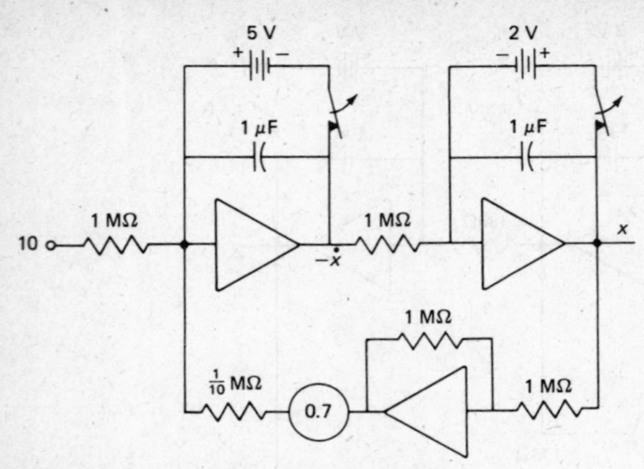

Fig. 13-45 Final simulation for Example 13-8.

Example 13-8
Find the simulation of

$$\ddot{x} + 7x = 10 \qquad \text{for} \quad \begin{matrix} \dot{x}(0) = 5 \\ x(0) = 2 \end{matrix}$$

Solution

$$\ddot{x} = 10 - 7x \qquad \text{and} \qquad -\dot{x} = -\int (10 - 7x)\, dt$$

See Figs. 13-44 and 13-45.

13-6 SIMULTANEOUS EQUATIONS:

In many instances the dynamics of a system are described by two or more simultaneous equations. Naturally as the number of equations increases, the analog simulation becomes more difficult because the number of amplifiers increases. Conceptually, though, there is no more difficulty in simulating simultaneous equations than single equations.

Example 13-9
Consider the equations

$$\ddot{x} + 2\dot{x} + 3y = 2 \tag{1}$$

$$\ddot{y} + 2y + x = 8 \tag{2}$$

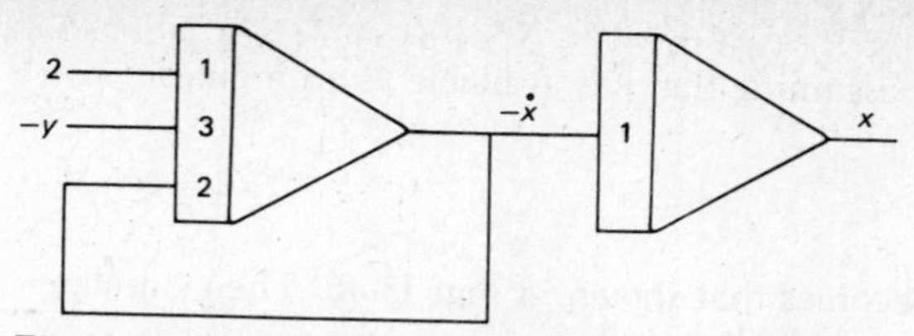

Fig. 13-46 Simulation of x in Example 13-9.

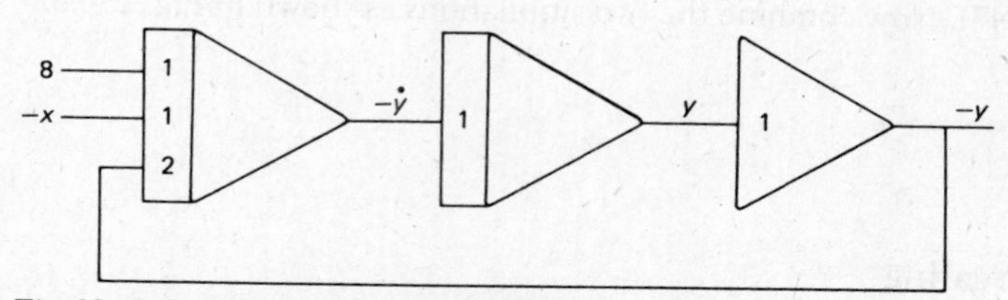

Fig. 13-47 Simulation of y in Example 13-9.

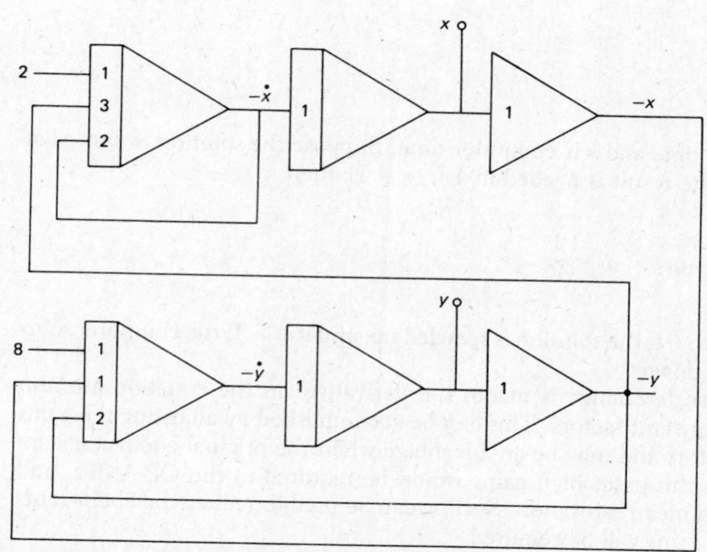

Fig. 13-48 Final simulation of Example 13-9.

Solution

First simulate Eq. (1) by assuming that y is available as an input or

$$\ddot{x} + 2\dot{x} = 2 - 3y$$

so that the simulation becomes that shown in Fig. 13-46. Then simulate Eq. (2) assuming that x is available or

$$\ddot{y} + 2y = 8 - x$$

(see Fig. 13-47). Now combine the two simulations as shown in Fig. 13-48.

13-7 SCALING

13-7-1 Time Scaling

In many simulations the response of the system may be required over long periods of time. To avoid tying up the computer, a change is made from real time to a speeded-up version of time; let

$$t = \frac{\tau}{a}$$

where t is real time and τ is computer time. Suppose the solution is desired at $t = 10$ h but the result is needed in 1 h ($\tau = 1$); then

$$10 = \frac{1}{a} \qquad \text{or} \qquad a = \frac{1}{10}$$

Therefore, if $a < 1$, the solution is speeded up, and if $a > 1$, the computer solution is slowed down.

When a scale change is made, the derivatives in the equation are also changed by constant factors. This can be accomplished by adjusting the gains of the integrators and may be an advantage when the original coefficients are very large. In this case, high gains would be required in the OP-AMPs, and high gains can mean saturation. Scaling can be used to reduce the coefficients so that lower gains will be required.

13-7-2 Amplitude Scaling

If the equation being simulated is not in terms of voltage, scale factors must be established which relate the voltages at the amplifier outputs to the variables being calculated. In addition, to avoid saturation, the scale factors are chosen so that when the variable reaches maximum value, the amplifier voltage is less than or equal to the specified maximum (its supply voltage).

As an illustration, suppose the equation for the temperature in a chemical process is given by

$$\frac{dT}{dt} + 0.4T = Q$$

where T is the absolute temperature and Q is the rate at which the chemical is mixed.

Let $Q = 8 = \text{const}$ and $T(0) = 10$

The steady-state or maximum value of T is 20 ($Q/0.4 = 8/0.4 = 20$). Further, assume that the maximum allowed amplifier output voltage is 10 V. The scale factor for T becomes

$$K_T = \frac{V_{\max}}{T_{\max}} = \frac{10}{20} = 0.5$$

Without scaling (1 V = 1°) the schematic is shown in Fig. 13-49, and the scaled program becomes that in Fig. 13-50. Then

$$T = \frac{V}{0.5}$$

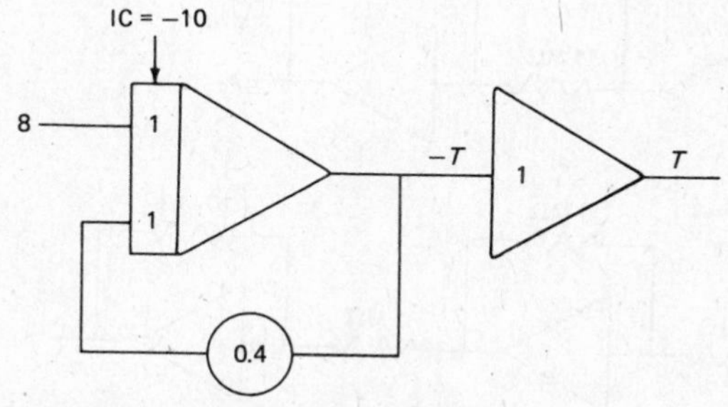

Fig. 13-49 Simulation without scaling.

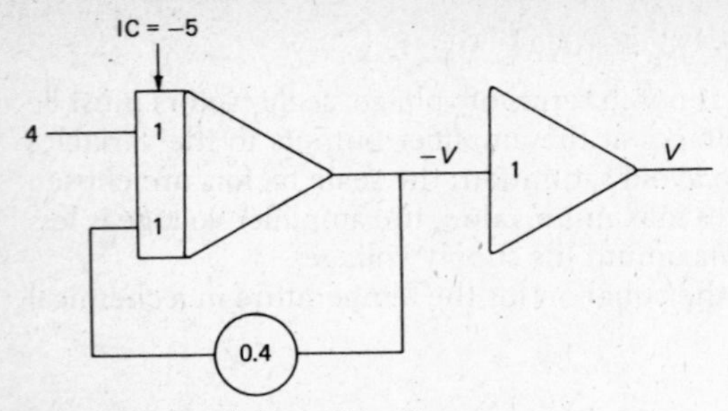

Fig. 13-50 Simulation with scaling.

An easy way to verify this scaled program is to write $V = 0.5T$. Then multiply both sides of the original equation by 0.5 (which does not alter the equation)

$$0.5\left(\frac{dT}{dt} + 0.4T\right) = 0.5Q \qquad \text{or} \qquad 0.5\frac{dT}{dt} + 0.4(0.5)T = 0.5Q = 0.5(8)$$

$$\frac{d(0.5T)}{dt} + 0.4(0.5T) = 0.5(8)$$

$$\frac{dV}{dt} + 0.4V = 4 \qquad \text{and} \qquad V(0) = 0.5[T(0)] = 0.5(10) = 5$$

PROBLEMS

13-1. Find e_o for the OP-AMP circuits in Fig. P13-1.

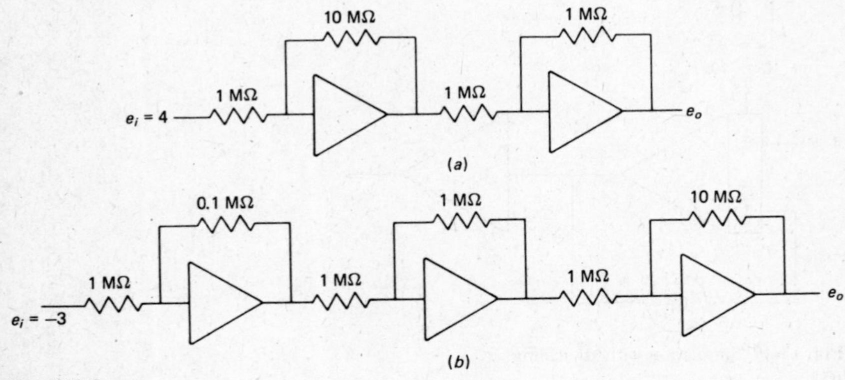

Fig. P13-1

13-2. Find e_o for the OP-AMP circuits in Fig. P13-2.

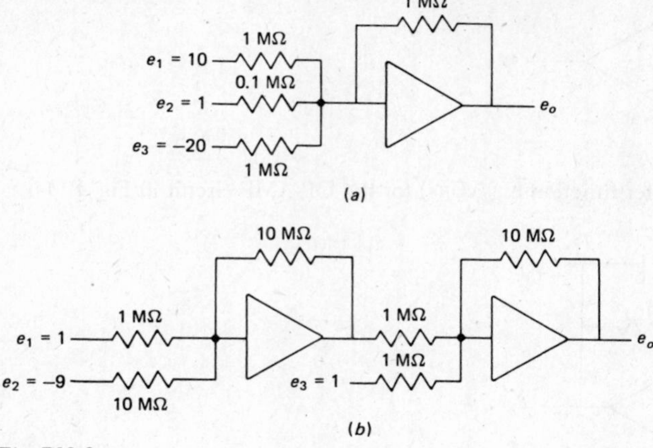

Fig. P13-2

13-3. Using the simplified schematics of the OP-AMP summers in Fig. P13-3, find e_o.

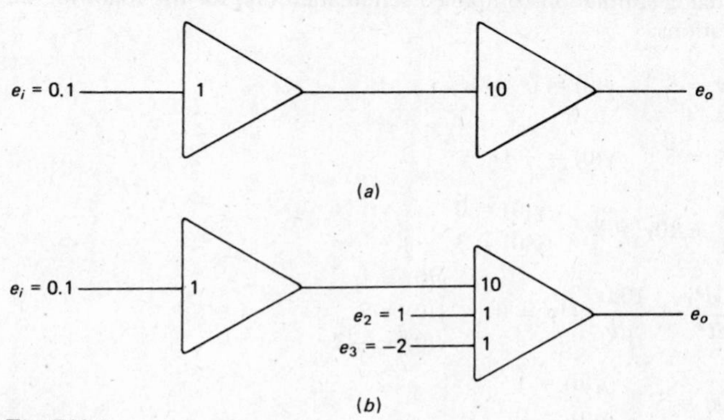

Fig. P13-3

13-4. (a) Sketch e_o for the OP-AMP integrator shown in Fig. P13-4.
(b) What is the value of e_o at $t = 0.2$ s?

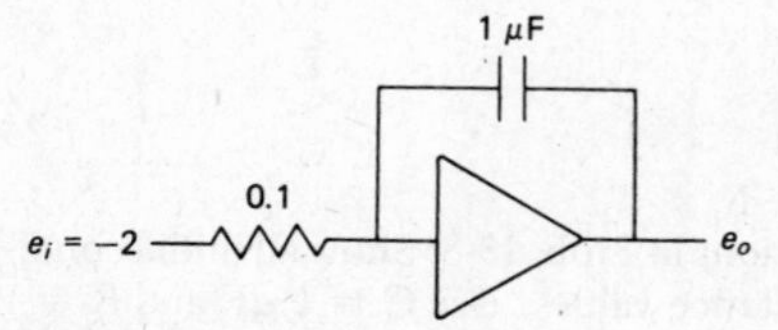

Fig. P13-4

13-5. Find e_o at $t = 0.5$ s for the summing integrator in Fig. P13-5.

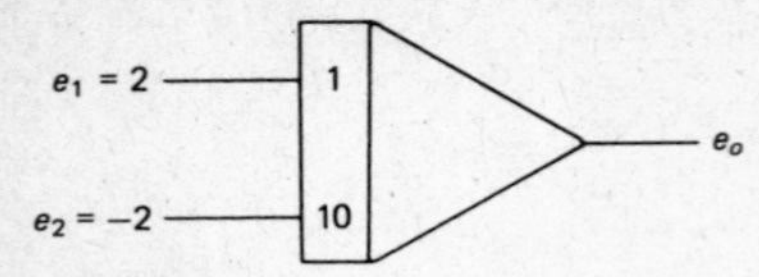

Fig. P13-5

13-6. Find the transfer function $E_o(s)/E_i(s)$ for the OP-AMP circuit in Fig. P13-6.

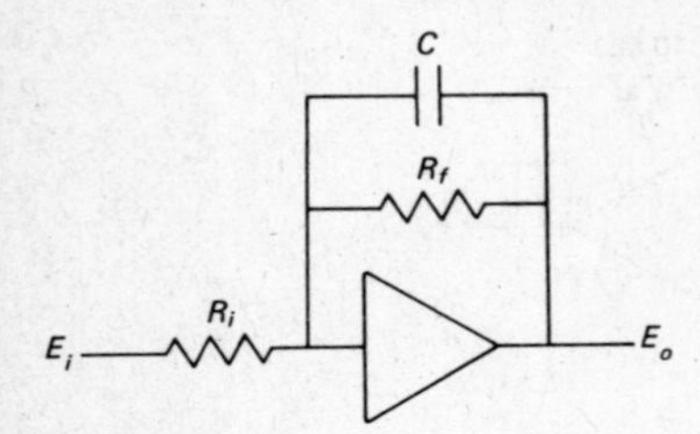

Fig. P13-6

13-7. Set up the analog simulation (simplified schematic only) for the following differential equations:

(a) $\dfrac{dv}{dt} + 10v = 5 \qquad v(0) = 0$

(b) $\dfrac{dv}{dt} + 10v = 5 \qquad v(0) = -3$

(c) $\dfrac{d^2y}{dt^2} + \dfrac{dy}{dt} + 10y = 8 \qquad y(0) = 0, \quad \dot{y}(0) = 3$

(d) $\dfrac{d^3y}{dt^2} + \dfrac{10d^2y}{dt^2} + \dfrac{10dy}{dt} + y = 0 \qquad y(0) = 1, \quad \dot{y}(0) = 2, \quad \ddot{y}(0) = -3$

(e) $\dfrac{d^2q}{dt^2} + 10q = 5 \qquad q(0) = 1, \quad \dot{q}(0) = 0$

13-8. Set up the analog simulation (simplified schematic only) for the following differential equations. Only gains of 1 and 10 are permitted. Use potentiometers where required.

(a) $\dfrac{dv}{dt} + 7v = 8 \qquad v(0) = 1$

(b) $\dfrac{d^2x}{dt^2} + \dfrac{5dx}{dt} + 0.6x = 3 \qquad x(0) = 0, \quad \dot{x}(0) = -1$

13-9. Set up the final simulation for the equations in Prob. 13-8. Show all initial conditions, potentiometer settings, and resistance values. Use C = 1 μF and R = 1 MΩ or 0.1 MΩ.

13-10. Set up the simplified analog simulation to determine the current in Fig. P13-10. Use gains of 1 and 10 only and show initial conditions.

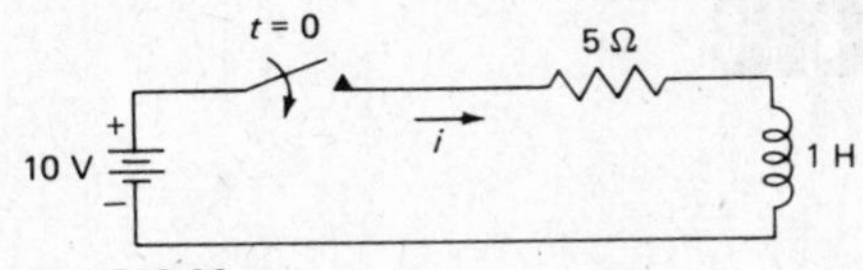

Fig. P13-10

13-11. Write the differential equation corresponding to the analog simulation in Fig. P13-11.

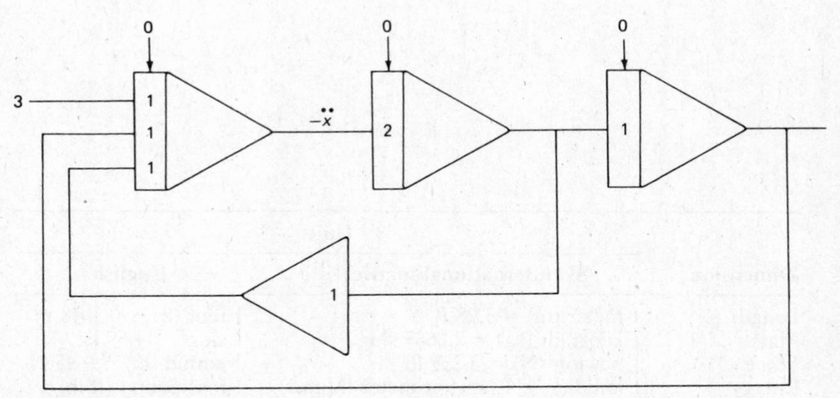

Fig. P13-11

13-12. Set up the simplified analog simulation to solve for i_1 and i_2 using the following simultaneous differential equations (use gains of 1 and 10 only):

$$\frac{2di_1}{dt} + 10i_1 = 10 + 10i_2$$

$$\frac{5di_2}{dt} + 20i_2 = 10i_1 \qquad i_1(0) = i_2(0) = 0$$

13-13. Scale the following differential equation for a maximum of 5 V. Then prepare the simplified analog simulation for the scaled equation.

$$\frac{dp}{dt} + 0.2p = 10 \qquad p(0) = 8$$

appendix A

Conversions Table

Dimension	Units	
	SI (international metric)	**English**
Length	1 meter (m) = 3.28 ft	1 foot (ft) = 0.3048 m
Mass	1 kilogram (kg) = 0.0685 slug	1 slug = 14.59 kg
Force	1 newton (N) = 0.225 lb	1 pound (lb) = 4.45 N
Energy	1 joule (J) = 1 newton-meter (N·m) = 0.738 ft·lb	1 foot-pound (ft·lb) = 1.36 N·m
Power	1 watt (W) = 1 joule/second (J/s) = 0.738 ft·lb/s	1 foot-pound/second (ft·lb/s) = 1.36 W
Torque	1 newton-meter (N·m) = 0.738 ft·lb	1 ft·lb = 1.36 N·m
Moment of inertia	1 kilogram-meter^2 (kg·m^2) = 0.737 slug·ft^2	1 slug·ft^2 = 1.357 kg·m^2
Velocity	1 meter/second (m/s) = 3.28 ft/s	1 ft/s = 0.3048 m/s

appendix B

Laplace Transforms: Theorems and Pairs

The following properties hold for Laplace transforms:

$$F(s) = \mathscr{L}[f(t)] = \int_0^\infty f(t)e^{-st}\,dt$$

$$\mathscr{L}[f_1(t) + f_2(t)] = F_1(s) + F_2(s)$$

$$\mathscr{L}[af(t)] = aF(s)$$

Two theorems help us determine the initial and final values of a function of time, the initial-value theorem

$$f(0+) = \lim_{t\to 0} f(t) = \lim_{s\to\infty} sF(s)$$

and the final-value theorem

$$f(\infty) = \lim_{t\to\infty} f(t) = \lim_{s\to 0} sF(s)$$

In solving differential and integral equations the following identities can be used:

$$\mathscr{L}\left[\frac{df(t)}{dt}\right] = \mathscr{L}[\dot{f}(t)] = sF(s) - f(0+)$$

$$\mathscr{L}\left[\frac{d^2f(t)}{dt^2}\right] = \mathscr{L}[\ddot{f}(t)] = s^2F(s) - sf(0+) - \dot{f}(0+)$$

$$\mathscr{L}\left[\int f(t)\,dt\right] = \mathscr{L}[f^{-1}(t)] = \frac{F(s)}{s} + \frac{f^{-1}(0+)}{s}$$

Table B-1 lists some of the Laplace transform pairs commonly used in control systems.

TABLE B-1

Time function $f(t)$	Laplace transform $F(s)$
$Ku(t)$ where $u(t) = \begin{cases} 0 & t < 0 \\ 1 & t > 0 \end{cases}$	$\frac{K}{s}$
Kt	$\frac{K}{s^2}$
Kt^n	$\frac{K(n!)}{s^{n+1}}$
Ke^{-at}	$\frac{K}{s + a}$
Kte^{-at}	$\frac{K}{(s + a)^2}$
$Kt^n e^{-at}$	$\frac{K(n!)}{(s + a)^{n+1}}$
$K \sin \omega t$	$\frac{K\omega}{s^2 + \omega^2}$
$K \cos \omega t$	$\frac{Ks}{s^2 + \omega^2}$
$Ke^{-at} \sin \omega t$	$\frac{K\omega}{(s + a)^2 + \omega^2}$
$Ke^{-at} \cos \omega t$	$\frac{K(s + a)}{(s + a)^2 + \omega^2}$
$\frac{K}{\omega^2 + a^2} - \frac{Ke^{-at} \sin (\omega t + \psi)}{\omega\sqrt{a^2 + \omega^2}}$ $\quad \psi = \tan^{-1} \frac{\omega}{a}$	$\frac{K}{s[(s + a)^2 + \omega^2]}$

appendix C

Nonlinearities

The analysis of second-order systems was performed assuming that the system components were all linear. Linearity implies a proportionality between the input and the output (in the steady state) for any transfer function. Nonlinearities manifest themselves in practically every control-system component. Minor nonlinearities have virtually no effect on the analyses of second-order systems, but large nonlinearities cannot be ignored. This appendix deals only with the effects of amplifier saturation, gear backlash, nonlinear friction, and nonlinear speed-torque characteristics of servomotors.

C-1 AMPLIFIER SATURATION

All amplifiers saturate at a particular output-voltage level. Further increases in input voltage produce no increase in output voltage once saturation is reached. Figure C-1 illustrates the effect of saturation. As long as the input signal is kept below $e_{i,\max}$, the amplifier gain is constant. If larger signals are applied, the gain is effectively reduced. This reduction in gain reduces the speed of response of a servo.

C-2 GEAR BACKLASH

Backlash is caused by the failure of mating gears to mesh precisely, as shown in Fig. C-2. As long as the gears are rotating in one direction, the teeth of the gears remain in contact. Upon a change in direction, the output gear loses contact with the input gear momentarily due to backlash. If the backlash is large enough, a nonlinear oscillation known as *chattering*, *hunting*, or a *limit cycle* can occur.

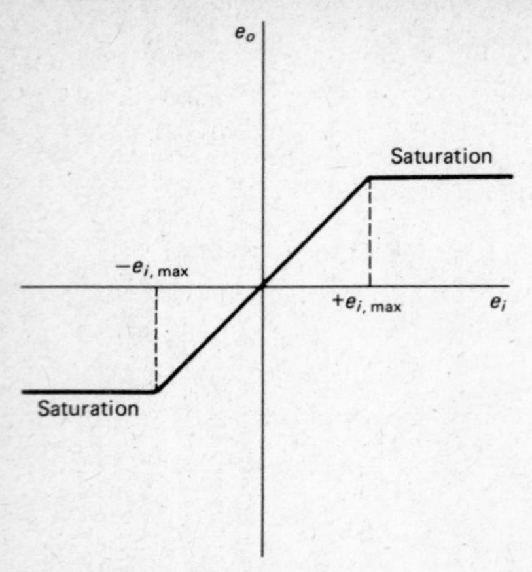

Fig. C-1 Amplifier saturation.

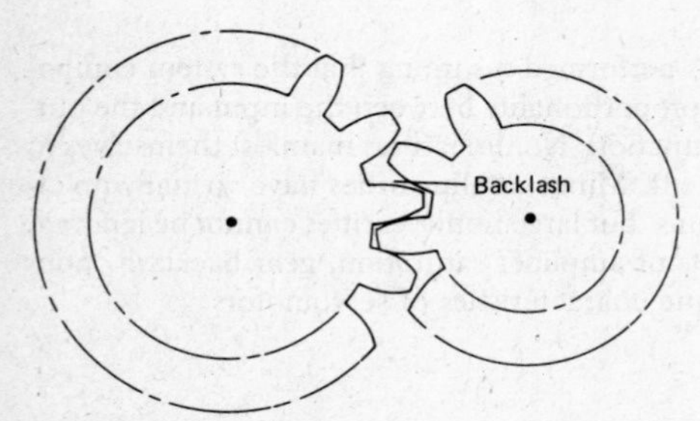

Fig. C-2 Gear backlash.

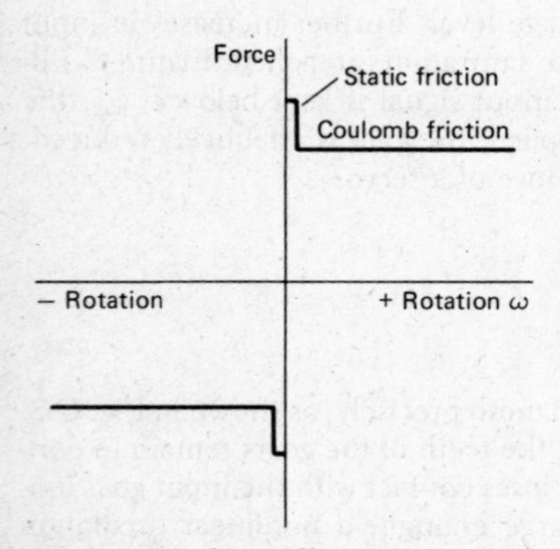

Fig. C-3 Nonlinear friction.

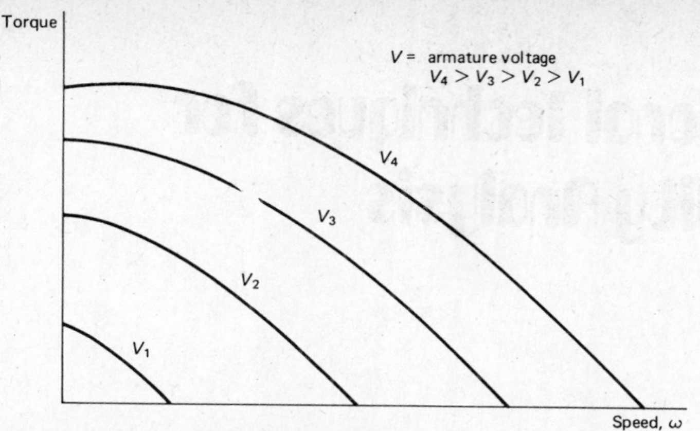

Fig. C-4 Nonlinear motor torque-speed characteristics.

C-3
NONLINEAR FRICTION

The viscous friction described in Chaps. 2, 4, and 5 is a linear effect because the frictional force is proportional to speed. For example, it is assumed that the force of friction doubles if the speed doubles. However, two forms of nonlinear friction are generally found in servosystems, static friction and coulomb friction. Static friction is eliminated once the system is in motion. As long as the motor delivers sufficient torque at start-up, this type of friction presents no problem. Coulomb friction is caused by sliding motion. The friction force is constant and opposite to the direction of motion. Figure C-3 illustrates the combined effect of static and coulomb friction. Coulomb friction has a tendency to stabilize a system. In a system with backlash, the coulomb friction can reduce or virtually eliminate chattering.

C-4
NONLINEAR TORQUE-SPEED CHARACTERISTICS OF SERVOMOTORS

Figure C-4 represents the typical torque-speed curves for a dc servomotor. The curves are not straight lines, nor are they equally spaced. This means that the back-emf constant (viscous-friction coefficient) and the motor torque constant are not really constant for all values of torque. If the linearized values are actually lower than the values of the constants over the entire response range, the true output will be somewhat slower than the predicted output.

appendix D

Additional Techniques for Stability Analysis

The general second-order system characteristic equation is

$$s^2 + 2\zeta\omega_n s + \omega_n^2 = 0$$

Solving this equation for the roots, we obtain

$$s = -\zeta\omega_n \pm \omega_n \sqrt{\zeta^2 - 1} \tag{D-1}$$

Making note of the fact that ω_n (natural frequency) must always be positive, we can make the following points about the two roots:

1. If $\zeta > 1$ (overdamped), the roots will be negative real numbers. For very large values of ζ, one root approaches zero while the other root approaches $-2\zeta\omega_n$.
2. If $\zeta = 1$ (critical damping), the roots will be equal negative real numbers ($s = -\omega_n$).
3. If $0 < \zeta < 1$ (underdamped), the roots will be complex, i.e., containing a real and imaginary part. The real part will be negative, as in cases 1 and 2 above. The imaginary parts will be $\pm\omega_n \sqrt{\zeta^2 - 1}$. The roots are said to be *complex conjugates with negative real parts*.
4. If $\zeta = 0$ (unstable with sustained oscillations), the roots will be $s = \pm j\omega_n$. Note that the roots are purely imaginary with a zero real part.
5. If ζ could be made negative* (absolutely unstable), the roots would have positive real parts.

* Note that $-1 < \zeta < 0$ corresponds to a divergent response with oscillations and $\zeta < -1$ corresponds to a divergent response without oscillations.

From items 1 to 5 the following conclusion can be drawn: as long as the characteristic equation has roots with negative real parts, the system will be stable. If the roots have positive real parts, however, the system will be absolutely unstable. While the above discussion does not constitute a proof, the conclusion drawn is correct and applicable to higher-order systems.

D-1
ROUTH STABILITY CRITERION

The Routh criterion is a mathematical method used to determine the stability (or instability) of a system. It involves setting up an array of numbers based on the coefficients of the s terms in the characteristic equation. For every change in sign in the first column of the array, a root of the characteristic equation exists with positive real part. This in turn would indicate an unstable system.

Consider the characteristic equation of an nth-order system

$$a_1 s^n + a_2 s^{n-1} + a_3 s^{n-2} + a_4 s^{n-3} + \cdots + a_{n+1} = 0$$

The following array is set up:

$$\begin{bmatrix} a_1 & a_3 & a_5 & \cdots \\ a_2 & a_4 & a_6 & \cdots \\ b_1 & b_2 & b_3 & \cdots \\ c_1 & c_2 & c_3 & \cdots \\ d_1 & d_2 & d_3 & \cdots \end{bmatrix}$$

where $b_1 = \dfrac{a_2 a_3 - a_1 a_4}{a_2}$ $\qquad b_2 = \dfrac{a_2 a_5 - a_1 a_6}{a_2}$

$$c_1 = \frac{b_1 a_4 - b_2 a_2}{b_1} \qquad c_2 = \frac{b_1 a_6 - b_3 a_2}{b_1}$$

$$d_1 = \frac{c_1 b_2 - c_2 b_1}{c_1} \qquad d_2 = \frac{c_1 b_3 - c_3 b_1}{c_1}$$

The criterion will be applied in the following example.

Example D-1

A fourth-order system has the following characteristic equation. Using the Routh criterion, determine the stability (or instability) of the system

$$s^4 + 2s^3 + 4s^2 + s + 6 = 0$$

Solution

$a_1 = 1$	$a_3 = 4$	$a_5 = 6$
$a_2 = 2$	$a_4 = 1$	$a_6 = 0$
$b_1 = 3.5$	$b_2 = 6$	$b_3 = 0$
$c_1 = -2.43$	$c_2 = 0$	$c_3 = 0$
$d_1 = 5.997$	$d_2 = 0$	$d_3 = 0$

From the first column, it is clear that there are two changes in sign, from b_1 to c_1 (plus to minus) and c_1 to d_1 (minus to plus). Since there are two changes in sign, the characteristic equation has two roots with positive real parts. The system is therefore absolutely unstable. It should be noted that if a fifth row consisting of e_1, e_2, . . . , etc., were calculated, it would consist of all zeros. For this reason the d row was the last one calculated.

D-2 ROOT-LOCUS TECHNIQUE

In general, a transfer function can be expressed as the ratio of two polynomials in s

$$T(s) = \frac{N(s)}{D(s)}$$

When $N(s)$ is set equal to zero and the roots of the equation $N(s) = 0$ are determined, the resulting values of s are called the *zeros* of the transfer function $T(s)$. Hence a zero is a value of s which when substituted into the transfer function makes it zero. When $D(s)$ is set equal to zero and the roots of the equation $D(s) = 0$ are found, the resulting values of s are called the *poles* of the transfer function $T(s)$. Hence a pole is a value of s which when substituted into the transfer function makes its denominator equal to zero, causing $T(s)$ to become infinite.

Example D-2

Find the zeros and poles of the transfer function

$$\frac{C}{R}(s) = \frac{10(s + 5)}{s(s^2 + 10s + 21)}$$

Solution

To find the zeros, set the numerator equal to zero:

$$10(s + 5) = 0$$
$$s + 5 = 0$$
$$s = -5$$

To find the poles, set the denominator equal to zero:

$$s(s^2 + 10s + 21) = 0$$
$$s(s + 3)(s + 7) = 0$$

This has three roots;

$$s = 0 \qquad s = -3 \qquad \text{and} \qquad s = -7$$

The transfer function $C/R(s)$ has one zero at $s = -5$ and three poles, $s = 0$, $s = -3$, and $s = -7$.

Classically, a root locus is a plot of the roots of the characteristic equation of a system as the loop gain is varied from zero to infinity. Since some roots are complex numbers, they will have both real and imaginary parts. The roots are plotted on a graph where the ordinate represents the imaginary part and the abscissa represents the real part of the root. The graph is called the *complex-frequency plane*.

Example D-3

Plot the root locus of a control system with the open-loop transfer function

$$HG(s) = \frac{K}{s(s + 6)}$$

where K is an amplifier gain.

Solution

First, the characteristic equation is obtained:

$$1 + HG(s) = 0$$

$$1 + \frac{K}{s(s + 6)} = 0$$

$$s^2 + 6s + K = 0$$

The gain K will now be varied from 0 to ∞. This in effect is the same as varying the loop gain from 0 to ∞. For each value of K the roots will be determined. They have been shown in Table D-1 for some values of K. The roots in Table D-1 are plotted in Fig. D-1, called the *root locus* of the system. The following points should be noted:

1. The roots corresponding to $K = 0$ are marked with a cross (at 0 and -6). Note that these roots correspond to the poles of the open-loop transfer function. This will always be the case. On the real axis $\zeta \geq 1$.
2. As the gain K is increased, the roots move along the real axis until they meet. At this point ($K = 9$) the roots are equal (and $\zeta = 1$). As the gain is increased fur-

TABLE D-1
Tabulation of Roots of the Characteristic Equation

K	Root 1	Root 2
0	$s = 0$	$s = -6$
5	$s = -1$	$s = -5$
8	$s = -2$	$s = -4$
9	$s = -3$	$s = -3$
13	$s = -3 + j2$	$s = -3 - j2$
18	$s = -3 + j3$	$s = -3 - j3$
25	$s = -3 + j4$	$s = -3 - j4$
73	$s = -3 + j8$	$s = -3 - j8$
∞	$s = -3 + j\infty$	$s = -3 - j\infty$

ther, the roots break away from the real axis and become complex-conjugate pairs.

3. For every pole of the open-loop transfer function, there will be a locus.
4. In this example, the root locus goes to infinity as K goes to infinity. This is because the open-loop transfer function $HG(s)$ does not have zeros. If zeros are present, the root locus will end up at the zero locations when K is infinite. For cases with one zero, one root will end up at the zero location and the other will go to infinity. The number of zeros must always be less than or equal to the number of poles in the open-loop transfer function.
5. The distance of the root from the origin corresponds to the natural frequency ω_n at the particular gain. In this example for a gain of 25, $\omega_n = \sqrt{3^2 + 4^2} = 5$.
6. If a line is drawn from the origin to any point on the locus, it makes an angle θ with the real axis. The damping ratio is equal to the cos θ. In this example at $K = 25$, $\zeta = \cos\theta = \frac{3}{5} = 0.6$. Note that reducing the gain K in this example decreases θ. When $K = 9$, $\theta = 0°$ and $\zeta = 1.0$ (critical damping). If K is reduced further, the system becomes overdamped.
7. The real part of any complex root (the distance from the imaginary axis) is equal to $\zeta\omega_n$. The time constant associated with a root is equal to 1/(real part). The overall system response will be governed by the largest time constant. In this example, for $K = 8$ the system has two time constants associated with its response. From the plot, they are found to be $\frac{1}{4} = 0.25$ s and $\frac{1}{2} = 0.5$ s. The system examined here will therefore have a time constant of 0.5 s for a gain of 8 and will be overdamped ($\zeta > 1$).

This example illustrated the methods used to extract information from a system's root locus. The information obtained could have been obtained using the techniques presented in Chap. 5 more easily than the root-locus technique because the system considered here was only a second-order system. When the system becomes complex (higher-order) and there are perhaps 10 roots which are affected by a gain's variation, the root locus becomes a valuable analytic tool. The root locus can be used not only for analysis but also for system design and compensation. Classically, the root locus has

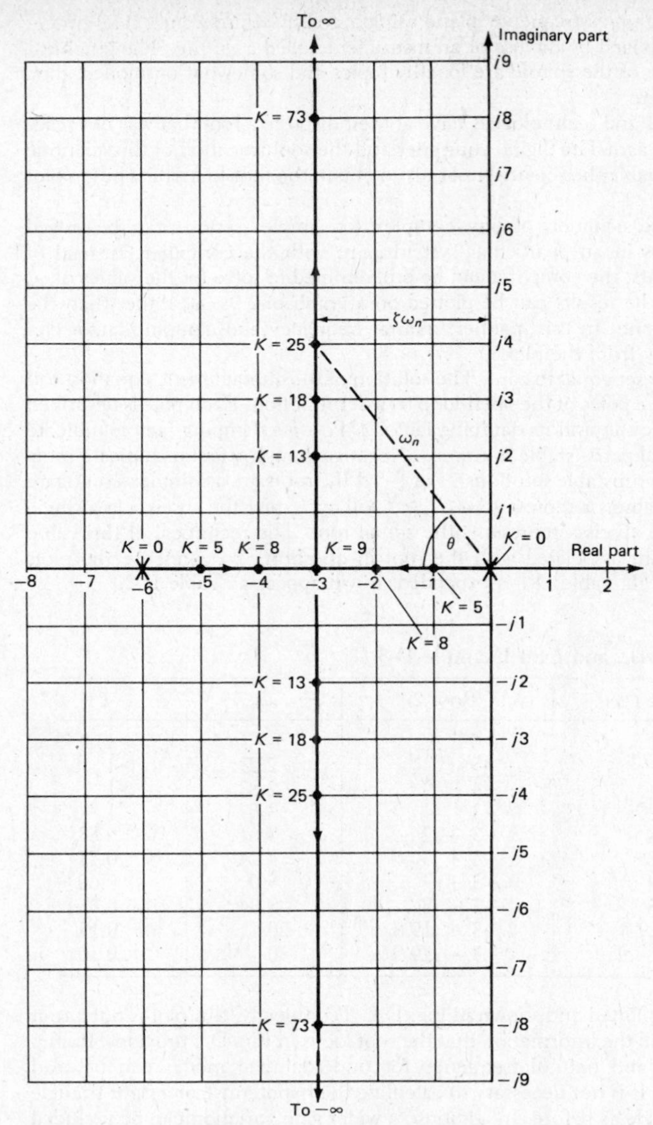

Fig. D-1 Root locus for system of Example D-3.

been plotted in the complex-frequency plane using a graphical procedure. The procedure is somewhat simplified by the use of an instrument called a *spirule*. The graphical procedure and the use of the spirule are lengthy topics and somewhat outmoded; they will not be covered here.

Today's engineers and technologists have at their disposal a tool that was not readily accessible 25 years ago. The digital computer and the sophistication of the scientific languages available enable the scientist not only to obtain the root-locus data but to plot it simultaneously.

The characteristic equation of any complex (or simple) system can be solved directly, for variations in any particular system gain, with the computer. Instead of solving just for the roots, the computer can be programmed to solve for the values of ω_n and ζ of each root. The results can be plotted on a graph of ζ vs. ω_n rather than the complex-frequency plane. In this manner, natural frequency and damping factor can be determined directly from the plot.

The gain K is first set equal to zero. The solution of the characteristic equation will yield the location of the poles of the open-loop transfer function. Each pole is described by its natural frequency ω_n and its damping factor ζ. Positive damping factors indicate roots with negative real parts (stable solutions). Negative damping factors indicate roots with positive real parts (unstable solutions). For $\zeta < 1$ the roots are a complex-conjugate pair. In this case, the abscissa represents ω_n. $\zeta \geq 1$ will appear on the abscissa as a single root. In this case, the abscissa represents the actual root. The reciprocal of this value gives the time constant associated with the root in question. To condense the plot, semilog paper is used. If Table D-1 is expanded, it will appear as Table D-2.

TABLE D-2
Tabulation of Roots, ω_n, and ζ for Example D-3

K	Root 1	Root 2	ω_n	ζ
0	0	−6	—	>1
5	−1	−5	2.23	>1
8	−2	−4	2.83	>1
9	−3	−3	3.0	1
13	$-3 + j2$	$-3 - j2$	3.61	0.83
18	$-3 + j3$	$-3 - j3$	4.24	0.71
25	$-3 + j4$	$-3 - j4$	5.0	0.60
73	$-3 + j8$	$-3 - j8$	8.54	0.35
400	$-3 + j19.8$	$-3 - j19.8$	20	0.15
900	$-3 + j29.9$	$-3 - j29.9$	30	0.10

The root locus is plotted and shown in Fig. D-2. Technically, this plot is not a root locus, but it provides all the information that the root locus in Fig. D-1 provides. In Fig. D-2, damping factor and natural frequency for underdamped modes can be read directly from the plot. It is not necessary to calculate the hypotenuse of a right triangle or the cosine of an angle as before. In addition, a wider gain variation can be included on a graph of a given size. This alternate approach becomes even more powerful as an analytical and design tool when higher-order systems are examined.

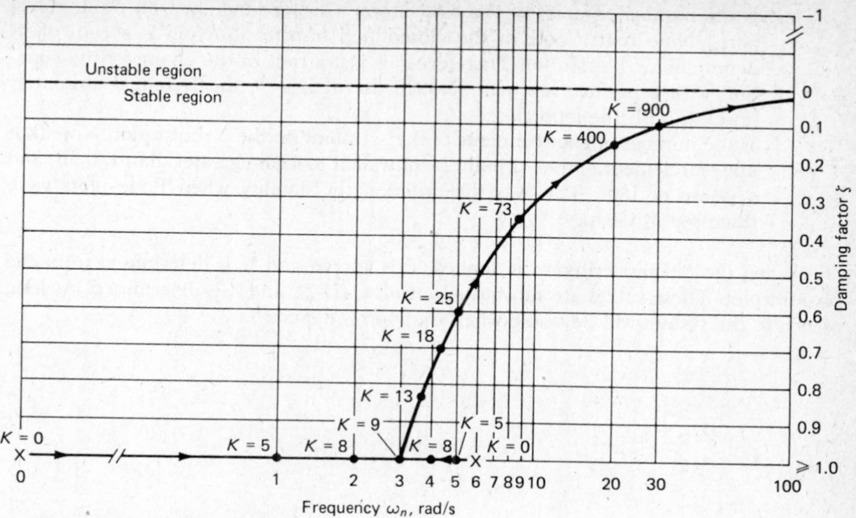

Fig. D-2 Alternate form of root locus.

D-3 NYQUIST CRITERION

In Chap. 7 a very general treatment of the Nyquist stability criterion was given. It will be explained here in more detail. The Nyquist plot, like a Bode plot, enables one to determine the stability of a system. Closed-loop stability can be determined with experimental data from the open-loop system or analytical data from the open-loop transfer function. The two plots show essentially the same thing. The Nyquist criterion, however, is more complete; if it indicates an instability, the system is absolutely unstable. There are some instances when a Bode plot will indicate an instability when in reality the system is stable.

A Nyquist plot is a graph of the open-loop transfer function $HG(s)$ in the complex-frequency plane as a function of ω. The Nyquist stability criterion states that

$$N = Z - P \tag{D-2}$$

where P is the number of poles of the open-loop transfer function with positive real parts. Note that if there are poles with positive real parts, it would indicate that the system with the loop open is unstable. This is sometimes the case; when the loop is closed, the system may become stable. Usually the open-loop system is stable and $P = 0$. It must be checked, however.

Z is the number of poles of the closed-loop transfer function with positive real parts. Note that a pole of the closed-loop transfer function is a zero of its denominator $1 + HG(s)$. Therefore Z is just a root of the characteristic equation with a positive real part. Notice that if $Z \neq 0$, the system is absolutely unstable with the loop closed.

N is the number of encirclements of the -1 point on the Nyquist plot. Note that an encirclement of the -1 point is equivalent to a gain greater than 1 (0 dB) and a phase of 180°. This was the criterion for stability when Bode plots were discussed in Chap. 6.

When the Nyquist criterion is applied, P is known and N is determined from the Nyquist plot. These values are substituted into Eq. (D-2), and Z is determined. As long as $Z = 0$, the system will be stable when the loop is closed.

appendix E

Control-System Analysis of The Phase-Locked Loop

A phase-locked loop (PLL) is essentially a frequency control system. It has an input signal, output signal, and a feedback signal. When the components and their respective transfer functions are known, the system can be analyzed using the techniques presented in this book. Figure E-1 shows the general PLL configuration.

The appearance of an input signal ω_i creates an error (or actuating signal). After filtering, the error signal represents the control signal to the voltage-controlled oscillator (VCO). The VCO output ω_o is the controlled variable of the system. The output frequency can be fed back to the input through a counter. By performing frequency division the counter provides a feedback gain less than or equal to 1. The phase detector compares the input frequency with the feedback frequency ω_f and produces an error signal proportional to the difference in phase between the signals. Since negative feedback is used, the error signal is driven to zero or a small value. When this occurs, the PLL is said to be in *lock*. The phase detector actually produces an output which is made up of the sum and difference of its input frequencies. The detector output is a mix of

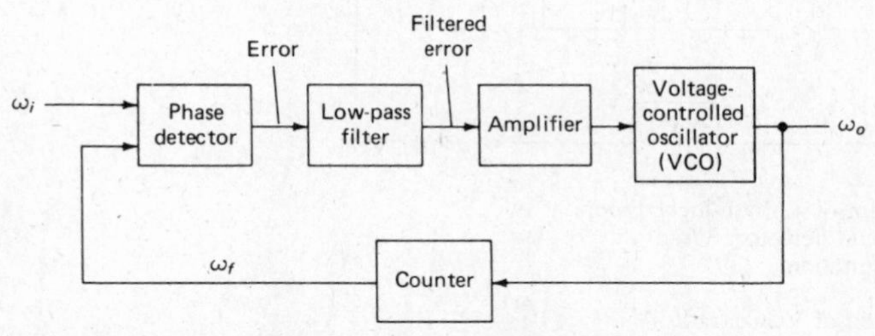

Fig. E-1 General configuration of a phase-locked loop.

$\omega_i + \omega_f$ and $\omega_i - \omega_f$. Since ω_f is driven to approximately ω_i, the detector output is made up of a high-frequency component ($\approx 2\omega_i$) and a low-frequency component (dc to $\approx$ 100 kHz) depending on the input step size. The low-pass filter passes the dc (low-frequency) signal and blocks the high-frequency signal. In addition, the filter rejects high-frequency interference.

In actual operation, the VCO without any input to the system produces a free-running frequency ω_o'. As the input frequency appears in steps, the VCO output changes accordingly to produce a system lock.

The PLL cannot acquire lock with any step change in input frequency. The *capture range* is the range of input frequencies around ω_o' over which the PLL can acquire lock. Once the PLL is in lock and the input frequency varies, the loop will remain in lock as long as the input stays within the *lock range* of the PLL. The lock range is always greater than the capture range. The *lock-up time* is the time required for the PLL to acquire lock in response to step changes in the input. A more detailed block diagram can now be drawn (Fig. E-2).

If the filter is a simple *RC* low-pass filter, its transfer function is given by

$$T_f(s) = \frac{a}{s + a} \tag{E-1}$$

where the 3-dB frequency a is given by $1/RC$.

Substituting Eq. (E-1) for $T_f(s)$ in Fig. E-2, we can write for the loop gain

$$HG(s) = \frac{K_d a K_A K_V}{Ns(s + a)} \tag{E-2}$$

We find the characteristic equation as follows:

$$1 + HG(s) = 0$$

$$1 + \frac{K_d a K_A K_V/N}{s(s + a)} = 0$$

$$s^2 + as + \frac{K_d a K_A K_V}{N} = 0 \tag{E-3}$$

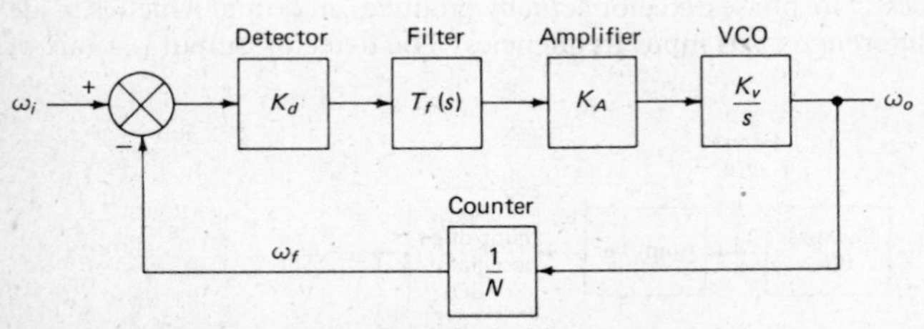

Fig. E-2 Block diagram of a phase-locked loop;
K_d = gain of the phase detector, V/rad
$T_f(s)$ = filter transfer function
K_A = amplifier gain
K_v/s = transfer function of VCO, (rad/s)/V
N = division in frequency performed by counter

From Eq. (E-3)

$$\omega_n = \sqrt{\frac{K_d a K_A K_V}{N}} \tag{E-4}$$

and $$\zeta = \sqrt{\frac{aN}{4K_d K_A K_V}} \tag{E-5}$$

Example E-1

A PLL has the configuration shown in Fig. E-2. The gains are $a = 3000$ rad/s, $N = 50$, $K_d = 0.1$ V/rad, $K_A = 1$, and $K_V = 5 \times 10^6$ (rad/s)/V. Determine ω_n, ζ, the percent peak overshoot, and the lock-up time to ±3 percent of steady-state frequency.

Solution

From Eq. (E-4)

$$\omega_n = \sqrt{\frac{0.1(3000)(1)(5 \times 10^6)}{50}} = 5477.2 \text{ rad/s}$$

From Eq. (E-5)

$$\zeta = \sqrt{\frac{3000(50)}{4(0.1)(1)(5 \times 10^6)}} = 0.274$$

From the curves given in Chap. 5 for a second-order-system step response, the percent overshoot for $\zeta = 0.274$ is about 40 percent. The time constant τ is given by $1/\zeta\omega_n$:

$$\tau = \frac{1}{0.274(5477)} = 0.67 \text{ ms}$$

After 4 time constants the output oscillation will have decayed to about 2 percent of the final value. Therefore the lock-up time will be approximately $4\tau = 2.67$ ms.

The filter break frequency a is a critical parameter. It is desirable to make a low in order to get good noise and interference rejection. However, as a decreases, both ω_n and ζ decrease. This in turn gives rise to a larger overshoot and longer lock-up time. In addition, if a is made too small so that it is less than the difference frequency $\omega_i - \omega_f$, it will block the actuating signal and the PLL will not capture.

An error analysis can also be made for the PLL. The equations derived in Chap. 5 for steady-state errors can easily be applied to Fig. E-2.

The emphasis in this appendix has been to analyze the PLL as a control system and to note the effects of the filter bandwidth on some critical system parameters. Depending on the particular application, the PLL will be used in different ways. Regardless of the application, its dynamic response and static accuracy can always be examined as they were here. Among the applications of the PLL are AM and FM demodulation, frequency-shift keying (FSK), frequency synthesis, frequency synchronization, and signal conditioning.

appendix F

Solutions to Numerical Problems

Chapter 1

1-5. $\dfrac{\theta_o}{\theta_i} = \dfrac{KK_aK_eK_p\tau}{1 + aKK_aK_eK_p\tau}$

1-6. a

Chapter 2

2-1. (a) 0.5% (b) 1.0%
2-2. (a) 0.01 (b) 0.005
2-3. (a) 2.86 V/rad, 0.05 V/deg (b) 1.43 V/rad, 0.025 V/deg (45 V)
(c) 0.72 V/rad, 0.0125 V/deg (22.5 V)
2-4. (a) 9.67 (b) 0 to 1.5 turns, 9.67 V/turn (c) 0 to 0.15 turns, 9.67 V/turn
2-5. (a) 10 (b) 15 ± 0.3 (c) 14.3 (d) 10
2-6. (a) 0.382 (b) 0.467 (c) 0.36
2-7. (a) 0.764 (b) 0.0133 (c) 0 ± 0.24
2-10. (a) 0.035 V (b) 1 V (c) 36 rad/s, 2062.8°/s
2-11. 0.0167 V/(rad/s), 0.003 V/(deg/s)
2-12. (a) 2 (b) 80 (c) 60 (d) 2 (e) 2.5
2-13. (a) 8 rad ccw, 2.67 rad cw (b) 2 rad/s ccw, 8 rad/s cw
(c) 5 lb · ft, 7.5 lb · ft
2-14. (a) 9 slug · ft^2, 0.6 lb · ft · s (b) 30 lb · ft (c) 26 lb · ft
2-15. (a) 40 rad/s
2-16. (b) 0.2
2-17. (a) 10 rad/s (b) 0 rad/s (stalled condition)
2-18. 0

2-19. 20
2-20. 4
2-22. 5
2-23. $t = 120$ s
2-24. 0
2-25. 6
2-26. 2.376
2-27. 1.25 V/deg
2-28. (a) 84.57 V (b) 88.63 V

Chapter 3

3-1. (a) $\frac{7}{s} + \frac{5}{s^2}$ (b) $3\left(\frac{1}{s} - \frac{1}{s+1}\right)$ (c) $\frac{2}{s^3} + \frac{8}{s^2 + 16}$ (d) $\frac{-5(s+3)}{(s+3)^2 + 25}$

3-2. (a) $\frac{1}{s} + \frac{2}{s+1}$ (b) $\frac{-4}{s+2} + \frac{9}{s+3}$ (c) $\frac{1}{s} + \frac{2}{s+1} + \frac{3}{s+2}$ (d) $\frac{\frac{1}{16}}{s} - \frac{\frac{1}{16}}{s+4}$

3-3. (a) $2 - 2e^{-3t}$ (b) $\frac{20}{9} + \frac{5}{18}e^{-9t} - \frac{5}{2}e^{-t}$ (c) $3 \cos 3t$ (d) $5 \sin t$
(e) $1 - \cos 2t - \sin 2t$ (f) $e^{-t} \cos t - e^{-t} \sin t$

3-4. $2 - \frac{20}{\sqrt{50}} e^{-5t} \sin (5t + 45°)$

3-5. (a) $\frac{\frac{5}{4}}{s} - \frac{\frac{5}{4}}{s + \frac{4}{3}}$ (b) $\frac{1/(s+5) + 6s/(s^2 + 49) + 4s + 7}{2s^2 + 3s + 4}$

3-6. (a) $4(1 - e^{-2t})$ (b) $2 + e^{-3t} - 3e^{-t}$ (c) $\sin 3t$
(d) $1 - 1.05e^{-t} \sin (3t + 71.57°)$

3-7. (a) 0 (b) $\frac{5}{s+1}$ (c) $-\frac{s}{3(s^2 + 9)}$

3-8. $4(1 - 3^{-1.5t})$
3-9. $3e^{-t} - 3e^{-2t}$
3-10. $e^{-t} \sin 3t$
3-11. (a) 0 (b) 3

Chapter 4

4-1. 1.5

4-2. $e_o = \frac{ak_l m}{k_x}$

4-3. 40s

4-4. (a) $\frac{\dot{\Theta}(s)}{V_m(s)} = \frac{K_m}{J_m s + F_m}$ (b) $\frac{\dot{\Theta}(s)}{V_m(s)} = \frac{K}{s + 1/\tau}$

4-5. $\frac{\dot{\Theta}(s)}{V_m(s)} = \frac{25}{s+2}$

4-6. $\dfrac{50}{s + 0.625}$

4-8. (a) $\dot{\Theta}_m = 700$ rad/s, $\dot{\Theta}_L = 350$ rad/s (b) 800 in · lb

4-9. (a) 0.167 s (b) $\dfrac{100}{s + 6}$ (c) $\dfrac{100}{s(s + 6)}$

4-10. (b) 0.2

4-11. (a) $\dfrac{K}{1 + K_1K}$ (b) K (c) KK_1 (d) $\dfrac{K}{1 + K_1K}$

4-12. $\dfrac{AB}{1 + ABC}$

4-13. (a) KK_1K_3 (b) K_1K_2 (c) $\dfrac{KK_1K_3}{1 + K_1K_2}$ (d) $\dfrac{C}{K_3}$

4-14. 48

4-15. (a) $\dfrac{KK_1}{1 + K_1}$

4-16. (a) $\dfrac{3}{s(s + 4)}$ (b) $\dfrac{4}{s^2 + 4s + 3}$ (c) $s^2 + 4s + 3 = 0$

4-17. (a) $\dfrac{14}{(s + 7)(s + 0.6)}$ (b) $\dfrac{14}{s^2 + 7.6s + 18.2}$ (c) $s^2 + 7.6s + 18.2 = 0$

4-18. (a) $\dfrac{300}{s(s + 5)}$ (b) $\dfrac{2000}{s^2 + 5s + 300}$ (c) $s^2 + 5s + 300 = 0$

4-19. (a) $\dfrac{5000}{s(s + 30)}$ (b) $\dfrac{5000}{s^2 + 30s + 5000}$ (c) $s^2 + 30s + 5000 = 0$

4-20. (a) $\dfrac{10}{s(s + 45)}$ (b) $\dfrac{100}{s^2 + 45s + 10}$ (c) $s^2 + 45s + 10 = 0$

4-21. (a) $\dfrac{192{,}000}{s^2 + 1230s + 200}$ (b) $\dfrac{240{,}000}{s^2 + 1230s + 192{,}200}$ (c) $s^2 + 1230s + 192{,}200 = 0$

4-22. (a) $\dfrac{80}{s(s + 2)}$ (b) $\dfrac{200(s + 4)}{s^2 + 2s + 80}$ (c) $s^2 + 2s + 80 = 0$

4-23. $\Theta(s) = \dfrac{ACDR_1(s) + CDR_2(s)}{1 + DE + BC + ACD}$

Chapter 5

5-2. 0
5-3. (a) $1 + 2e^{-3t} - 3e^{-2t}$ (c) 2.5 s (d) overdamped
5-4. (a) $1 - 1.414e^{-t} \sin (t + 45°)$ (c) 5 s (d) underdamped
5-5. (a) 3 (b) 0.1 (c) 2.98 rad/s (d) 3.33 s (e) 70%
5-6. (a) 0.1 rad (b) $\zeta = 0.5$, $\omega_n = 5$ (c) underdamped (d) 0.354

5-7. $\zeta = 0.5$, $\omega_n = 2$
5-8. (a) 0.1 rad (b) -0.5 rad
5-9. (a) 0.01 rad (b) 0.09 rad
5-10. $\zeta = 0.4$, $\tau = 1.5$ s, $\omega_n = 1.66$, $\omega_d = 1.52$
5-11. (b) 964 (c) $\omega_n = 288$, $\zeta = 0.007$
5-12. (a) 0.2 (b) 10 (c) 9.8 (d) 50%
5-13. (a) 5 (b) 0.6 (c) 4 (d) 10%
5-14. (a) 20% (b) 6.25 (c) 0
5-17. 0
5-18. (a) 5 (b) 0.5

Chapter 6

6-5. $\frac{27}{s}; \frac{5}{s}$

6-6. $\frac{16}{s^2}; \frac{81}{s^2}$

6-7. $\frac{512}{s^3}; \frac{8}{s^3}$

6-8. $\frac{s}{2}; \frac{s}{4}; \frac{s^2}{9}; \frac{s^2}{49}; \frac{s^3}{64}; \frac{s^3}{256}$

6-14. $\frac{10}{s} \frac{s/5 + 1}{(s/30 + 1)^2}$

6-15. $\frac{10(s/50 + 1)^2}{(s/10 + 1)(s/1000 + 1)}$

6-16. $\frac{100(s/100 + 1)(s/20{,}000 + 1)^2}{(s/10 + 1)(s/1000 + 1)}$

6-17. $\phi_{PM} = -45°$, GM $= \infty$
6-18. (b) GM $= -4$ dB, $\phi_{PM} = +2°$ (c) No
6-19. (b) GM $= -10$ dB, $\phi_{PM} = +25°$ (c) No
6-20. 4.47
6-21. $5 < K < 9.92$

Chapter 7

7-1. 0.89
7-2. 0.15
7-3. $K_t = 0.0142$
7-6. Uncompensated: GM $= -4$ dB, $\phi_{PM} = +5°$ compensated: GM $= +16$ dB, $\phi_{PM} = -60°$
7-11. Uncompensated: GM $= 12$ dB, $\phi_{PM} = -35°$ compensated: GM $= 32$ dB, $\phi_{PM} = -50°$

Chapter 8

8-4. (b) $\frac{X_A}{R} = \frac{20}{10s^2 + 10s + 20}$

8-5. $K_1 = 5$ in^3/(s · mil) $K_2 = 10^{-4}$ in^3/(s · lb · in^2)

8-6. (b) 0.5 s

8-7. Underdamped

8-10. (a) 100 mV (c) 80 mV

8-11. (a) 200 mV (c) 160 mV

Chapter 9

9-2. (a) 1.8° (b) 150 r/min (c) 15.7 rad/s

9-4. 130

9-5. 8-step sequence

9-6. (a) 200 (b) 1.8° (c) 600 r/min (d) 62.8 rad/s

9-7. 50 pulses at a rate of 500 Hz

9-8. 180°

9-9. $R = 6\ \Omega$, $L = 30$ mH

9-10. (a) 19 Ω (b) 6 ms

9-11. 60.26 in · lb at 500 steps/s

9-12. 55.26 in · lb at 1000 steps/s

9-13. 811.6 oz · in at 180 steps/s

9-14. 831.6 oz · in at 90 steps/s

9-16. 010000, 5.625°

9-17. 135 to 157.5° ccw, 22.5°

Chapter 10

10-3. (a) 0.39% (b) 0.039 V (c) 0.664 V (d) 11000000 (e) 9.96 V

10-4. (a) 1% (b) 10 mV (c) 0.82 V (d) MSD = 0011, LSD = 0111
(e) 0.99 V

10-5. (a) 1.56%, 0.1 V (b) 6.4 V (c) 2.8 V (d) 2.8 ± 0.05 V

10-6. (a) 10 mV (b) 0.38 V

10-7. (a) 10 mV (b) 0.38 V

10-8. (a) 1.875 V (b) 6.25% (c) 0.3125 (d) 0.15625 V (e) 4.6875 V

10-9. (a) 1% (b) 0.05 (c) 3.6 V (d) 4.95 V

10-10. (a) No (b) No (c) 0.48 V/step

10-12. (a) 0.2 V (b) ≈0.19 V/step (c) Yes (d) 1.4 V

10-14. (a) 1.41 mA (b) 21.2 mA (c) 15.51 ± 0.71 mA

10-15. Nonmonotonic, not linear

10-17. 10 bits

10-18. (a) 2.925 V (b) ≈108 μs (c) 9259 conv/s (d) 4.725 V (e) 3 (or 000011)

10-19. (a) 9.75 V (b) ≈108 μs (c) 9259 conv/s (d) 15.75 V (e) 10(or 001010)

10-20. (a) 0.286 V (b) ≈145 μs (c) ≈75 μs (d) 6897 conv/s

10-21. 10.43 V
10-22. (a) 8 μs, 125,000 conv/s (b) 0.125 V
10.23. (a) 50 μs, 20,000 conv/s (b) 0.1 V
10-24. 2.5 μs, 400,000 conv/s
10-25. (a) 8 μs (b) 0.5 μs (c) 125,000 conv/s
10-26. (a) 4 μs (b) 250,000 conv/s
10-27. (a) 15 (b) 0.375 V
10-29. (a) 150 Hz (b) 187.5 to 300
10-35. 1.6 kHz (i.e., 1600 channels/s)
10-36. (a) 200 Hz (b) 100 Hz (c) 5 ms (d) 1 ms
10-37. $A = 1, B = 0$
10-38. −71.44 dB
10-39. 50
10-42. 8 poles
10-43. 3 poles

Chapter 11

11-4. Originate MODEM 1 = 1270 Hz, 0 = 1070 Hz; answer MODEM 1 = 2225 Hz, 0 = 2025 Hz
11-12. 2
11-13. (a) 2 (b) 1
11-14. Conversion time + acquisition time < 0.31 ms
11-15. (a) conversion time + acquisition time < 1.56 ms
(b) conversion time + acquisition time < 62.5 μs
11-16. (a) $t_c < 31$ μs (b) $t_c < 1.24$ μs

Chapter 12

12-11. 927 r/m
12-12. 117.21 V

Chapter 13

13-1. (a) 40 (b) 3
13-2. (a) 0 (b) 0
13-3. (a) 1 (b) 2
13-4. (b) 4
13-5. 9
13-6. $\dfrac{-R_f/R_i}{R_f sC + 1}$
13-11. $\ddot{x} + 2\dot{x} + 2x = 3$

Index

Index